Geology Illustrated

JOHN S. SHELTON

Formerly Associate Professor of Geology, Pomona College

Geology Illustrated

Drawings by HAL SHELTON

W. H. FREEMAN AND COMPANY

 San Francisco and London

A SERIES OF BOOKS IN GEOLOGY

Editors: *James Gilluly, A. O. Woodford*

Preface

This book is addressed to thoughtful and observant people who enjoy the outdoors: especially to those with some curiosity about the origins and meanings of familiar rocks and landscapes. It presupposes no particular acquaintance with science and will, I hope, be intelligible to readers of many different ages and backgrounds.

Most elementary geological truths are best discovered and explored where geology is—in the field—while looking at the evidence. On the pages that follow we take a step toward this ideal by using photographs of localities that might be visited and, so far as practicable, treating these scenes as prime sources of information. No photograph can replace reality, but what is lost is partly offset by the freedom to examine areas that would be out of reach in any reasonable program of field trips.

The photographs are arranged in sequences that develop, through this observational approach, some of the main principles of geology. The majority of the illustrations are drawn from the American West and Southwest both because many geological features are grandly displayed there, relatively unobscured by vegetation, and because this is the region I know best.

As a means of communicating geological concepts, the pictures are fully as important as the words that accompany them. On most pages the photographs represent the facts, the words supply the interpretation. Many of the illustrations will, therefore, repay a little of the kind of attention that would be accorded the real feature in the field. In keeping with this, almost no identifying marks have been placed on the photographs and very few on the drawings. The text (which almost invariably concerns an illustration on the same or a facing page) serves as an expanded legend for the picture; if, while reading it, it is necessary to look more than once to identify some feature with certainty, this is no more than Nature asks of those who contemplate her unlabelled cliffs and hills.

Many of the scenes include examples of geological features or processes other than the one discussed in the text adjacent to them. The index will assist the reader in making fuller use of these examples; under each entry, such as *alluvial fan* or *erosion*, are listed all the figures in which these features or processes may profitably be recognized, whether or not they are mentioned in the accompanying discussion.

Most of what might be considered conventional textbook style and arrangement has been deliberately avoided. There are no summaries or review questions, and few suggested readings. Definitions are woven into the text rather than set apart as phrases to be memorized. The six major parts into which the book has been divided constitute a somewhat unconventional approach that cuts across the customary divisions of physical and historical geology. Yet step by step, largely through examples, the book develops most of the concepts that lie near the heart of traditional geology.

I am more interested in engaging the reader in the quest for better understanding of earth's record in rock than in setting forth the achievements of geology. Accordingly, the geological interpretations outlined herein are presented more as samples of method than as *faits accomplis;* indeed, if the contents of this volume should lead anyone to formulate different explanations, as good or better than those offered here, I would count the book a real success.

As a teaching aid, the book is a point of departure rather than something to lean on. I would urge that wherever possible, teachers and readers expand and supplement the topics introduced here through field trips and analysis of whatever local geology is accessible to them. We learn best from what is real, and teach best from what we know through firsthand experience.

A word about units of measurement: At present the majority of us who live in English-speaking countries prefer to use inches, feet, miles, pounds, and degrees Fahrenheit, although nearly all of our scientists have adopted the metric system and the Centigrade (or, more properly, Celsius) scale of temperature. Recent events indicate that we are almost surely entering a transition period during which most of us will need to be familiar with both systems and no choice of units will please everybody. In writing for a general audience at this time it seems reactionary to ignore the metric system and presumptuous to adopt it completely. My conservative compromise between good communication and good sense has been to use millimeters and Centigrade but keep inches, feet, miles, pounds, and tons, instead of adopting centimeters, meters, kilometers, grams, and kilograms. Simple devices for quick estimation of equivalents will be found at the back of the book, just preceding the index.

It is not often that a geologist and artist can work closely with each other, unhurriedly probing, comparing, and blending their sometimes very different insights about the same scene. The drawings in this book are the product of such a joint effort, during which my brother and I found increased respect for each other's discipline and pleasure in this opportunity to create together. Except as otherwise credited in the legends, the photographs are my own.

Inevitably the author of a work of this kind draws heavily upon the current pool of knowledge in its field. I am acutely aware that I have done this, both consciously and unconsciously, and that I am indebted to many indirect contributors. Specifically, I acknowledge generous assistance on individual subjects from Raymond M. Alf, Edward C. Beaumont, C. Wayne Burnham, Doak C. Cox, J. S. Creager, Bruno D'Argenio, Joseph Ernst, J. H. Feth, Carl Fries, Jr., Mason L. Hill, Carl L. Hubbs, Charles B. Hunt, Douglas L. Inman, Richard H. Jahns, Vincent C. Kelley, Robert L. Kovach, Chester R. Longwell, J. Hoover Mackin, John H. Maxson, Richard Merriam, Jack C. Miller, Thane H. McCulloh, Donald B. McIntyre, Edwin D. McKee, Berlen C. Moneymaker, George E. Neff, Carleton H. Nelson, Antonio Parascandola, Robert P. Sharp, Eugene M. Shoemaker, L. T. Silver, and J. W. Wilt.

Helpful correspondence and discussions have been conducted with John H. Aldrich, Clarence R. Allen, Charles A. Anderson, William S. Cooper, Wakefield Dort, Jr., Richard F. Flint, George M. Stanley, William D. Thornbury, and C. A. Whitten.

Special thanks are due Robert Frampton and Paul Ulmer, upon whose darkroom wizardry I have leaned heavily, and Joan Kemp, who patiently and skillfully adapted her fine calligraphy to a number of special needs. Mrs. Ruth Douglas typed the manuscript with meticulous skill.

Parts of the manuscript have been read and helpfully commented upon by members of my family and by Frederic W. Dundas, Chester R. Longwell, Howard E. Norris, and Robert M. Walker and family.

Most of all, I am indebted to Professor Alfred O. Woodford, who as teacher, colleague, and friend has supplied inspiration and encouragement during more than thirty years of stimulating and happy association, and to Chester R. Longwell for memorable times in both classroom and field and for generous permission to include some unpublished data.

The ideas expressed on the pages that follow do not necessarily have the approval of all these good people; any weaknesses and errors reflect my ignorance and stubbornness in the face of good counsel.

Claremont, California
December, 1965

CONTENTS

Introduction

Geology is the science of the earth. It is based primarily on the study of rocks exposed at the earth's surface and is therefore an outdoor science in the sense that most of its fundamental data must be gathered in the open and most of the information acquired in the laboratory must ultimately be evaluated in the field.

But a major goal of geology is knowledge of conditions, relations, processes, and events that in large part cannot be directly observed, either because they are out of sight beneath the earth's surface or because they belong to times that predate human history. Much of geology is therefore concerned with indirect evidence and with working out possible combinations of processes and events which, taking place in and on the crust during past eons of time, might reasonably have produced the present scene. In short, geology probes downward to depths far below what we can see and backward to times long before there was anyone to look.

Speculation about what is out of sight underfoot must be as old as thinking man; it certainly is a principal part of the motivation of geologists, all of whom have continuing interest in such questions as: What is the earth's crust? How thick is it? What lies beneath it? What causes such phenomena as volcanism and earthquakes that occur within it? Furthermore, what is out of sight underfoot is often of considerable economic importance. Modern geology was hardly fifty years old when, in 1856, R. A. C. Godwin-Austen predicted that coal could be found in the Weald lowlands, southeast of London, more than a hundred miles from the nearest place where coal was visible at the surface. Such a notion must have seemed almost incredible to most of his contemporaries, but his reasoning was beautiful in its simplicity; he knew the sequence and thickness of the layers overlying the coal north and west of London and recogzined that they were also present, very gently domed, in the Weald. If coal occurred beneath them in one place, it was probable that it also did in another. In 1890 a borehole proved him correct and a mine was established there. Today our industrial civilization depends in part on the ability of trained geologists to predict where such hidden wealth as petroleum and uranium and nearly all the ores that sustain the machine age can be found. The examples and analyses on the pages that follow should lead to increased understanding of the relations between what we see at the surface and what lies beneath.

The unending effort to reconstruct conditions of the distant past that is such an important part of geological research may also have economic as well as scientific value. For example, the search for uranium deposits in the Colorado Plateaus region in the late 1940's developed into a problem in reconstructing the details not only of the paths but even the flow-patterns of rivers that existed about 150 million years ago—because the patterns of sand they deposited had obviously been factors that influenced the localization of the ore. Much of the world's petroleum is associated with shallow-water marine sediments whose delineation involves the location of ancient shorelines now deeply buried under younger deposits. The more accurately the position of these ancient features can be determined, the more effectively the uranium and oil can be located and recovered.

Very often, however, the geologist is driven by plain curiosity. How old are the various rocks of the earth's crust? If they are continuously disintegrating in high places and being washed down to low places, why are there any mountains left? Where and how are rocks like granite produced? What keeps rivers flowing during long seasons of no rain? How do we know there was an Ice Age? Have the continents always had their present sizes, shapes, and positions? Can we accurately locate prehistoric mountains and seas? What was happening in Pennsylvania when any particular layer in the Grand Canyon was being deposited? Indeed, the same curiosity can be aroused simply by pausing over the view from any hilltop and asking: Has it *always* looked like this?

The geological answer will always be no, and the logical next question, What *was* it like?, can be asked again and again for the successive chapters of earth history in any given place. Numerous examples of how the search for the answers to this question is conducted will be found throughout this book.

The quest for knowledge about what is out of sight underfoot and backward in time begins with what we can see now. We must reason from what we observe in the present to what we postulate at depth or in the past. This kind of extrapolation, so often dramatized in detective stories, is one of the main sinews of science. Geology provides a rich field for the exercise of this technique. Much of the language it uses is relatively nontechnical. The starting points, the observable data, are relatively easy to grasp. Many good questions can be asked without special knowledge of facts and processes. What was the geography west of Rapid City, South Dakota, before the Black Hills came into existence—or have they always been there? Possible answers readily come to mind as visions of deserts, volcanoes, plains, tropical forests, an ice cap, or a shallow sea. Most people will have less difficulty examining these critically, and deducing the consequences of each, than they would in formulating and choosing among alternative explanations for the properties of electricity, or the atom, or the role of catalysts in chemical reactions. In geology most investigations begin with visible things known by relatively familiar names.

This book will have fulfilled its purpose if it helps anyone to visualize what lies below the earth's surface, and understand some of the changes that have occurred over the vast reaches of geologic time. To discover this added meaning in the land we live on is to attain *geological insight*.

I Materials

A brief look at the rocks that compose

the accessible part of the earth's crust.

1 A FIRST LOOK

Beneath the Surface

Upon returning home from a long journey we often comment about how good it feels to be on familiar ground again. Yet for most of us this familiarity is at best skin deep. We know the shape of the land surface, the relative positions of many hills and valleys, cliffs and streams, highways and towns. But most of us know very little about what lies beneath these familiar features. Is our town built on old lava flows? River deposits? Glacial deposits? Marine deposits? An old lake bed? An old landslide? Granite? Each of these would mean a drastically different geological story embracing the events of the thousands, or millions, or even hundreds of millions of years, that are recorded in the rocks at and beneath the surface. How can we distinguish between these possibilities—or recognize still others?

A good first step would be to go out and look more closely at the familiar scene. In some areas we might see nothing but *soil*—loose tillable material derived principally from the decomposition of rocks. Except in regions of rugged topography, soil covers most of the land. But our main interest lies in what is underneath, and important though they are, soils conceal more than they reveal regarding the *bedrock*, the underlying solid rock that has not lost its structure and character through decomposition. Our quest leads, therefore, to *outcrops*—places where undisturbed bedrock shows through the blanket of soil or is exposed in stream banks, cliffs, road cuts, or excavations.

Suppose, for example, that our town is on fairly flat ground, a short distance from the base of steep mountains. On the outskirts is the large gravel pit whose walls are shown in the photographs below. There are other such pits between the town and the mountains. Exposed here are pebbles, cobbles, and boulders embedded in fine gravel and sand, the whole exhibiting a crude but unmistakable layering or *stratification*. Each pebble and larger piece is composed

Fig. 1. Wall of gravel pit just outside Claremont, California, looking north toward the San Gabriel Mountains in the background.

of hard rock, yet the edges and corners of all are smoothed and rounded, as though worn by abrasion. (See Figures 3 and 4 for a closer view.) Abrasion and wear indicate movement; how far have these rock fragments travelled? In what direction? Were they moved by waves, by streams, or in some other way? Where and what was their source? Why are the pebbles and cobbles in crude layers? Do the layers extend under the mountains, or lap up against them? Already the simplest observations have led us to a series of specific queries. Can such questions be answered? To find out we need more facts.

Every few decades heavy rains, sometimes aided by melting snow, create torrents in the mountain canyons that are visible in the background of the upper view, a little over two miles away. Flood waters pour out of the canyon mouths, carrying mud and sand and rolling cobbles and boulders from the canyons onto the adjacent flatter ground. As the flood subsides it leaves an irregular and thin deposit of this material. Historical records show that three or four such floods have occurred in the past 100 years. From Figures 1 and 2 it is clear that the surface of the ground, and therefore all such new layers, are exactly parallel to the stratification below. The ground surface and the gravel layers slope away from the mountains, at an angle of less than 2 degrees near the gravel pits, increasing to more than 3 degrees at the mountain front. The size and proportion of boulders also increases toward the mountains, both in the surface material and the layers in the pits.

It is highly improbable that the newest layers of gravel would so exactly parallel and resemble the material beneath them, and that both would be parallel to the present land surface, unless all were products of the same continuing process. Indeed, the most conspicuous variation in the walls of the pits is the change from fresh gray gravels near the top to rust-stained and slightly decomposed ones near the base—just what would be expected if they are progressively older downward.

For these reasons we will tentatively conclude that all of the visible gravels were washed out of the mountains, that the whole lowland region is underlain by gravel and sand, and that the top of each deeper gravel layer is a buried former land surface.

Fig. 2. Looking west at the gravel pit wall shown in the foreground of Figure 1. Loose material at the bottom has fallen from the pit wall.

Rocks Are Not All Alike

The tentative conclusion that all the gravels were washed out of the mountain canyons might be tested by comparing the transported cobbles with the available source rocks in the mountains. If the same kinds of rock occur in both places, and especially if there are some unusual varieties among them, the hypothesis would be greatly strengthened.

If we examine a few dozen stones we discover that there are many different kinds—in fact it is difficult to find two exactly alike. Looking even more closely, as in Figure 4, we see that all the differences of color and texture are caused by differences in the kinds and shapes of particles that make up each rock, and in the way these particles are arranged. The particles are mineral grains and the proportions of the different kinds constitute the mineral composition of the rock. Whether the grains are large or small or variable in size; rounded or irregular or geometric in shape; intergrown, cemented together, randomly distributed, or in a layered arrangement—these determine the *texture* of the rock. It is on the basis of these two fundamental attributes, mineral composition and texture, that almost all rocks are most reliably distinguished.

Among the varieties of rock shown in Figure 4, all of which (and more) are present in larger masses in the mountains, several are distinctive. One (marble, second from left in bottom row) is plentiful in several parts of the mountains but rare as pebbles because, being soft and somewhat soluble, it does not survive much travel. Several others are so rare that pieces could not be found where this picture was taken.

All the facts about the composition of the gravels are consistent with the hypothesis that they are deposits left by repeated outpourings from the mountains, and that the crude stratification reflects successive floods. But might the gravels not have been worked over and modified by the waves of an ancient sea? We know this is not true of those at the present surface, which is more than 1,000 feet above sea level. It is also highly unlikely for the deeper layers, both because they resemble the surface deposits and because a few fossil bones of land animals, but no marine shells, have been found in them. In short, these are land-laid or continental, rather than marine, sediments. Further, to answer the question raised on page 3, they could

Fig. 3. Closer view of the wall of the gravel pit shown in Figure 1: note sorting into layers, finer near the middle, coarser near the top and bottom.

not possibly extend under the mountains because they were derived from them.

But we have taken only the first step. Can it be assumed that these gravels have been accumulating ever since the earth was born? Probably not, for if they had, the mountains might well be eroded away by now. Using a little disciplined imagination, let us undo the one step we are sure of by putting all the pebbles back where they came from. This should more than fill the canyons in the mountains, but poses some new problems: What have we uncovered by removing the gravel deposit; what kind of basin floor was it laid on? (Information from deep wells might be helpful here.) And, to take a further step back in time, what is the origin of the mountains (that shed the stones that made the gravel that buried the floor and built the plain that provided the site for the town that we built)? This is a double problem, for we want to know not only what the mountains are composed of, but how they came to stand so high above their surroundings. If they are old volcanoes, we may solve both problems at once. But what if they are made of rocks that can be formed only on the sea floor, or only deep in the earth's crust?

To make even a small start toward answering these questions we must know how to recognize rocks of different origin. Then we can see whether this second major step back in the geologic history of the area will lead us to a time of volcanic activity, will require us so to reconstruct the geography that the sea could have occupied the region, or will force us to find a means of bringing to view a deeper part of the crust—to name but a few possibilities.

Naturally, the geologist must be able to tell one rock from another, just as the botanist must be able to distinguish different plants and the zoologist different animals. It is a basic skill at which he works all his life, improving with experience. But as the problem of the gravels demonstrates, the geologist's primary interest in rocks is their origin; he wants to know where and how they were originally produced, and how they arrived at the places where we find them. Fortunately, although there are many thousands of varieties of rocks and hundreds of rock names, there are only a few fundamentally different origins. It is far more useful to be able to recognize these origins than to be able to determine the specific name of a rock.

The next fifty pages are therefore devoted to exploring, by means of examples, the origins of some common types of rocks. In Parts II and III we will investigate some of the ways in which they are moved.

Fig. 4. A few pebbles and cobbles from the wall shown at the left, selected to illustrate variety in texture and composition.

2 THE ORIGIN OF IGNEOUS ROCKS

The Birth of Volcanic Rocks

Almost every rock has two things to tell. First, it has within it a record of the conditions under which it formed, whether under the sea or on land, in a volcano or deep in the earth's crust. Second, it commonly records some of the things that have happened to it since it was formed— bending, breaking, squeezing, partial recrystallization, or merely the fragmentation and surface rounding exhibited by the gravels just considered.

Among the important rock types, lavas have the most obvious origin. Hundreds of thousands of people have witnessed eruptions, and millions have seen motion pictures of such active volcanoes as Mauna Loa, Vesuvius, or Parícutin. In simplest terms, molten material wells up, usually accompanied by some gas, flows out on the surface and congeals. In a few hours, or at most a few days, it is hard rock, though it may take much longer to cool completely.

Some of the broader aspects of this process can be examined in the scene at the right. The tongue of black lava, which stopped in the foreground, apparently issued from the base of the cone at its far end (cf. Fig. 62). Fresh lava, or the newly exposed surface of old lava that has been broken open, is generally glistening black. By contrast, the surfaces of flows that have been exposed to the weather for many centuries are usually rusty and may have decomposed enough to form a thin soil which, with its vegetation, is normally lighter colored. Evidently the conspicuous dark part of the tongue in Figure 5 is the most recent outpouring. But the tongue is not all black; scattered spots, a large area near the middle, and some of the margins, are brownish and thinly covered with volcanic ash and grass. Their distribution suggests a slightly earlier flow from the same vent that received a thin blanket of ash before it was largely covered by the fresher lava. In the background are cones in various stages of destruction by erosion, surrounded by lava and ash almost completely obscured by soil; these evidently belong to still older periods. Volcanic activity must have been going on here for some time.

The mountain on the skyline is also composed entirely of volcanic material, and there are a great many more volcanic piles nearby, but outside the area encompassed by this photograph. In short, we are looking at part of a *volcanic field*. Some of its boundary can be traced in the photograph: the darker volcanic rocks, with their generally more rounded forms, lie on top of the lighter-colored layered rocks; the distinct horizontal ledges of the latter represent flat-lying beds of sedimentary material (cf. Fig. 19).

The surfaces separating any distinguishable rock body from its neighbors are its *contacts*. The greater part of any contact surface is, of course, underground. But if the rock is exposed, the edge of the contact surface is visible on the ground as a line along the rock's boundaries— between it and the rocks below, beside, or above it—and may therefore be plotted on a map. Points on the contact between the two recent lavas, and along their contact with the sedimentary rock, are indicated in the drawing at the right.

By following the contact between these lavas and the adjacent nonvolcanic rocks on the ground we could obtain an accurate outline of the volcanic field and a look at all the visible older rocks on which it rests. Plotted on a map, this outline would show a volcanic field covering more than 1,000 square miles (less than one-fifth of which is shown here), everywhere resting on the same sedimentary layers. Clearly, then, these volcanoes must have been fed by conduits that bored up through the sedimentary plateau. The fact that small fragments of sandstone and other nonvolcanic rocks of the plateau are abundant in the lava and ash strengthens this deduction; they must have been torn from the walls of the conduit during the eruption. It follows that, unlike the mountains in the background of Figure 1, these mountains are younger than the sedimentary deposits around them.

Fig. 5. Looking south into the San Franciscan volcanic field on the Colorado Plateau in northern
Arizona. Flagstaff is just behind Humphreys Peak, the large mountain on the skyline.
The lava flow in the foreground is 4.5 miles long and about 100 feet thick.

Lava and Ash as Rocks

The two important ingredients in all volcanic eruptions are the molten rock (called *magma* as long as it is still liquid underground and *molten lava* when it reaches the surface) and the accompanying gases (mostly water vapor). The gases are dissolved in the magma and held there by the pressures existing within the earth's crust. As the magma rises in the throat of the volcano, this confining pressure is reduced, until the gas—like that in a bottle of carbonated drink that has been uncapped—forms bubbles within the magma that rush to the surface, in a process that somewhat resembles violent boiling.

If the amount of gas is relatively small, the lava pours out fairly quietly, perhaps forming a few lava fountains (e.g., Fig. 70) or tossing an occasional hot plastic blob of lava into the air. If the proportion of gas is relatively high or the magma viscous there may be numerous explosions. In extreme cases much or all of the molten lava may be blown high into the air, sometimes accompanied by fragments from the walls of the volcanic conduit. By the time the mixture of lava and rock particles settles back to earth it is so dispersed as to consist almost entirely of small fragments, collectively known as *volcanic ash* when it first settles and as *tuff* if and when it later becomes consolidated into rock.

Since lava, ash, and tuff are "born of fire," coming directly from the molten state, they belong to the great group of *igneous* rocks (from Latin: *ignis*, fire; compare ignite). Igneous rocks are almost surely more abundant than any other kind in the crust of the earth, and those that erupt on the earth's surface, the volcanic rocks, are an important variety.

In the view below can be seen some of the products of a series of eruptions that ended in 1851. In the foreground and forming the cinder cone in the distance are ash and associated explosion products. At the left is part of one of the lava flows; the loose blocks that cover it are remnants of its outer crust, which hardened and broke up while the interior of the flow was still molten and moving. The source of the flow was near the cinder cone in the background.

But if we are to use rocks to read the history of the earth we must be able to recognize lava or tuff when there is no volcano nearby—when they occur between layers of other rocks in a road cut or beach cliff, or as isolated pebbles in a stream bed or gravel pit. As a first step in developing this skill let us look more closely at some samples whose origin is beyond doubt.

Fig. 6. View southwest toward Cinder Cone in the northeast corner of Lassen Volcanic National Park. Blocky lava flow at the left; volcanic ash underfoot.

Figure 7 shows a small piece of this lava at natural size. The individual mineral grains are too small for identification with the naked eye: most conspicuous are numerous small cavities whose irregular shapes, many of them a bit squeezed, reflect their origin; they are bubble holes recording the last unsuccessful efforts of gas to escape after cooling had increased the viscosity of the lava. If the lava was not flowing at this stage these cavities or *vesicles* tend to be open; if it was oozing along they may become somewhat flattened (as here) or closed.

In the next view (Fig. 8) we are looking through a thin slice of the same rock, about 0.03 mm thick and magnified 10 times by a microscope. The vesicles now appear as large irregular white patches, and the rock can be seen to be composed mostly of minute clear lath-shaped crystals (which look white in the figure) scattered through a dark *groundmass* that consists chiefly of natural *volcanic glass*. The latter is a true silicate glass which, owing to the presence of finely divided dark and opaque ingredients, is commonly no more than slightly translucent; it can, however, often be identified by holding the thin edge of a fresh chip up to a bright light and examining it with a magnifying lens.

These two characteristics—the presence of vesicles and a texture of very fine grains embedded in glass—are earmarks of lava. They can be found in the smallest piece and therefore serve to identify pieces that may be found far from any other evidence of volcanic activity.

The microscopic features of volcanic ash and tuff also reflect their origin. Ash from the foreground of the scene at the bottom of the facing page is shown in Figure 9, and a threefold enlargement of part of this sample is shown in Figure 10. The ash is made up mostly of small fragments of highly vesicular rock resembling cinders or slag, though a few solid glassy pellets may be seen in both views. Not visible here is a fine dusty component that includes minute shards of glass. When viewed under the microscope some of these shards can be seen to retain telltale curved shapes that clearly relate them to the breaking up of clusters of bubbles. Larger fragments of this solidified glassy froth, known as *pumice*, are often recognizable in both ash and tuff. All this fragmental material differs from the lavas only in having been blown into the air instead of flowing out on the surface; the high gas content that was necessary to accomplish this is recorded in the abundant fragments with spongy texture.

Fig. 7. Hand specimen of the lava shown in Figure 6.

Fig. 9. Sample of volcanic ash in foreground of Figure 6; standard pin for scale.

Fig. 8. The same lava seen under the microscope, ×10.

Fig. 10. Closer view of the ash in Figure 9.

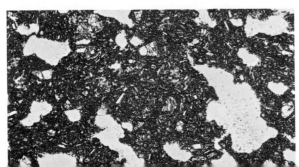

A Word About Textures

All volcanic rocks can hardly be expected to look alike. The color, texture, and mineral composition of any sample are sensitively related both to the composition of its parent magma and to the way it was erupted and cooled.

The chemical compositions of lavas—or of all the common igneous rocks, for that matter—show less variation than one might expect. Oxygen and silicon are the chief elements in all of them, combined with varying proportions of aluminum, iron, magnesium, calcium, sodium, and potassium, plus a trace of water. In the rocks virtually all of these are locked up in less than a dozen minerals, most of which are usually easy to recognize after a little practice. Because the minerals are what one actually sees when examining a rock, distinctions between rocks are, as stated earlier, generally based on mineral composition and texture, both of which can usually be identified with relatively little special equipment.

Texture reveals much more about the origin of most rocks than does mineral composition. Such fundamentally different rocks as sandstone and granite, for example, may have virtually the same mineral composition, but are easily distinguished from one another by their textures. These textures faithfully record their very different origins: the mineral grains of the granite are intergrown crystals that could only have formed at high temperature and pressure; those of the sandstone are rounded from wear and are held together by a natural cement, showing that they accumulated as sediment.

Because our interest in rocks is more in knowing their origins than in being able to name varieties (many of which come with the same histories anyway), we shall concentrate on textures here.

The textures of igneous rocks are produced during solidification. Through the study of active volcanoes and by melting rock samples in the laboratory and then cooling them under controlled conditions, much has been learned about the factors that influence texture. Most lavas are molten at about 1,100°C. As the temperature falls, submicroscopic crystals begin to grow. This process of crystallization can be stopped at any time by sudden cooling (quenching), which halts crystal growth and converts the still molten material into glass made murky by the dispersed embryonic crystals it contains (e.g., obsidian). If not thus interrupted, however, the crystals continue to grow, drawing their needed constituents from the diminishing liquid fraction surrounding them.

Low viscosity (high fluidity) and time favor the growth of mineral grains by permitting the ions (charged atoms) needed for each crystal to migrate easily and in quantity to their proper niches in the crystalline structure of the growing grain. In the course of laboratory experiments with rocks it has been learned that the increase in viscosity one would expect as the melt cools does not always occur. Instead, the small amount of water vapor and certain other volatile constituents present in almost every magma are largely rejected by the growing crystals and thus constitute an increasing proportion of the residual liquid. These volatiles lower the melting point (thus postponing total solidification), greatly decrease the viscosity, and thus promote crystal growth. However, it requires high confining pressures to keep them from escaping; if the pressure is released the magma suddenly boils and thickens, and growth of the crystalline mineral grains is slowed or stopped.

From this it is obvious that a fully crystalline coarse-grained rock can be produced only by slow cooling under pressures adequate to seal the magma chamber and to hold in the volatiles—in other words, well down in the earth's crust. The geological occurrence of such rocks fully bears this out. Volcanic rocks, on the other hand, are fine grained or glassy because eruption at the earth's surface brings about rapid cooling and permits escape of most of the volatiles. It is this escape, of course, that explains the explosiveness of most eruptions and the vesicular nature of most lava.

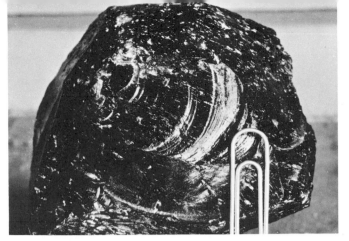

Fig. 11. Obsidian from Mammoth Lakes, California.

Fig. 12. The same obsidian seen in thin-section, ×10.

There are many reasons why, despite their common eruptive origin, different volcanic rocks have different textures: (1) The mineral grains normally start to grow at different times and grow at different rates, depending primarily on their composition; (2) cooling may occur in several stages differing in temperature, viscosity, and duration; (3) volatiles may be released in different amounts at different times; (4) crystals grown during one stage may be partly or wholly redissolved during another; (5) eruption may end the whole process at any stage.

The accompanying pairs of photographs of four different volcanic rocks illustrate some of the evidence upon which these generalizations are based. In each pair, the one on the left shows a natural surface as it might appear to someone scrutinizing a piece held in the hand, and the one on the right a *thin-section* (slice 0.03 mm thick) of the same rock, seen through a microscope at tenfold magnification.

The first pair (Figs. 11 and 12) are of typical natural volcanic glass, or *obsidian*. Although the rock contains a very low proportion of dark minerals, the specimen is almost black because these consist of small particles so uniformly dispersed that they color the whole mass. The fine-grained varieties of any rock are always darker than coarse-grained ones of the same composition. The coarse-grained equivalent of obsidian, for example, is granite, a rock whose overall appearance is generally very light gray to almost white. (See Figs. 21 and 29.)

Under the microscope the obsidian is seen to be thoroughly streaked out from flowage when the glass was molten and viscous. A few small dark crystals are visible, but the myriads of minute ones, whose concentration largely accounts for the darker streaks, are much too small to show at this magnification. In some obsidian the lighter streaks, especially the larger ones, are produced by concentrations of microscopic bubbles.

A slightly more advanced stage of crystallization is illustrated by the next pair (Figs. 13 and 14). The weathered surface shows a crude horizontal lamination resulting from flow when the lava was molten. This is even clearer in the thin-section; notice that the dark glass is charged with swarms of needle-like or lath-shaped crystals of feldspar that tend to line up and to wrap

Fig. 13. Weathered surface of fine-grained basalt porphyry, San Jose Hills, California.

Fig. 14. The same basalt seen in thin-section, ×10.

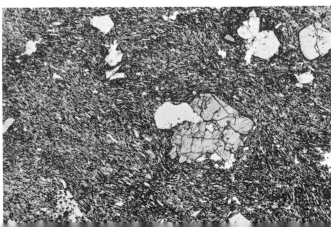

Fig. 15. Biotite porphyry from Ship Rock, New Mexico.

Fig. 16. The same biotite porphyry seen in thin-section, ×10.

around the few large crystals. A fine-grained igneous rock such as this in which a few scattered grains have grown to significantly larger size than the rest is called a *porphyry* and the individual large grains are *phenocrysts* (Greek: *phainein*, to show).

This porphyry has had a more complex cooling history than the obsidian. The phenocrysts probably represent minerals which began to crystallize long before eruption and under relatively stable conditions. At a later stage, possibly during and after eruption, the small feldspar crystals began to grow, but eruption and consequent chilling ended the process and converted the remaining liquid to glass, in which they are fixed like a school of frozen minnows. The irregular clear-white areas are vesicles.

In the rock shown in Figures 15 and 16 the process of crystallization has been carried one small step farther. The specimen, from one of the dikes shown in Figure 19, is also a fine-grained porphyry. Some of the phenocrysts can be seen in the hand specimen as well as in the thin-section. Flow texture is present but much less pronounced than in the preceding examples, as might be expected of molten material that filled a crack instead of flowing freely onto the land surface. The absence of vesicles is a result of solidification under pressure great enough to prevent bubbles from forming. The still smaller proportion of glass reflects the slower cooling and higher degree of crystallization fostered by these conditions.

The last specimen (Figs. 17 and 18) is deceptive and points out the critical value of texture in deciphering the origin of rocks. At first glance the unmagnified specimen seems to be a porphyry studded with phenocrysts of quartz, but as the magnified thin-section shows, most of the quartz grains are broken pieces, embedded in dark glass. Furthermore, the glass is not homogeneous; within it are light-colored pieces with shredded ends that are bent and squeezed by the larger grains, and fragments that differ from one another in frothiness. A few contain microscopic crystals. Such an intimate association of small pieces of glass, pumice, and lava with differing cooling histories could not be the simple product of solidification of a single extrusion. The magma must have been chilled and quenched at different stages, probably at different levels in the throat of the volcano.

The kind of volcanic activity that best accounts for these attributes is an explosive eruption. This rock is a *welded tuff*—formed by ash that was so hot when it accumulated that some of the still-soft particles of pumice and glass adhered to each other and to the mineral grains, binding the whole together. This "welding" is most likely to take place in the deeper parts of thick accumulations, and the resulting rock, after the associated loose ash has been removed

 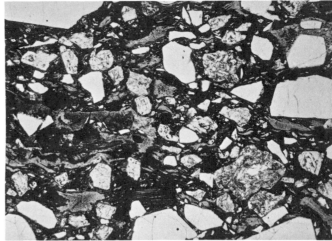

Fig. 17. Welded tuff from near Current, Nevada.　　　　Fig. 18. The same welded tuff seen in thin-section, ×10.

by erosion, may easily be mistaken for porphyritic lava. Embedded, flattened pieces of pumice are a good clue to welded tuff.

The logical next question is: What about the magma that does not reach the surface? When a volcano permanently ceases its activity the throat is not empty; it is filled with volcanic rock. Near the surface some of this rock may be broken lava and ash that have slumped from the inside walls of the crater, but deeper down it should be solidified magma that might extend down to the chamber from which it came. Are such deeper rocks ever visible? If so, what would they look like? On the next few pages we will explore the question of igneous rocks that solidified underground.

The Volcanic Neck, a Shallow Intrusive Igneous Body

It would be interesting to compare erupted lavas with the rocks deep in the throat of the volcano that produced them; the two would be, respectively, the *extruded* and *intruded* products of the same magma. Such a comparison is seldom possible, however, since by the time the deeper rocks are uncovered the ejected material has usually been completely removed by erosion.

We can, nevertheless, find volcanoes in all stages of destruction by erosion; some early stages were seen in Figure 5. By carefully comparing such remnants it is possible to recognize a *volcanic neck*, the body of igneous rock that solidified in the throat, or pipe, that once fed a volcano.

Ship Rock, shown at the right, is probably such a neck. Although the surrounding region consists almost entirely of nearly horizontal sedimentary layers (note outcrops in these views), Ship Rock is composed almost entirely of volcanic material that was evidently emplaced under such pressure that radial cracks formed to accommodate the expansion. Magma was apparently injected into the cracks as fast as they opened, producing the radiating *dikes* seen here.

Did this intrusion reach the surface of its time and actually feed a volcano? The surviving evidence provides some clues:

1. The intrusion is one of more than a dozen within a radius of twenty miles whose uniformly erect shapes are most simply explained as filled vertical pipes and whose distribution is characteristic of a volcanic field (cf. Fig. 5).

2. Ship Rock contains many small pieces of older rocks that must have been torn from the walls of its pipe. These include fragments of sedimentary rocks lying beneath those exposed in this view and occasional pieces of granite (inset, Fig. 19), which is known, from nearby drilling for oil, to occur only at depths more than 9,000 feet below the present summit. Blocks of foreign rocks carried two miles vertically upward imply vigorous upward flow; although such currents might result from convection, they would be a near-certainty in an open pipe.

3. The rock exposed in Ship Rock is mostly tuff, cut by a few dikes. Such fragmental material is usually produced by the copious or explosive escape of gases; this can take place only upon release of pressure, which can most easily occur at or near the surface.

4. We may also reason that it is unlikely that molten rock under pressure great enough to produce and fill the radiating cracks that surround the neck could exist within a short distance of the surface without breaking through.

Although such arguments do not prove that Ship Rock fed a volcano they favor that interpretation and have been taken into account in the reconstruction of probable earlier conditions shown in the lower view. The drawing also shows an additional thickness of sedimentary rock, since removed by erosion, through which the vent probably bored and on top of which the ash and cinders probably accumulated. The assumed funnel shape of the old crater is compatible with late-stage collapse around the vent, which in turn may explain some of the tuff in the upper part of the neck.

The dike is shown filling a fracture that did not reach the surface. We reason that if it had broken through, the escaping material would probably have reamed out a vertical pipe; the uniform width of the dike argues against this. It is also likely that the dikes were injected before the main vent broke out at the surface, because once the vent was opened extrusion should have been easier than intrusion. This inference is supported by the nonvesicular texture of the dikes, shown in Figures 15 and 16.

Note that according to this analysis the present height of Ship Rock is entirely due to its passive resistance to the erosion that removed the surrounding sediments; it does not represent any kind of upward thrusting of a column of hard rock.

Fig. 19. *Upper:* Looking north at Ship Rock in northwestern New Mexico. The monolith is about 1,400 feet high and the dike in the foreground is 5 miles long. *Inset:* Granite inclusion in tuff at base of Ship Rock. *Lower:* The same scene with a cut-away restoration in the background showing probable earlier conditions.

The Stock, A Larger And Deeper Igneous Intrusive

Volcanoes can be observed in action, and many volcanic necks can be recognized beyond any reasonable doubt. But from here on, the trail toward an understanding of deeper igneous intrusions becomes dimmer and the evidence less direct.

The same general range of compositions is found among common igneous rocks of all types, regardless of where they cooled. But their textures, especially the range in size of the crystalline grains, tend to differ according to the size, shape, and situation of the whole rock body. This observation implies that the environment in which the rock cools usually leaves its mark on the rock, and that from this imprint we can therefore often learn something of the early history of the rock mass. What effects should we expect from greater depth at time of solidification and greater size of the intrusive body? Both depth and size tend to retard temperature change, so such bodies should show evidence of much slower cooling, in many cases probably extending over thousands of years. Depth also imposes higher confining pressures, which should inhibit the escape of water vapor and other gases. Under these conditions all of the magma is likely to become visibly crystalline and there should be no glass.

The smaller bodies of such originally deep-seated rocks, ranging from a few thousand feet to about ten miles across, are *stocks*. Many stocks have a porphyritic texture (e.g., Fig. 20, inset photo at the right) intermediate between those found in volcanic and granitic rocks, though nearly complete gradation with both groups can be found.

Consider the example shown below. The stock proper, about 3,000 feet in diameter, is surrounded by a wide zone of intensely shattered sedimentary rocks into which magma has been injected. Farther out this gives way to less-disturbed *country rock* (a general term for the older host rock surrounding any intrusive igneous body) containing scores of smaller intrusions. These intrusions include both cross-cutting dikes and *sills*—sheets injected between the layers of sedimentary rock. Many of the dikes and sills are interconnected, and the whole intrusive complex—stock, shatter zone, and retinue of dikes and sills—occupies the center of a bulge, or dome, in the country rock.

Fig. 20. *Left:* Looking southwest at Mt. Ellsworth in the Henry Mountains of southern Utah. The dark summit of igneous rocks is 2,000 feet above and 3 miles beyond the canyon in the foreground.

Putting all this together, we conclude that the magma forced its way up from below, doming the older sedimentary layers as it did so. Whether or not it broke through to the surface, the accompanying stresses produced a surrounding zone of shattered country rock. The inner part of this zone became intimately mixed with the molten porphyry (Fig. 20, *right*), and farther out magma was injected along cracks and between layers. There are four similar stocks nearby, three of them larger than this one, around which the offshoot intrusions extend from three to six miles, as presently exposed. Each cooled and crystallized under conditions that yielded a rock of intermediate texture, which later erosion has brought to view.

How do we know the conditions in the core of the mountain, several thousand feet below the surface, as they are shown in the drawing? The first step is to climb all over the mountain, carefully plotting the distribution of the different rock types on a map. Fuller sequences of the sedimentary layers are extensively exposed a few miles away (e.g., Fig. 88) and since the strata are relatively uniform throughout this region their thickness and character there are a reliable guide to what is below the surface here. For the igneous rocks, the height of the mountain itself, and the deep canyons on its flanks, provide information through about 3,000 vertical feet. Furthermore, some of the others in this group of five similar stocks are more deeply dissected by erosion than the one shown here and it is reasonable to use them as guides to the obscured part of this one. Such considerations provide some control over how the visible relationships should be extended downward. Obviously the probable accuracy of the extrapolation diminishes with depth.

Some of the layers that have been eroded from the vicinity of the stocks are present nearby. By combining their thickness with that of the upwarped strata adjacent to the stocks it is estimated that there were roughly one to two miles of rock on top of these intrusions at the time of their emplacement. Perhaps the fact that the pressures were sufficient to raise this stratified cover at least 3,000 feet helps to explain the unusually extensive shattered zones around the stocks.

Fig. 20. *Right:* A quarter has been cut away to show the internal structure. *Inset:* A sample of the porphyritic texture found in the Mt. Ellsworth stock, as seen under the microscope (*Courtesy of Charles B. Hunt*).

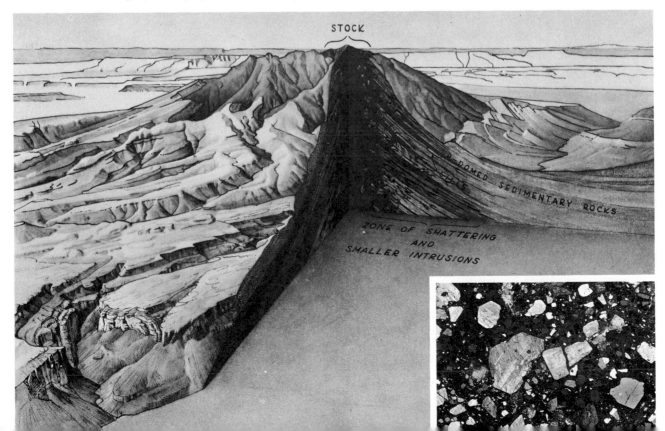

The Batholith, Largest and Deepest of the Intrusive Igneous Masses

On all the continents there are areas in which coarse-grained igneous rocks are exposed over hundreds or thousands of square miles of the land surface. Some, like the Sierra Nevada shown here, are mountainous; others are lowlands. All observations indicate that these rock masses are many miles thick and so deeply embedded in the earth that their bases have never been exposed to view by erosion. Such large igneous intrusions are *batholiths* (Greek: *bathys*, deep + *lithos*, stone); some of them are the largest known bodies of intrusive igneous rock.

The rock of a typical batholith has a rather uniform texture, like that shown in Figure 21.

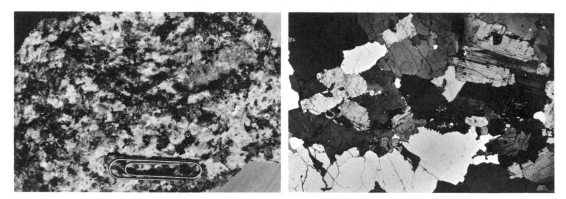

Fig. 21. Specimen (*left*) and thin-section, ×10 (*right*) of granite from Sentinel Dome near Yosemite Valley, California. Dark tones in latter are produced by polarized light; most of the grains are actually colorless. (*Specimen collected by D. B. McIntyre.*)

It consists entirely of intergrown crystalline mineral grains averaging about 5 mm across; there is no residual glass. A few grains have one or more straight sides marking crystal faces, but the majority have irregular shapes that interlock with those of their neighbors. The interpenetration along crooked boundaries implies that the different mineral grains were all growing at about the same time. This texture indicates conditions more favorable to crystal growth than any we have yet examined; i.e., still slower cooling and greater retention of volatile constituents. This is what we should expect in enormous masses that had cooled at great depths, where the possibilities of a sudden change in temperature, position, or composition would be minimal.

In most batholiths, and in all the very large ones, light-colored minerals are much more abundant than dark ones. This, and the characteristic texture of these rocks, gives them the speckled or overall light-gray appearance that is familiar to all who have hiked in the Sierra Nevada, visited Mt. Rushmore in the Black Hills, Stone Mountain in Georgia, or almost any cemetery. These are *granitic* rocks.

In the upper view at the right the foreground ridge of light granite is capped by dark rocks, more of which can be seen on top of the granite in the background. In the scene below the same granite, 40 miles farther north, is seen partly enveloping a mass of striped, darker rocks that are embedded in it from above. In both areas dikes and irregular protrusions of granite invade the dark rocks, thereby indicating that these must have been there first, as part of the country rock into which the granite was intruded. Because most such visible remnants of these older rocks are associated with low places in the top of the batholith—sags in the cover beneath which the batholith solidified—they are called *roof pendants*.

The exposed parts of batholiths are often more irregular in shape than stocks. In the Sierra Nevada, the southern part of the great batholith has been completely unroofed by erosion, exposing over 10,000 square miles of granitic rocks, but the northern part is still partly obscured by patches of older overlying rocks. In the latter area, which lies in and beyond the background of the upper view at the right, some of the places where the batholith shows through this cover have the size and character of stocks; indeed, it is probable that many stocks are actually upward protrusions on deeper batholiths, the "size" of either depending partly on the extent to which it has been uncovered by erosion. In view of this relationship geologists often lump stocks and batholiths together as *plutons* (from *Pluto*, Greek god of the nether world). All igneous rocks may, then, be referred to two great groups, *volcanic* and *plutonic*, whose typical textures and occurrence are highly distinctive, although of course there are gradations between them.

Fig. 22. Looking northwest at roof pendant along crest of Sierra Nevada batholith near Split Mountain, 24 miles south of Bishop, California.

Fig. 23. Looking west at roof pendant exposed north of Mt. Morgan, 25 miles northwest of Bishop, California.

Fig. 24. Dark dike about 3 feet wide in contact with granitic rocks, east central Alvord
Mountains, Mojave Desert, California.

Some Small-Scale Features of Igneous Rocks

Dike with Chilled Margins

The righthand half of a vertical dike is shown in Figure 24. If we ignore the cracks and super-ficial crumbling of some of the rocks, significant differences in texture become apparent. These differences occur in three principal vertical bands. At the far right is the rough surface of the coarse-grained granite into which the dike was intruded. At the far left is the central part of the dike, which is almost equally coarse, though its weathered surface is less rough. Between these, the texture of the dike rock becomes progressively finer toward the right, where its sharp contact with the granite is marked by a shadowed recess in the cliff. To the left of the pencil, for example, the mottled surface clearly discloses many grains 3 mm and more across; to the right of the pencil there are none of these, and close to the contact with the granite the dike rock is so fine grained it appears homogeneous.

These differences, even more clearly discernible under the microscope, are probably attrib-utable to the cooling history of the dike. The surrounding granite must have been solid at the time the dike was injected, and almost certainly cooler than the invading magma. Thus the incoming magma was probably chilled along its contacts with the granite walls and crystallized relatively rapidly there, as a mass of small crystals. The central part of the dike remained hot longer and therefore its grains grew larger. If this analysis is correct we see here, in the space of a few inches, an illustration of the effect of cooling on texture.

For a chilled margin to form it is obviously necessary that the dike remain essentially un-disturbed during its solidification. The fact that such dikes are not very common probably means that in many dikes there is some circulation of the fluid magma, and consequent mixing, even in late stages of crystallization.

Fig. 25. Quarry face in the Harding pegmatite, Taos County, New Mexico. The box and gallon jug give scale. (*Photograph by R. H. Jahns*).

Pegmatite and the Growth of Giant Crystals

Irregular dikes of light-colored rock with abnormally coarse crystalline texture occur in and near granitic rocks all over the world. The rocks that constitute such dikes, known as *pegmatites*, sometimes contain crystals of unusual chemical composition and exceptional beauty.

The view in Figure 25 includes the upper half of an almost horizontal body of pegmatite exposed in a quarry. The dark enclosing country rock is visible in the upper-right part of the cliff face, below which the lighter pegmatite appears in broad bands of contrasting texture and tint which slope gently down to the right. The whole pegmatite body is about 50 feet thick, measured across the layers, and has been traced for more than 350 feet along the outcrop and 650 feet by drilling into the mountain.

Like most pegmatites, this one is composed mostly of quartz and feldspar, but special interest centers in its less common constituents. Most spectacular here is the zone 10 to 15 feet thick that passes just above the old tunnel entrance and consists of a tangle of fence-post-size crystals of the lithium-bearing mineral spodumene. Above this are zones rich in beryllium, phosphorus, and fluorine, and the zone below it was once the largest source of the relatively rare element tantalum in the United States.

Why should such a small igneous body include such giant crystals? The answer is only partly understood. Crystals are distinguished from other solids by the highly ordered three-dimensional patterns of their constituent atoms. A crystal grows only by the addition of the proper atoms at the proper places. The growth of crystals several feet long obviously requires either plenty of time or great mobility of the available atoms in order that, in enormous numbers, they may find their particular niches in the appropriate growing framework. A rock body as small as the one shown here could hardly have remained fluid long enough to account for such giant crystals by time alone; the largest crystals in huge batholiths, whose cooling must have

23

taken very much longer, seldom exceed an inch or two in length (see Fig. 21). The explanation must lie in the other factor, mobility of the atoms.

Laboratory experiments with natural melts of granitic composition have shown that in the temperature range where most of the crystallization takes place (roughly 600° to 700°C) viscosity is more sensitive to small changes in water content than to changes in temperature. For example, a drop in temperature of about 100°C produces only a ten-fold increase in viscosity whereas a reduction in water content from 8% to 2% increases it about a thousand times. A high water-vapor content in pegmatites is thus strongly implied, but it is probably not the only factor influencing their texture. The largest crystals in pegmatites are typically compounds rich in such elements as lithium, beryllium, boron, and fluorine. These and the hydrogen and oxygen of water vapor are all very light elements, and it seems likely, therefore, that all of them contribute to reducing the viscosity of molten rock. Apparently when they are present in above-average amounts, the high chemical mobility necessary for the growth of crystals can exist even at relatively low temperatures and crystals grow large despite the fact that they probably have less time in which to do it than do the minerals in a batholith. Pegmatites thus represent an extreme case of the influence of chemical composition on texture.

Cooling Cracks

After a body of igneous rock solidifies, it must cool several hundred degrees to reach thermal equilibrium with its surroundings. Extruded rocks cool most rapidly, of course; the centers of lava flows lose their excess heat in a few years—commonly less than twenty. Large intrusive bodies, on the other hand, probably require thousands of years to cool, with variations depending on depth, size, and shape.

Like most solids, igneous rocks shrink slightly as they cool. In deep-seated rocks the confining pressure of overlying rocks inhibits the formation of cooling cracks, but in lavas and shallow intrusives they are sometimes strikingly developed. A thin body like a dike loses the greater part of its heat to its walls. But these surfaces, often acres in extent, cannot possibly shrink toward one center; to do so would mean dragging the extremities inward many inches, and the tensile strength of the rock is not sufficient for this. Instead, as the dike cools, tension develops fairly uniformly over the entire surface that is in contact with the cold wall rock. Under favorable conditions the result is shrinkage about many equally spaced centers. But such contraction means tension between the centers, which results in a system of short straight cracks separating each center of contraction from its neighbors (Fig. 26). The inevitable result is a polygonal pattern of cracks, ideally hexagonal, but often departing from this because of irregularities in such factors as the thermal conductivity. The most familiar analogy is the pattern of mud cracks formed on the floor of a dried-up pond (see Fig. 81).

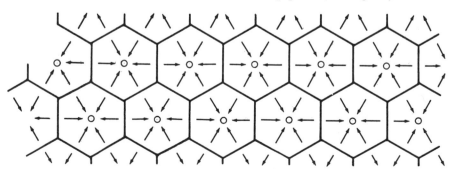

Fig. 26. Diagrammatic representation of shrinkage attending cooling or drying of a homogeneous surface. O = uniformly spaced centers of contraction; arrows indicate direction of shrinkage; lines, the resulting tension cracks.

Fig. 27. *Left:* Andesitic dike about 50 feet wide cutting volcanic breccias, exposed on U.S. Highway 101, Conejo Grade, 3.5 miles east of Camarillo, California. *Right:* Devils Post Pile National Monument near Mammoth Mountain on the east side of the Sierra Nevada. (*Photograph by R. C. Frampton*).

As cooling progresses into the igneous body, contraction advances with it and each polygon formed on the surface becomes the end of an inward-growing column. These columns may grow until they reach similar prisms begun at the opposite side (e.g., Fig. 316), or they may curve in response to variations in the temperature gradient.

Figure 27, *left*, shows a vertical dike about 50 feet wide intruded into darker volcanic rocks and exposed in a highway cut about 200 feet high. Notice the crude horizontal columns, perpendicular to the walls of the dike. Figure 27, *right*, shows part of a thick basaltic lava flow on the upper San Joaquin River in the eastern Sierra Nevada. Conditions must have been almost ideal here, for some of the columns are 60 feet long and the majority are beautifully regular in size and shape. Over half of them have six sides and most of the rest are five-sided. Their average diameter is about two feet. Note the changing orientation and curvature of some of the columns which must reflect uneven cooling rates, probably caused by irregularities in the ground over which the lava flowed. (For a closer view of the tops of these columns, see Figure 209.)

Figure 28 shows a block of volcanic rock embedded in tuff produced about 20 million years ago. (The methods of establishing such time intervals are explored in Part IV.) The radial

Fig. 28.
Andesite block embedded in andesitic tuff breccia. Glendora volcanics, west of San Dimas Canyon in San Gabriel Mountains, California.

shrinkage cracks in this rock must have formed after the block and its associated ash were deposited, for otherwise it would have fallen apart during the explosive eruption required to produce and distribute the accumulation. (It is now breaking apart under its own weight.) We can only conclude that it cooled in its present position. There are many other blocks like it scattered through great masses of tuff here and nearby—which is not only evidence that this ancient eruption was explosive but also that the materials traveled so fast and accumulated so rapidly that they were still hot when they came to rest, even though several miles from the eruptive source. Such eruptions, though not common, have been observed in historical times; one of the best known took place on the flanks of Mont Pelée on the island of Martinique in May, 1902, when 28,000 people were fatally smothered by the intensely hot ash in the town of St. Pierre, 6 miles from the vent.

Inclusions

It is partly because rocks like granite are so homogeneous throughout such enormous volumes that special interest attaches to any irregularities in them.

Figure 29 shows a typical dark *inclusion* in granitic rock. It is obviously of finer texture than its host, and much richer in dark minerals. Parts of some batholiths contain great numbers of these, which can be seen scattered a few yards apart across road cuts or other fresh exposures.

It is often difficult to choose among plausible explanations for such inclusions. Are they (1) undigested chunks of country rock torn from the walls of the magma chamber by an actively migrating magma; (2) unmelted relics of country rock that was being quietly fused to form magma; (3) residual chunks of country rock that was being converted to granitic magma by chemical diffusion; or (4) segregated clots of early-formed dark minerals that crystallized from the magma itself? Perhaps some combination of such mechanisms is involved. The example shown here is believed to be a relic of country rock rather than a segregation because similar larger inclusions nearby clearly match the country rock. The frayed end (near the point of the pencil) probably means that the dark rock was in the process of being assimilated by the magma when complete solidification halted further changes.

Inclusions have a bearing on the larger problem of the origin of granite, discussed on pages 20–21, for in many cases they are the chief visible evidence by which we may hope to distinguish

Fig. 29.
Close-up of dark inclusion in granitic
rock exposed on California Highway 78,
0.6 mile southeast of Santa Ysabel
in San Diego County.

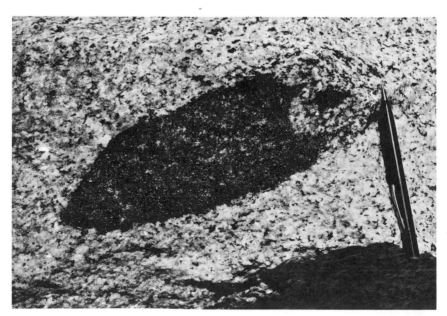

Fig. 30. Two views of textural variations in granitic rock exposed along upper Rock Creek, eastern Sierra Nevada.

granite formed by selective fusion, granite emplaced by forceful injection, and granite formed in place by chemical alteration of older rocks.

Flow Structure

A close look at good exposures of fresh granite will often disclose interesting variations in texture. One of the most common is a simple alignment of the minerals like that shown at the left in Figure 30. Notice that on the top surface of the block, behind the knife, the texture is typically granitic—a random distribution of light and dark minerals—but that on the front face there is a distinct vertical alignment that is especially noticeable in the dark minerals. Such alignment is almost certainly the result of movement in the rock during the late semi-molten stage—after some crystals were well formed but before the mass had become completely solid. This is sometimes referred to as the "crystal-mush" stage. Presumably the fluid filling the space between the crystals was highly viscous, and the crystalline plates and prisms, like matchsticks in honey, were oriented by slow currents in the gooey mass. The currents may have arisen from active intrusion of the magma, from crustal movements that changed the shape of the magma chamber, from convection produced by local differences in temperature and density, or from combinations of these.

The righthand view is a close-up of a granite outcrop showing a small shear zone that appears to have developed just before complete solidification. Above the left end of the knife the texture is quite normal, but at the right the minerals have been dragged into a narrow zone of alignment along which the right side moved down relative to the left. Beyond this (out of sight at the right) the texture returns again to normal. If the movement had occurred when the rock was completely solid and brittle we would expect a clean break with perhaps some crushed rock along the zone of movement. But here the zone fades gradually into normal rock on either side, which would seem to require that the rock be very nearly solid but, at least in this zone, plastic enough to drag the crystals into alignment, grinding some of them into finer sizes in the center of the shear zone. Such plasticity almost surely exists at some stage during solidification, and it probably can also develop later if deformation takes place under high confining pressures. Distinguishing between such events, many millions of years after they took place, is a typical problem facing the geologist trying to understand the history of such a rock body.

3 THE ORIGIN OF SEDIMENTARY ROCKS

Introduction

After lavas, the rocks with the most obvious origin are *sedimentary* rocks. The commonest of these are compacted and cemented accumulations of gravel, sand, mud (silt and clay), or chemical precipitates—often containing shells, bones, or vegetal remains—that were deposited by water, wind, or ice.

Because the accumulation of sediment is almost never a steady process, most sedimentary rocks are stratified. Individual layers may be tens of feet thick, as in the Alaskan coal and sandstone deposit in Figure 31, or less than an inch thick, like those in the recently deposited sands on the reservoir floor shown in Figure 32. A rock composed almost entirely of extinct marine shells, shown in Figure 33, is a sample of the more unusual compositions found among sedimentary rocks.

The ultimate resting place of most sediments is on the bottoms of the world's oceans, seas, and bays. Enormous quantities accumulate at the mouths of large rivers (probably a cubic mile every twenty years or so off the Mississippi, for example) and waves and currents spread sediment almost continuously along all coasts and on the floors of such natural basins as Hudson Bay or the Mediterranean Sea. Lesser quantities accumulate in river valleys, desert basins, and along the bases of steep mountains (as in the gravel pit shown in Figs. 1 and 2). When such sediments become so compacted and cemented that they break rather than crumble when struck with a hammer, or produce pebbles and boulders that remain coherent instead of disintegrating when moved, they are sedimentary *rocks*. Conditions intermediate between loose sediment and hard sedimentary rock are common; the coherence often varies greatly within a single rock mass.

Marine sedimentary rocks are far more common and widespread on land today than all other kinds of sedimentary rocks combined. This is one of those simple facts that fairly cry out for explanation and that lie at the heart of man's continuing effort to understand more fully the changing geography of the geologic past.

Fig. 31. Black layers of coal interstratified with light sandstone and exposed on Lignite Creek near Healy, Alaska. (*Photo by Bradford Washburn.*)

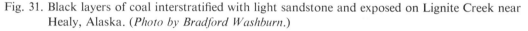

Fig. 32. Stratified sand deposited during a few years on the floor of San Gabriel Reservoir, California. The white scale is 7 inches long.

All sediments may be divided into three groups. (1) *Clastic* (or mechanical or detrital): accumulations of flakes and grains and pieces of weathered rock, such as silt, sand, and gravel. (2) *Chemical:* natural precipitates, such as rock salt and gypsum. (3) *Organic:* accumulations of organic remains, such as coal or shells. Clastic sediments are about three times as abundant as chemical and organic, and since the stages in their production and accumulation are more easily seen and photographed they will receive the major emphasis in what follows.

We will begin by looking at sedimentary transport, the middle stage in the production of sedimentary masses, because it is the most obvious, and then turn to the more subtle beginning (derivation of sediment) and end (accumulation of sediment).

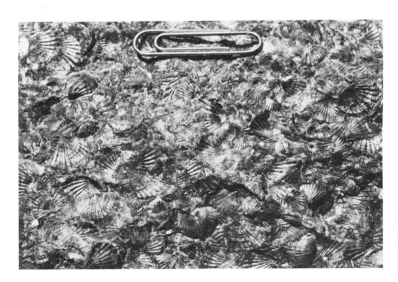

Fig. 33.
Sample of a sedimentary rock composed almost entirely of extinct marine shells (*Eocoelia* [*Coelospira*]), near Rochester, New York.

Transportation, the Middle Stage of the Process

The narrow suspension bridge shown in these views crosses the Colorado River in the deepest part of the Grand Canyon. The rocky inner gorge of the Canyon is 1,000 feet deep here and in many places the river is confined between nearly vertical walls that are only 400 feet apart at the water's edge. The muddy, swirling river seethes and churns; a pervasive, distant roar rises out of the turbulence; and the wet rocks and occasional muddy sand banks fill the damp air with a fresh earthy smell. Relentless power pervades the scene; the river knows no indecision, it always flows in the same direction, always laden with mud and sand. Its rocky walls, and the boulders it rolls along its bed during flood stages, are etched and worn by the fine particles suspended in the current. No rock in the river bed can escape this grinding fluid.

This is a river at work. Similar activity prevails throughout the lengths of tens of thousands of streams all over the world, some long, some short, some lazy, some vigorous. Multiply these by millions of years and the total amount of work accomplished is beyond comprehension—the greatest hauling job of all time, still going on with no hint of an end.

How much mud and rock is the river carrying? Where does it all come from? Where is it going? Is it even possible to find the answers to such questions?

The small encircled white object beneath and beyond the left half of the bridge in the lower view is an observation car hung from a taut cable across the chasm, just upstream from the bridge. From it, government observers make daily measurements of the depth and velocity of the river and take samples for determining its *suspended load* of clay, silt, and sand, and its *dissolved load* of salts carried in solution. Records kept for many years at hundreds of such gaging stations in all parts of the United States disclose the seasonal habits of streams and make it possible to plan the use of water and the control of reservoirs, as well as to forecast floods in time to give warning to threatened areas.

Two impressive facts emerge from the figures obtained at this Grand Canyon station. The first is that the suspended load carried by the river is prodigious. For the water year ending September 30, 1957, this load averaged more than 425,000 tons a day, or almost five tons each second. To move earth at this rate would require the continuous operation of 73 power shovels, each capable of picking up a two-ton scoop every 30 seconds, day and night. (Average figures for a 50-year span are even higher—about 550,000 tons per day, equal to the capacity of 95 such shovels.)

The second notable fact is that there is great variation in the amount of water discharged and in the load that it carries. In this same twelve months the discharge under the bridge varied from 2,870 cubic feet per second (October 1, 1956) to 124,000 cubic feet per second (June 13, 1957); the second of these rates is more than 42 times the first. (Sometimes the discharge doubles in twenty-four hours.) On October 1, the river carried a load of about 500 tons; on June 13, about 2,000,000 tons. This increase of 4,000-fold demonstrates a fundamental principle: the capacity of a stream to transport sediment increases in more than direct ratio to its discharge. For example, figures like those just cited show that on the average, each time the flow of water in the Colorado doubled, the load increased more than four times. Obviously a stream may be expected to do most of its work during floods. (This idea is examined further on page 130). Undoubtedly the largest boulders, like those in the foreground of the lower view, are moved only in periods of highest discharge.

The load carried by the Colorado River is on its way to becoming sedimentary rock. To complete the picture, we must go upstream to see where the sediment comes from, and then downstream to see where it is deposited.

Fig. 34. Aerial view of the inner gorge of the Grand Canyon of the Colorado River, looking upstream. Mouth of Bright Angel Creek at left and Kaibab suspension bridge just beyond.

Fig. 35. The Colorado River, looking upstream from the mouth of Bright Angel Creek toward the Kaibab suspension bridge.

Feeding the Streams

Where does the Colorado River pick up these millions of tons of sediment that it moves each year?

A small part is scraped up from the bed of the river by the abrasive action of the stream and its load, aided by water-induced decomposition of the rock in the streambed. (The universal effectiveness of this process is proved by the depths of the gullies, canyons, and valleys in which streams are always found.) But if this were the only source of sediment, the Grand Canyon should be about the same width at the top as at the bottom, perhaps with one wall overhanging. In reality, however, the chasm is more than ten miles from rim to rim, although the average width of the river itself is only about 500 feet.

Evidently the sides have fallen in—certainly not all at once, but bit by bit through slumping and downwash as the river gradually cut deeper. The amount of material thus contributed to the river is more than forty times as much as that scraped from its bed: most of the stream's load has come from its banks rather than its bottom.

This process of feeding streams comprises two important stages; first, the partial disintegration and decomposition of the rocks in the valley walls by *weathering*, and second, the *downslope movement* of this material under the influence of gravity. Weathering and downslope movement will be considered in greater detail later, but the scene at the right illustrates how they work. The stream is flowing in a canyon cut through almost horizontal layers of sedimentary rock. In the foreground is a heap of blocks derived from sandstone cliffs higher up the slope. The beds in the lower part of the slope, shown here, are thinner and finer grained; they are also less resistant to weathering and produce a gentler slope, which in most places is obscured by debris from above. At the left this covering debris may be seen as a chaotic accumulation of blocks of the upper sandstone embedded in decomposed rock. The tumbled positions, angular shapes, wide range of sizes, and irregular distribution of the blocks is characteristic of the mixture of *rubble* and soil, often called *colluvium*, that is found on the sides of all valleys and canyons, though of course differing combinations of climate and rock type produce variety in the details. Part of this cover (at the right) has been undercut by the current and has slumped into the stream, producing a natural exposure of the underlying bedrock and showing in several places that the colluvium is only five or six feet thick.

Geologically this scene is one of action. The blocks of sandstone are caught, as in a single frame of a motion picture, in various positions as they twist, creep, and slide down the slope. Successive photographs of the block marked X show that in five years it rotated about 30° and moved down the slope about fifteen feet. The blocks embedded in the colluvium also move, although less rapidly than this, because they are surrounded by material containing minute flaky grains of silt and clay. These very fine ingredients tend to expand and become slippery when moistened, so that with each cycle of wetting and drying, as well as of freezing and thawing, the whole permeable colluvial mass swells and contracts a little. Under the constant tug of gravity on a steep slope the material settles back a little lower each time; a net shift downward is inevitable.

In this way rock debris is removed from the valley wall and fed to the stream—and the valley widened as a consequence. The mechanism can be likened to a railroad system in which the streams, like endless freight trains, haul away the weathered material brought to both sides of their tracks by downslope movements.

Fig. 36. Slumping of right bank of Eagle River, 21 miles above its junction with the Colorado River.

The Deposition of Sediment in Quiet Water

The actual accumulation of sediment under water is relatively difficult to observe. Some ocean currents can move some sediment (page 77)—but is that the whole story? The completion of Hoover Dam in 1935 created a lake more than 80 miles long in the path of the Colorado River and provided an unusual opportunity to study what happens when a stream carrying 200,000,000 tons of sediment a year enters a body of standing water. Scientists were interested in the processes of sedimentation that would be demonstrated, and there were strong engineering and economic reasons for observing and measuring the accumulation because the life and the management of the reservoir would depend in part on what became of this load.

The photograph below shows the visible part of the process. The water from the canyon at the left is colored light tan by its suspended load, but the lake is dark and clear because, upon entering tranquil waters, the sediment begins to sink and in a mile or so is out of sight. The identity of the river is lost. Its velocity is absorbed, and the turbulence which was so apparent under the bridge in the Grand Canyon, and which is chiefly responsible for keeping the sediment in suspension, now all but disappears. Consequently most of the load settles to the bottom. Part of the *delta* thus deposited in the upper end of the lake is visible in this view.

But this is not all that happens. Soon after the lake was formed, regular measurements and sampling disclosed that once or twice a year a tongue of clay-laden water crept along the reservoir floor all the way to the dam, where its arrival was recorded in a rapid rise of the bottom, sometimes of more than 30 feet in a few weeks, followed by slow settling until the deposit was about two-thirds its original thickness. These changes quite regularly followed peaks in the suspended load of the Colorado River (as determined at the station shown in Fig. 35), and were approximately proportional to them.

These observations demonstrated something that had been predicted by engineers working with laboratory models: Sometimes, when maximum loads are delivered to the upper end of the lake, although sand and silt are deposited on the delta the finer clay (possibly joined by mud stirred up through slumping of the soft bottom) forms a cloud or tongue of suspended

Fig. 37. Muddy Colorado River water issuing from the Grand Canyon at the left and entering the upper end of Lake Mead at the right. Looking southeast.

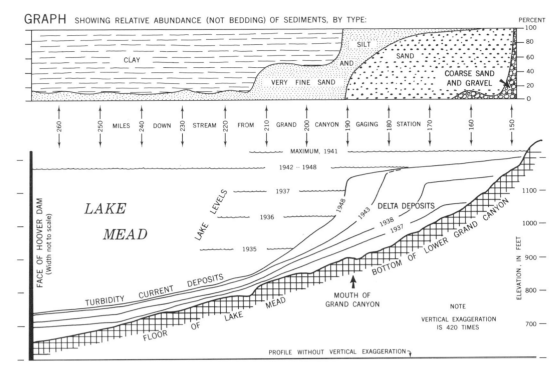

GRAPH SHOWING RELATIVE ABUNDANCE (NOT BEDDING) OF SEDIMENTS, BY TYPE:

Fig. 38. Successive profiles along the Colorado River channel during the first eleven years after Lake Mead was formed showing growth and composition of delta and turbidity-current deposits. (*Compiled from W. O. Smith et al.*, U.S. Geological Survey, Professional Paper 295, *1960*.)

matter that creeps along the bottom of the reservoir until stopped by the dam. Such an underflow of mixed water and sediment is a common form of *turbidity current*. As the name implies, there is enough turbulence within the mass to keep the mixture of sediment and water from separating, and, since it is heavier than clear water, gravity pulls it downslope. In Lake Mead such flows are several feet thick and travel a few inches per second, taking about a week to cover the 70 or more miles from the steep front of the delta to the dam—a route along which the average slope is less than 5 feet per mile. The subsidence that follows each arrival at the dam is undoubtedly a result of settling and compaction of the loose sediment.

The diagram on this page shows, with enormous vertical exaggeration, successive profiles across the delta and along the reservoir floor to the dam: the size distribution of sediments is indicated by the graph above the profiles. It is clear that the coarser sediment accumulates on the delta and that successive blankets of clay have been spread between it and the dam. (The useful life of the reservoir was estimated in centuries, however, even before Glen Canyon dam was completed upstream.)

Geologists were quick to explore this method of underwater transportation of sediment. Experiments showed that on steep slopes silt and sand and even pebbles and cobbles will move as part of a turbidity current. A growing body of evidence now indicates that this mechanism may be as important as ocean currents in distributing sediment on the sea floor.

The amount of material delivered to a stream, and the load it can carry, fluctuate seasonally and even daily. Turbid flows and underwater slumping are occasional events. And in the ocean, most near-shore currents change with the direction and intensity of the wind. Normally therefore, the delivery of sediment to its final resting place is not a steady process, and stratification, or *bedding*, is a universal characteristic of clastic sedimentary rocks. The layers, or *strata*, differing from each other in composition and grain size, are the physical record of this nonuniform accumulation.

Chemical and Organic Sedimentary Rocks

Mud (composed of silt and clay) is by far the most abundant sediment; it is the chief ingredient of about half the total known volume of sedimentary rocks. Sand is second in abundance, accounting for more than half of the rest. Coarse materials such as rubble and gravel are minor contributors, though they are important in many localities. Altogether, these siliceous clastic (Greek: *klastos*, broken) sediments account for about 80% of all sedimentary rocks. The rest are composed of a variety of sediments, chiefly chemical precipitates or of organic origin. These are *non*clastic; most of their ingredients are produced at the site of accumulation instead of being transported as fragments from other sources.

Organic and chemical sediments vary greatly in composition. *Coal* is an altered and condensed derivative of vegetation preserved in swamps. *Rock salt* and *gypsum* are formed from chemical precipitates produced when shallow lagoons or lakes become natural evaporation pans. *Limestone*, by far the most abundant nonclastic sedimentary rock on the continents consists mainly of calcium carbonate. We use the term limestone loosely here, as is frequently done, to include any solid sedimentary rock composed dominantly of calcium carbonate, whether produced through chemical precipitation, secretion by algae, or the accumulation of shell fragments and grains of older carbonate rocks. Many limestones were formed by a combination of these processes (see Fig. 57, *left*) and thus are partly clastic in origin.

Nearly all limestones are marine and most contain well-preserved fossils. Many also contain considerable amounts of calcium magnesium carbonate; if this predominates the rock is technically *dolomite*. Many limestones contain irregular nodules and thin lenses of *chert* (cryptocrystalline silica) which, being insoluble and very hard, protrude from weathered outcrops (e.g., Fig. 274). The flint extensively mined south of London for use in early firearms, and some of that used by primitive man for implements, are varieties of chert. Limestone had been much used as a building stone and is the chief source of lime as well as the main ingredient of commercial Portland cement.

The upper view here shows a typical exposure of limestone in a dry climate. Note that stratification is only moderately prominent. This is characteristic of the nonclastic sediments and follows from their mode of origin, which is relatively independent of the fluctuations in deposition that cause thin bedding in many clastic sediments.

The lower view is a close-up of the same limestone outcrop showing the homogeneous fine-grained texture and rough surface characteristically developed in fairly dry climates; the roughness results from the solvent action of rainwater (see also Fig. 229). In humid climates the solubility of limestone prevents it from even showing through the soil over large areas (Figs. 131 and 232).

There is, of course, no reason why clastic and nonclastic sediments may not be deposited more or less simultaneously in the same place. Limestones are frequently interbedded with shale or contain admixed clay, silt, and sand. Rocks representing almost every conceivable mixture of these four materials can be found in nature and are referred to as silty limestone, calcareous shale, and so on.

Fig. 39. General view of the Redwall limestone in the Grand Canyon, Arizona. Total thickness
of beds shown is about 350 feet.

Fig. 40. Closer look at an outcrop of Redwall limestone in the Grand Canyon, Arizona.
Height of view, about 5 feet.

Textures of Some Sedimentary Rocks

The most widely used and easily applied basis for distinguishing different types of sedimentary rocks is the size, shape, and composition of their constituent fragments or particles. Let us examine samples of a few common types.

Fig. 41.
Sedimentary breccia of sandstone
and limestone fragments exposed on
lower Kaibab trail, Grand Canyon.
Width of view about 5 feet.

Breccia (Italian: fragments of stone; pronounced brech'ia) is sedimentary rock composed of cemented angular rock fragments. As ordinarily used, the word implies a coarse accumulation of rubble like that shown here, typically produced as colluvium (Fig. 36), landslide debris (Fig. 306), talus (Fig. 117), or broken lava (volcanic breccia). Notice in the example shown here the chaotic unstratified arrangement of light, dark, and striped fragments from less than an inch to more than a foot across. These are embedded in a *matrix* of sand, and the whole is cemented together with calcium carbonate deposited by natural waters percolating through the accumulation.

This breccia, clinging to an alcove several hundred feet above the Colorado River in the Grand Canyon, contains fragments of most of the rocks found in the walls towering 4,000 feet above it. Breccias are always found near the source of the rubble, for transportation of the fragments tends to round off their corners and edges and to sort them according to size, resulting in a deposit more likely to produce conglomerate than breccia.

Conglomerate is the hard rock produced by cementation of gravel. Like breccias, conglomerates are typically coarse deposits (Fig. 42); most of the fragments are at least one-fourth inch across. Unlike breccias, conglomerates are composed of rounded pebbles and cobbles, and are commonly stratified. Most conglomerates include a good deal of sand, both as the matrix in which the pebbles are embedded and as associated beds and lenses of sandstone. In fact, since natural agents cannot ordinarily spread gravel over large areas, most conglomerates

Fig. 42.
Outcrop of conglomerate and
associated cross-bedded sandstone,
Puente Hills, California.

are of limited extent and grade into finer deposits nearby. In the example shown here the sandy matrix is loosely cemented with calcium carbonate.

The pebbles in a conglomerate become smooth and rounded through weathering and abrasion as they are tumbled along in the vigorous currents that are needed to move such material in quantity. Conglomerates are commonly made up of material from at least a few miles away, and some pebbles have been traced to source rocks a hundred or more miles distant.

Locating the sources of even a few of the pebble types in an ancient conglomerate often requires skillful geological detective work, but the rewards may be great. Few other rocks contain so good a record of their source and history, and this record may provide the clue to the location of such features as ancient mountains, valleys, and coasts. The special value of conglomerates in this regard is simply that pebbles are more complex and have more character than mineral grains. The more completely a sediment is broken down into its constituent mineral grains the more it looks like other sediments. Mica flakes, for example, can be derived from dozens of different kinds of rock in as many different places, but mica-bearing lava is a fairly unusual rock and pebbles of it can have come only from the vicinity of a correspondingly distinctive volcanic vent.

Because water can normally move gravel only with vigorous currents, early investigators concluded that practically all conglomerates were deposited either by swift streams or by waves along exposed coasts—the only two agents believed competent to spread pebbles and cobbles. Furthermore, since vigorous currents are characteristically variable in velocity and direction, some kind of stratification seemed inevitable.

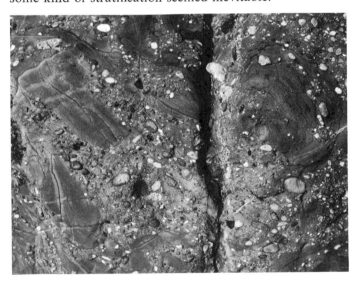

Fig. 43.
Sand-rich pebbly mudstone in sea cliff at Pigeon Point, California. Height of exposure is 15 inches. (*Photo courtesy of John C. Crowell.*)

Against this background, such completely unstratified and unsorted deposits as the one in Figure 43 pose a problem. Here, pebbles of different sizes are randomly distributed through a matrix which is itself a mixture of sizes from clay to coarse sand. Furthermore, some of the "pebbles" are chunks of sandstone that was evidently so soft that it was bent during transportation and deposition. (To distinguish these unsorted and unstratified deposits from normal conglomerates they are often referred to as *pebbly mudstones*.)

In investigating the origin of such rocks it was discovered that the strata next to many of them contain remains of marine organisms whose modern counterparts live at depths of several thousand feet, and that many pebbly mudstones and associated conglomerates contain the mingled remains of both shallow- and deep-water species.

Submarine slumping and turbidity currents provide a plausible answer to this problem. Marine pebbly mudstones are now generally believed to have been produced by flows of mixed clay, sand, pebbles, and water, which was set in motion by earthquakes or storm waves and crept and slid into deep water from accumulation sites near shore: the bent slabs of sandstone are probably pieces of partly consolidated deposits picked up from the sea floor during the descent, and the mixture of shallow and deep water fossils is a logical consequence of the whole process. Slumping and sliding on the steeper parts of the sea floor should produce the most chaotic mixtures; far-travelled turbidity flows would be more likely to lay down comparatively uniform blankets of graded sand and silt (see page 72). Probably all gradations between these two are possible.

Some nonmarine pebbly mudstones have been interpreted as glacial deposits and others as the work of mudflows.

Sandstone is cemented sand. The particles in most sandstones are dominantly quartz, the most abundant durable mineral at the earth's surface. In some regions, however, sandstones also contain conspicuous amounts of feldspar, fine rock fragments, volcanic debris, mica, or other distinctive ingredients. Viewed with appropriate magnification, almost any sandstone may be seen to contain occasional grains of durable and colorful gem-like minerals. A small amount of clay may be present as matrix between the sand-size grains. Most sandstones are cemented with some form of calcium carbonate or silica; a few extremely hard ones, usually composed of almost pure quartz sand, are cemented with silica and are sometimes referred to as quartzite. Most such cementing materials are deposited in the pore spaces of the rock by percolating groundwater (see page 241).

For most geologic purposes it is important to distinguish between sandstone derived from sand deposited under water and that derived from sand deposited on land by wind (dunes, for example). These two types record very different geologic events.

Figure 44 shows the surface of a piece of Coconino sandstone from a typical ancient eolian deposit (Latin: *Aeolus*, god of the winds) now exposed as a conspicuous gray-white cliff just below the rim of the Grand Canyon. The same rock, magnified ten times, is seen in the left half of Figure 45. (Some grains that are actually clear and glassy look dark here because polarized light was used to bring out the texture of the rock.) Notice particularly that the horizontal stratification visible in Figure 44 is produced by the sorting of fine and coarse grains into thin zones. Equally distinctive are the smooth, rounded outlines of most of the individual grains.

These features contrast sharply with those of the marine sandstone, shown in the right half of Figure 45, which was probably deposited by a turbidity current. (The origin of this bed is discussed on page 72). It represents the opposite extreme—a mixture of highly angular fragments in a wide range of sizes, some smaller and some larger than any in the eolian deposit. On a larger scale this sandstone shows a different kind of size-sorting, a gradual decrease in the size of the largest grains from bottom to top of each layer. This sorting is clearly visible in the outcrop of the bed from which this sample was taken (shown in Figure 73).

It is sometimes important to know whether a given water-land sandstone accumulated in the sea or in fresh water. The best evidence is provided by fossils buried with the sediment and now found in the rock, since the fossil remains from marine environments are readily distinguished from those of rivers and lakes.

Fig. 44.
Close view of vertical break across wind-deposited Coconino sandstone from upper wall of Grand Canyon.

Fig. 45. *Left:* Coconino sandstone under polarizing microscope, ×10. *Right:* Topanga marine
sandstone from near Duarte, California, shown at same magnification.

Commercially pure quartz sand is the principal ingredient of all glass, and well-cemented
sandstones are often used as building stones. Most of the world's petroleum is recovered from
the pore space of sandstones into which it probably migrated from associated shales (page 364).

Stratified deposits of the finest-grained clastic sediments—fine mud, silt, and clay—are
known as *shale* after compaction and cementation have hardened them into rock. Shale is the
most abundant sedimentary rock, probably accounting for about half of the total volume of
sediments on the continents and an even larger fraction of those on the sea floor.

Figure 46 shows a typical outcrop of shale. Because of its characteristically thin layers
(laminations) the rock often breaks down into flakes and platy chips on weathered surfaces.

Another shale, as seen in outcrop and under the microscope, is shown in Figure 47. The
magnification is again ten times, which permits direct comparison of grain size with the sandstones of Figure 45.

Shales are usually interbedded with minor layers of sandstone or limestone, or both. (Together, these three make up about 99% of all sedimentary rocks.) In some places where exposures are unusually extensive, as in the walls of the Grand Canyon, shale beds may be seen
grading laterally into limestone or sandstone within distances of a few hundred yards to a few
miles (e.g., Fig. 269).

Nearly all extensive shale bodies are of marine origin. A few very interesting ones, however,
accumulated in inland lakes and basins, as is shown by their fresh-water fossils and rhythmic
layering, the latter apparently related to seasonal changes in the lake and its tributary streams
(see Fig. 293).

Marine shale is probably the chief source of petroleum, which seems ultimately to be derived from microscopic marine organic remains that accumulated with the mud and were subsequently transformed by bacterial and chemical processes that may have taken millions of years.

Fig. 46.
Outcrop of Bright Angel shale
in Frenchman Mountain, southern
Nevada.

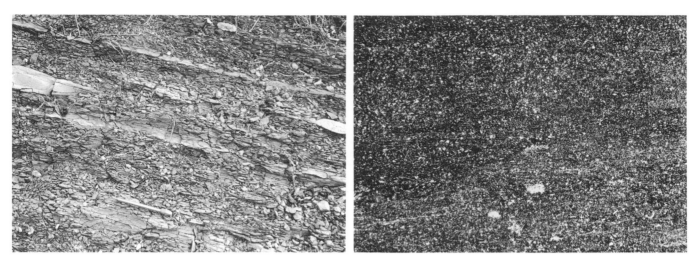

Fig. 47. Outcrop (*left*) and view through microscope, ×10, (*right*) of Holz shale,
Santa Ana Mountains, California.

Shale of certain kinds is a minor ingredient of Portland cement, and clay-rich sediments are, of course, the raw material of the world-wide and centuries-old pottery and ceramics industries.

One further and important attribute of common sedimentary rocks should be mentioned. For obvious reasons, it is almost exclusively in rocks of this class that the remains and impressions of land and sea life can easily become buried and preserved. Most of these represent, as do the samples illustrated here (Figs. 48 and 49), prehistoric shallow-water marine life. This is true even though they may be found thousands of feet above sea level and far inland from present shores—facts that have far-reaching implications regarding past positions of land and sea. Indeed fossils are not only a key to ancient geography; they are so different in rocks of different geologic age that a fair sample from one group of strata serves to establish its age relative to other fossil-bearing strata almost anywhere in the world. Such uses of fossils will be explored in Parts IV and V.

Figure 48 shows an outcrop of silty sandstone containing abundant shells of extinct relatives of living marine snails and clams; Figure 49, a layer of sandstone crowded with scallop shells and the internal casts of large smooth marine clams. Although extinct, these species are not very different from living forms.

These outcrops are unusually fossiliferous. Equally good ones can often be found in limestone or shale. Usually, however, the fossils are less concentrated and large thicknesses of many sedimentary rocks lack any such remains.

Fig. 48. Fossiliferous Williams sandstone, Santa Ana
Mountains, California.

Fig. 49. Fossiliferous unnamed sandstone near Turtle
Bay, Baja California, Mexico.

Fig. 50. Closer view of the Split Mountain roof pendant shown in Figure 22. About 24 miles south of Bishop, California.

4 THE ORIGIN OF METAMORPHIC ROCKS

The Country Rock Surrounding Stocks and Batholiths

Thus far we have examined two totally different kinds of rock; one (igneous) formed from molten material produced within the crust, the other (sedimentary) from fragmental material and chemical precipitates that were produced and accumulated on land or beneath the sea — i.e., at the earth's surface. If these were the only two kinds of rock, all sedimentary rocks should rest on igneous rock or on older sedimentary rocks, and all igneous bodies should be surrounded by sedimentary rocks or older igneous rocks. Even the most casual observations prove that this is not true—there is a third kind.

When uplift and erosion expose granite or other plutonic rocks at the surface, the surrounding rocks that are exposed in unbroken contact with them are the ones against which they cooled. All the reasons for believing that granitic rocks form within the crust (page 20) are also reasons for believing that these surrounding rocks have been at the same depths and, in fact, were there before the granite was. Furthermore, since they mark the limit of the once-molten mass, they must have been relatively solid when the plutonic magma intruded them.

In Figure 50, above, we are looking at a part of the steep east side of the Sierra Nevada in central California. This is a closer view of the roof pendant shown in Figure 22. The band of black rocks that can be followed from the left foreground around to the right as a capping on the light-colored granite below is a remnant of the country rock that surrounded the granite as it cooled and therefore merits special attention in the present context. An even closer view of the dark band and some pale streaks within it is shown on the next page.

Fig. 51. Detail of part of the Split Mountain roof pendant shown in Figure 50.

Figure 51 comprises about a thousand feet of the steep cliff a little to the left of the center of the scene in Figure 50. Across the middle, immediately above the light granite, is a zone of very dark rocks which, especially near the center of the view, has distinct lighter bands in its upper part. At the left this striped zone is bent upward and involved in some steep zig-zag folding. Above this zone, and forming the top of the cliff, there is a third rock of a uniform dark gray. Steep dark dikes (especially at the right), and almost horizontal thin white ones may be seen cutting through most of the cliff.

The dark, banded middle zone is composed of rocks unlike any others we have examined. Although the rocks are as obviously layered as most sedimentary rocks, they do not consist of grains derived from other rocks and cemented together, but are composed of solidly intergrown crystals. Yet their layers, which may differ sharply in composition and appearance (like those in Fig. 52) distinguish them from igneous rocks, which are essentially homogenous.

Furthermore, the chemical composition of some of the layers is such that they could have come only from sediments. Light-colored marble, for example, is a metamorphic rock which has virtually the same bulk composition as limestone (see Fig. 57).

Under the microscope one can see that such rocks consist of intergrown mineral grains, but well-shaped crystals are far less common than in igneous rocks; most of the grains are irregular in shape and have rather ragged outlines, producing a texture similar to that shown in Figure 53.

Fig. 52.
Detail of banding in metamorphic rock,
inner gorge of the Grand Canyon.

These facts are interpreted to mean that such rocks were once sediments. Where, as here, they are in solid, welded contact with granite (note, in Figure 51, the unevenness of the boundary surface between them) we know they were once at the same depth as granite and hence were subjected to very nearly the same temperatures and pressures. Parts of the original strata may have melted and been added to the granite, thus losing their identity completely. But the banded rocks did not melt, as shown by the relic bedding and ragged texture, both of which would have been lost if the rocks had become molten and then recrystallized from the liquid.

The essential change that has taken place, therefore, is *recrystallization in the solid state*. This has destroyed the distinction between clastic grains and cementing material, yielding a wholly crystalline rock—but one that retains recognizable features inherited from its sedimentary ancestors.

Rocks like these belong to a third great group, the *metamorphic* rocks (Greek: *meta*, implying change; *morph*, form), so named because they have been changed, without melting, as a result of having spent some time deep in the crust. Such rocks are characteristically found around and embedded in batholiths.

It is useful to distinguish two general types of metamorphism: (1) That which takes place throughout large masses of rock that are not directly associated with igneous intrusions, and involves negligible change in the bulk composition of the rock; this is *regional metamorphism*. The new texture and any new minerals are formed from ingredients available in the original rock. (2) That which takes place close enough to igneous intrusions to show special effects produced by the magma. Although these sometimes stem only from the local rise in temperature, they are likely also to include some changes in composition arising from chemical reactions between the country rock and fluid constituents of the molten mass. This type, *contact metamorphism*, is most extensive around stocks and some batholiths and plays an important part in the production of some kinds of ore deposits. As might be expected, the two types of metamorphism grade into each other. The zone of contact metamorphism around a batholith normally fades outward into regionally metamorphosed country rock, the latter commonly exhibiting finer banding and less intense recrystallization.

The dark rock immediately above the granite in Figure 50 is a product of contact metamorphism. Note that in some places, dark protrusions have been enveloped by the light-colored granite, while in others narrow dikes of granite have invaded the older rock. The youngest rock in the view is that forming the nearly vertical dark dikes which cut through granite and overlying rocks alike, probably filling cracks that developed in the course of cooling or uplift of the batholith.

Fig. 53.
Texture of dark bands in metamorphic rock
of Figure 52 as seen under microscope, ×10.

Some Effects of Metamorphism
on Sedimentary Rocks

Among the many varieties of metamorphic rock, none provide us with more information about their origin than those which, like the ones just mentioned, are recognizably derived from sedimentary rocks—the *metasediments*. Even though these rocks have made a round trip from the surface to a depth of several miles and back again, accompanied by burial and exhumation, their original texture and composition can often be estimated rather closely, thus permitting some evaluation of the degree and kind of change they have undergone.

The upper photograph shows a block of metamorphosed conglomerate with a surface cut smooth to bring out details of the texture. Many variations among the original pebbles are still discernible; some are light, some dark, some fine-grained, some coarse, some homogeneous, some streaky. A few of the light-colored ones retain fairly typical pebble shapes and sharp outlines, but the majority are a little vague around the edges, and all are flattened. Note especially how the gray ones have been squeezed and dented by their neighbors and their edges frayed and drawn out into wisps. In general, plutonic and metamorphic rocks are but little affected by a return to conditions similar to those that produced them, whereas pebbles of sedimentary and volcanic rocks, having been produced under near-surface conditions, undergo considerable change in adjusting to a metamorphic environment.

It is particularly noteworthy that the matrix is a wholly crystalline mass of intergrown mica and quartz, representing the complete recrystallization of what was probably a mixture of clay, feldspar, quartz sand, and a water soluble cement such as calcium carbonate. That this recrystallization has taken place in the solid state is proved by the present texture of the rock; had the material melted it would have become homogeneous. We can only conclude that deep in the earth the atoms composing the minerals found here are able to rearrange themselves in relatively solid material stimulated, perhaps, by energy added to the rock in the form of heat from the metamorphic environment. We will return to this theme on following pages.

The lower view is an outcrop of metamorphic rocks in the Rocky Mountains. The irregular layering leads to the suspicion that the rocks may be metasediments, and closer examination supports this interpretation. The rough-looking layers that have been etched by weathering consist almost entirely of coarse, intergrown *calcite*, the mineral name for the crystalline form of calcium carbonate, the essential ingredient of limestone (page 36; also Fig. 57). Limestone that has become coarsely crystalline through metamorphism, like this example, is called *marble* in geology; in the building-materials industry "marble" is used more loosely and often includes limestones that may be satisfactorily cut and polished.

Most limestones are not pure; they are often intimately interbedded with shale and usually contain 10% to 30% of clay and fine sand as "impurities"; these represent mud that accumulated with them on the sea floor. It is not surprising, therefore, that in this outcrop the layers of marble alternate with layers consisting chiefly of quartz, mica, and feldspar, all of which could have been derived from such mud. The original proportions of the mud and limestone cannot safely be estimated from the surviving thicknesses, however, because under the pressures accompanying even mild metamorphism, calcite becomes so highly mobile that it is readily squeezed into nearby rocks, filling spaces in and around less plastic masses.

Fig. 54. Cut face on block of metaconglomerate from Panamint Range, California.
(*Specimen collected by A. K. Baird.*)

Fig. 55. Exposure of interbedded marble and gneiss (page 48) on U.S. Highway 40 near
east base of Rocky Mountains.

Mylonitization, the Grinding of Rocks Under Very High Confining Pressure

Because it takes place deep in the crust, metamorphism is always accomplished under pressure. This is not necessarily just simple all-sided confining pressure resulting solely from the weight of the overlying rocks. The very existence at different places of exposed metasediments of different ages implies otherwise. Each such rock must (1) have originally formed at the earth's surface, (2) subsided, become deeply buried, and there undergone metamorphism, and (3) later been uplifted and exposed by erosion. Since all metasediments were not created at the same time, the vertical movements of the crust required in these steps must have differed as to when, where, and in what direction they took place. To simple compression are thus added unbalanced or distorting forces. These tend to change the shape of a body by stretching, flattening, twisting, or by combinations of these in different proportions and directions. The inevitable result is *shearing* (sliding one part of a body past another), for it is impossible to change the shape of a body without moving some of its particles past others, whether the deformation takes place along definite slip planes or by plastic flow.

Rocks sheared under compression usually develop *foliation* (Latin: *folium*, leaf), a crude layering often accompanied by the tendency to split because of the parallel orientation of flat minerals like mica. Roofing slate is a very fine-grained, foliated metamorphic rock. More common are *schist* (Latin: *schistos*, stone that cleaves easily) and *gneiss* (old Saxon mining term, pronounced "nice" in English-speaking countries) which are, respectively, medium-grained, thinly foliated (<1 mm), and coarser-grained, more thickly foliated metamorphic rocks. Both schist and gneiss are purely descriptive words, used whether the foliation is inherited from bedding or was produced by metamorphism. The rock in Figure 55 is gneiss. The third specimen from the top in Figure 56, at the right, is a schist, as is the host rock for the crystals in Figure 58.

The rocks pictured at the right demonstrate an extreme case of shearing under high confining pressure. They were collected from different parts of a zone a few thousand feet thick in a mountain range composed entirely of crystalline rocks and are here arranged to show progressively increasing effects of granulation by shear (top to bottom). In each pair, the left one is the natural surface of the rock and the right a thin-section of the same specimen seen under the microscope at tenfold magnification.

The top pair shows a normal granitic rock in which the minerals are uniformly distributed. The thin-section shows the texture to be typically granitic; note the intergrown grains in random orientation, about half of them showing one or more straight edges corresponding to crystal faces.

In the second pair the rock has slight gneissic banding (vertical here), and some crushing is evident in the thin-section. Except for a few remaining large grains, rounded off and traversed by many cracks, the rock is now medium to fine grained and the texture somewhat streaky (diagonally from upper right to lower left as shown here). Note especially the finely crushed material between the large grains; there is nothing like this in the original rock.

In the more advanced stage shown in the third pair the rock is finely foliated (parallel to the paper clip). The dark minerals are reduced to thin laminae and only a few of the light ones remain as discrete rounded grains large enough to recognize. In the thin-section it is clear that most of the rock has been ground into a pasty mass; well over half of it consists of particles too small for visual identification. The swirls of this material around the well-rounded relics of larger grains show that the latter have rolled—further proof that shearing has taken place.

The bottom pair shows a black rock with an extremely fine, almost flinty texture. The faint streakiness parallel to the paper clip is very clear in the thin-section. Only a few minute relic grains can be seen, some with thin tails trailing off between the laminae of the pasty groundmass. The rock is so smeared out that it would be almost impossible to identify if it did not grade into the less extreme varieties pictured above it.

Such rocks, the result of metamorphism by grinding along shear zones under pressure sufficient to keep the product from becoming incoherent, are *mylonites* (Greek: *mylon*, mill). Many mylonites become partly recrystallized by later metamorphism of other kinds.

Fig. 56. Suite of four specimens (*left*) and corresponding photomicrographs, ×10 (*right*) showing progressive mylonitization of granitic rock in the southeastern San Gabriel Mountains, California. (*Three of the speciments were collected by R. M. Alf.*)

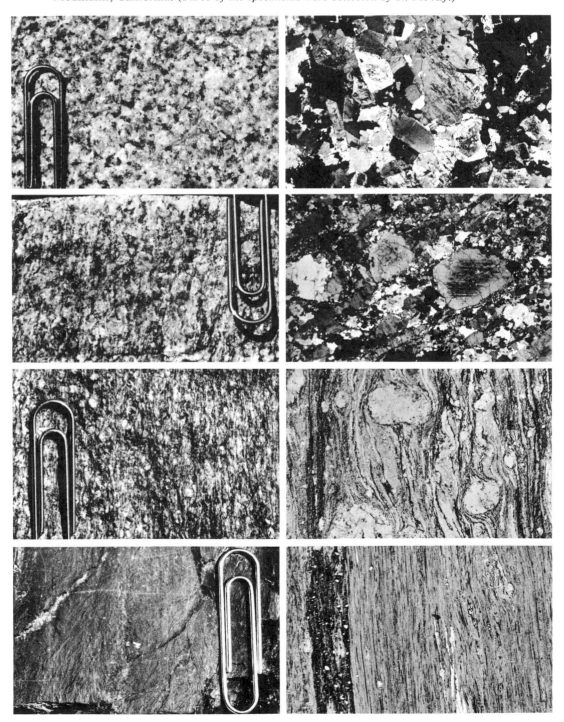

Growth of Mineral Grains in Solid Rock

From the examples on the last few pages we may conclude that earth movements carry rocks down into the crust and back up again, and that changes in texture, often accompanied by distortion, are the visible result. The most characteristic single feature of these textures, and of metamorphic rocks in general, is *recrystallization*—the enlargement of old mineral grains and growth of new ones within the unmelted rock while it is deep in the crust. (Unrecrystallized mylonites with small grains, like those in the lower part of Figure 56, are relatively rare and probably did not form at great depth.)

Like plutonic rocks, metamorphic rocks are composed of crystalline mineral grains—grains whose constituent atoms are arranged in the highly ordered patterns that distinguish crystalline from other (amorphous) substances. The "growth" of a crystalline grain requires the addition in three dimensions of layers of the proper atoms. For this to happen it is obviously necessary that the ingredients be free to move about between the grains. Everything about metamorphic rocks indicates that, deep in the crust, such mobility existed in rocks that recrystallized without becoming molten. The examples at the right illustrate some further evidence.

A typical limestone, as seen under the microscope, is shown in the left half of Figure 57. The large white grains with serrate margins and chambered interiors are fragments of shells of marine organisms. These are embedded in a very fine calcereous "mud" probably composed mostly of tiny pieces of broken shells and chemically precipitated calcium carbonate (with which sea water is very nearly saturated).

At the right in Figure 57, for comparison, is a typical recrystallized limestone, or marble, at the same magnification. The two rocks have almost identical composition, hardness, and color. The important difference between them is that in the marble the ingredients have been completely rearranged into large intergrown crystalline grains—a spectacular change in texture through recrystallization. The beginnings of such textures have been produced in the laboratory by subjecting fine-grained limestones under stress to temperatures of 500° to 800°C and pressures of 3,000 to 5,000 atmospheres for an hour or less. The experiments indicate that under natural conditions of deformation lasting centuries instead of hours, 300°C might be an adequate temperature, especially if the rock is under shearing stress. Such conditions undoubtedly exist eight or ten miles below the earth's surface.

The lower photo (Fig. 58) shows the weathered surface of a rock whose mineral composition and texture are both characteristically metamorphic. Dark brown prismatic crystals of the mineral staurolite are embedded in mica with a few small garnets and a little quartz. The mica, which constitutes nearly half the rock, includes both the dark variety (biotite) and the clear (muscovite) as closely packed thin flakes that give the finely foliated texture of a schist to the main part of the rock. This alone is good evidence of metamorphic origin, for igneous rocks seldom contain as much as 25% mica and intimate mixtures of the two kinds in the same plutonic rock are not common. Further, staurolite and certain kinds of garnet, including that found here, are so uncommon in igneous rocks that they may be considered quite reliable indicators of metamorphism. Staurolite typically occurs as well-shaped crystals embedded in mica schist derived from shale or shaly sandstone.

The rocks in Figures 57 and 58 were produced by regional metamorphism (page 45). The new minerals and enlarged grains grew in the solid rock, fed by atoms derived from the old ones and rendered mobile under metamorphic conditions. The bulk chemical composition is essentially the same after metamorphism as before. The staurolites embedded in mica, and the garnets in both, illustrate the general rule that when several ingredients are available among the mobile atoms, each growing crystal accepts from this supply only those that fit properly into its particular lattice. Thus each maintains its own composition and sometimes, as here, even its own shape.

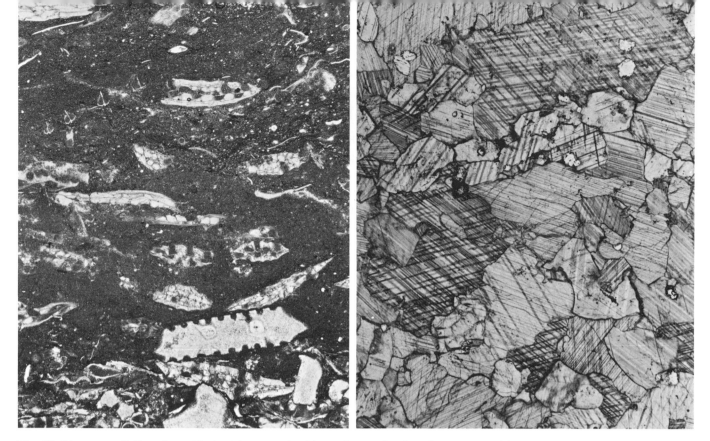

Fig. 57. Limestone (*left*) and marble (*right*) as seen in thin-section under the microscope, ×14.
(*Limestone photographed by D. B. McIntyre.*)

Fig. 58. Weathered surface of a boulder of garnet-staurolite-quartz-mica schist.
S = staurolite, G = garnet, M = mica.

Variety Among Metamorphic Rocks

A metamorphic rock can be produced from any rock that is carried into the crust deep enough to bring about changes (chiefly by recrystallization) without melting. The variety among metamorphic rocks thus stems basically from just two independent factors—the composition and texture of the original rock and the kind and degree of metamorphism to which it has been subjected.

In regional metamorphism some rocks are much more susceptible to a given metamorphic environment than others. For example, some combinations of heat and pressure that would convert limestone into marble or shale into slate would have little or no effect on schist or granite, which are already the products of more extreme conditions. What happens in any particular metamorphic environment depends largely on what rocks are put there.

On the other hand, a given rock may be converted into any of several different kinds of metamorphic rock by subjecting it to different metamorphic conditions. For example, one of the most easily traced progressions is from shale through slate, phyllite, and schist to gneiss, each of which represents a more advanced stage, or higher degree, of metamorphism (and coarser texture) than the last. Probably, other things being equal, each also represents the effects of higher temperatures and pressures than its predecessor, although greater lengths of time and increasing chemical change through mobility of rock ingredients at the atomic level may also be involved.

The same generalization applies to contact metamorphism, for when there is a chemical reaction between the country rock and magmatic fluids, the results again depend upon the composition of the surrounding rock and on the temperature and chemistry of the magmatic emanations. Sometimes the effect is drastic, converting the surrounding rocks to very different kinds for hundreds of feet from the contact; sometimes it is difficult to see any change even a few inches away from the intrusion.

It follows from these considerations that we are most likely to recognize a history of metamorphism in metasediments and metavolcanics because these rocks, produced near the surface, undergo the greatest changes when placed in a deeper environment. The metamorphism of a granite, on the other hand, if it takes place at all, may appear chiefly as streakiness arising from shearing. This may be difficult to distinguish from an original texture produced by slow currents within the batholith just before the rock became completely solid (see Fig. 30). Still more difficult to recognize is the remetamorphism of an already metamorphic rock like gneiss: the changes are subtle at best, but can often be identified, especially if the direction of shearing was different in each episode.

There is no reason why sediments that have descended, become schists, and then started toward the surface again, should thereupon cease being involved in crustal deformation. This means that we should be able to find some twice- and thrice-metamorphosed rocks as well as one-cycle metasediments and metavolcanics. We do, and one of the fascinating specialties of geology is the study of these superimposed episodes of metamorphism and the history they record; it requires careful work with microscopes specially equipped for measuring the exact orientations of individual mineral grains.

Some Perspective on Rocks in General

A few rocks have commercial value because of the ores they contain or their usefulness in construction. But all rocks have scientific value because of the information they contain and their usefulness in reconstructing the geologic events in which they have participated. This was demonstrated when we found that some of the most obvious questions about an area cannot be answered until we can tell one rock from another—especially those of different origins. Not until we can tell lava from limestone, shale from schist, or mylonite from granite, can we even begin to reconstruct the history of the mountains that shed the pebbles that make up the gravel exposed in the quarry of Figures 1 to 4.

When we contemplate the stuff of which the various rocks are made and the processes that produce them, we begin to see a kind of unending interplay in the origin of rocks. Briefly stated it is this: Sediments, and therefore sedimentary rocks, can be derived from any kind of rock exposed to weathering and erosion. Metamorphic rocks can be derived from any kind of rock that is buried deep enough in the crust to bring about changes without melting. Igneous rocks can probably be derived from the melting at depth of any common rock, sediments first being metamorphosed in the process. Almost all exposed rocks, therefore, have been derived from other rocks through transformations involving travel back and forth (during periods of tens or hundreds of million years) between the deep-seated realm of metamorphism and plutonism and the surface realm of weathering and sedimentation.

Since nearly all rocks are made out of other rocks, the chemical compositions of most fall within a remarkably small range. The chief exceptions are a few sediments like coal, limestone, quartz sandstone and rock salt, which are temporary concentrates produced by weathering and organic processes. But these are subject to later incorporation in other rocks through metamorphism or the production of magma, and in the long run the endless mixing and re-mixing tends to keep the composition of the majority, especially igneous and metamorphic rocks, close to the average composition of all.

It follows that, in broad terms, the most distinctive feature of any rock is more likely to be the imprint of the last process it went through—the particular circumstances that produced it—than the ingredients that went into it. The chemical composition, and even the minerals, of a feldspathic sandstone, a granite, and a gneiss may be almost identical; the great differences between them stem from the processes that produced them, not from their components.

This leads inevitably to a classification of rocks according to their mode of origin. To make this workable, we must attach prime importance to those particular attributes of any rock that reflect the conditions that produced it. With this in mind let us review the processes that produce rocks.

Those processes that operate at the earth's surface account for *sedimentary rocks*, which may be recognized by such distinctive features as stratification, cemented-grain textures, or their high content of water-soluble precipitates (limestone, gypsum, rock salt) or of organic material (coal and many limestones).

All other rock-producing processes must take place within the crust, below the surface. From observations in deep wells and mines, from study of volcanic eruptions, and from laboratory experiments, we know that here high temperatures and pressures are inescapable. We find it most practical to distinguish two kinds of rocks produced under these conditions.

First, those that melted, which we call *igneous rocks*. These are more homogeneous throughout greater volumes than either of the other kinds. Having passed through a liquid stage, they can retain no trace of structures or textures possessed by any ancestors.

Second, those that did not melt, which we call *metamorphic rocks*. Here we place all those rocks, whether previously igneous, sedimentary or metamorphic, that have been recognizably changed (but not melted) by conditions deep in the crust. They commonly retain traces of bedding or other structures, or show lamination or streakiness produced by shearing, or contortion from plastic deformation. Some have unusual compositions inherited from sedimentary ancestors.

Now, let us look again at the pebbles from the gravel pit (Fig. 4). This time we recognize that they are all igneous (mostly plutonic) and metamorphic types, and that they include metasediments derived from sandstone (metaquartzite, lower left corner) and limestone (marble, second from left in lower row). We are now in a position to follow again the geological trail we abandoned in a thicket of questions on page 4. Even without knowing that the metasediments occur as roof pendants in plutonic rocks, we can be sure that there were at least three stages in the development of the mountains shown in the background of Figure 1. These were: (1) The accumulation of sandstone and limestone at the earth's surface. (2) Deformation of the crust that carried these sedimentary rocks downward several miles, where they were metamorphosed. (3) Further deformation, which uplifted the metamorphosed rocks, bringing them so near the surface that subsequent erosion exposed them and some of their associated plutonic rocks. Exactly why this happened we do not know, but that it did happen is beyond doubt. These slow steps, probably requiring more than 100 million years, had to take place before there could be any pebbles to form the gravels on which the town is built.

The meager available evidence indicates that the full cycle from rock to sediment to rock again, or from rock to magma to rock again, has often taken place in a few hundred million years or less. Since the crust is estimated to be at least 4,000 million years old (page 309), it is rather unlikely that there exist today any large remnants of an "original crust" or "first rock" that we could readily distinguish from its descendants.

II Structure

The geometry of rock bodies, including the effects of bending and breaking by forces within the crust.

5 ON THE SIGNIFICANCE OF GEOLOGIC STRUCTURES

In Part I we learned where and how some common rocks are made, both upon and within the earth's crust. Among these were rocks composed of materials that originally accumulated on the surface, later became recrystallized at depths of several miles, and still later reappeared in mountains. Such changes require movement within the crust. What can be learned of these movements? Is there evidence of their magnitude, their direction, or of when they took place?

A large part of the answer is recorded in the architecture and arrangement of rocks— their *geologic structure*. This term embraces both the internal geometry of rock masses, such as stratification, and their external relations, including the kinds of contacts that separate them and whether one is found above, beside, beneath, or within another.

Few rocks are wholly undisturbed, and most of these are geologically very young. The deformation of all the rest, by uplift and subsidence, tension, compression, shearing, bending, and breaking, is caused by forces that seem to originate within the earth. The geologic structures thus produced are the best record we have of the restlessness of the crust during past hundreds of millions of years.

Obviously we need to know the original structure or anatomy of undeformed rocks in order to be able to recognize and evaluate deformation. Since volcanic eruptions and the deposition of sediment both take place at the surface of the earth, we can see most of the original structures that are built into these materials. We will begin, therefore, by examining the primary structures of volcanic and sedimentary rocks. Later we will sample the deformation many of them have subsequently undergone.

Since metamorphic and plutonic rocks are produced within the crust, many of their primary structures can only be surmised. Even if we knew them accurately it is unlikely that we would see many, for the crustal movements necessary to bring these rocks to the surface probably superimpose deformational structures on most of them. The interpretation of the resulting complexities is one of the oldest challenges in geology; it is being met with such tools as polarizing microscopes and X-ray spectrographs which can detect small changes in texture and composition, and with high-pressure–high-temperature apparatus that can simulate conditions deep in the crust.

6 ORIGINAL STRUCTURES OF VOLCANIC ROCKS

Introduction

Some volcanoes erupt mostly lava, others mostly ash. Lava that is very viscous may form a thick mass that moves only a few feet a day; lava of low viscosity may flow in a thin sheet at speeds up to 20 miles per hour. Not only do the eruptions of one volcano differ from those of another, but there may be differences between different eruptions from the same vent.

The variety thus created is preserved in such measurable quantities as the thicknesses and steepness of the flows, the presence or absence of a cinder cone, the relative amount of breccia, the proportion of ash or tuff and its distribution, and many other details. Taken together these determine the original structure of the volcanic pile. If we know how some of the variations come into being we can often use this structure to reconstruct the general type of eruption which produced ancient volcanic rocks as well as to tell something of their subsequent history.

Since the original form and structure of volcanic rock masses are determined by the kind of eruption that produced them, we may distinguish between *central eruptions*, in which the lava and ash escape from localized, often solitary, pipe-like vents, and *fissure eruptions*, in which lava wells up along one or more cracks, often several miles long, and there is usually little ash.

Central eruptions produce three common types of volcanoes. These are (1) the steep-sided *cinder cone*, many of which have lava flows emanating from the base (e.g., Fig. 5), (2) the majestic *stratovolcano*, a huge cone constructed of both lava and ash, and conforming to the popular concept of a volcano (e.g., Mt. Rainier in Washington, Mt. Hood in Oregon, Mt. Shasta in California, Fujiyama in Japan), and (3) the *volcanic dome*, a more or less bun-shaped protrusion of glassy lava, originally highly viscous, that congealed on top of its vent. Some domes are situated within an earlier crater and some are not (e.g., Fig. 67 and left foreground of Fig. 65). Cinder cones and domes are usually less than 1,000 feet high and are not likely to become well-known individually, although they are vastly more numerous than stratovolcanoes.

Among the products of fissure eruptions, two types of accumulation are easily distinguished, both formed from highly fluid lava that erupted with negligible explosiveness (hence very little ash). These are (1) the *shield volcano*, a broad, gently sloping pile of superimposed lava flows from recurring eruptions of magma that rose through more or less radially arranged cracks in the accumulation itself (e.g., the island of Hawaii), and (2) *plateau* or *flood basalts*, essentially horizontal accumulations of lava characteristically basaltic in composition (page 66), several thousand feet thick and covering tens of thousands of square miles (e.g., the Columbia River basalts of eastern Washington and adjacent parts of Oregon and Idaho, and similar outpourings in South America and western India). Plateau or flood basalts were fed through widely distributed fissures (seen today as dikes) by lava fluid enough to spread evenly over great distances. Fissure eruptions account for the greatest known volumes of erupted rock on the continents, and may be considerably more abundant on the floors of the ocean basins.

On the next few pages we will examine a cinder cone and its associated lavas, a stratovolcano, a group of domes, and a shield volcano. An area of plateau basalt will be considered later on pages 338–340.

Fig. 61. Parícutin in eruption at dawn, looking south on February 20, 1944. (*Photo by Tad Nichols.*)

Central Eruptions I:
Parícutin

Parícutin, which burst through a cornfield in Michoacán, Mexico, in 1943, is one of the best-known small volcanoes in the world. The view above (Fig. 61) shows its cinder cone on February 20, 1944, exactly one year after its birth. This 15-minute exposure at dawn records both the incandescent bombs tossed into the air and rolling down its slopes, visible best in the dark, and the general shape of the cone and its surroundings, seen best by day. The rim of the cinder cone is about 1,100 feet above the original cornfield, a height it reached in about eight months and never greatly exceeded. During its first year it erupted an estimated billion tons of ash, cinders, and bombs, mostly during the early months, plus about one-fifth this weight of lava, mostly during the last four months.

The upper view at the right (Fig. 62) shows an active lava flow emerging from the right base of the cone and advancing over older flows and ash. In this scene the volcano is 29 months old. Pine trees on the ridge at the left have been killed by falling ash.

The sequence of eruptions, in which explosive emission of ash dominated the early stages of activity, implies that compressed gases were concentrated near the top of the magma chamber. At this time bombs were observed to reach heights of 4,000 feet above the cone, probably requiring velocities of more than 375 miles per hour at the vent. Later the explosive activity

Fig. 62. Air view of Parícutin looking southeast, July 31, 1945. The light colored active lava flow
is about 3 weeks old. (*Photo by Tad Nichols.*)

Fig. 63. Cross section of 1944 lava flow exposed by storm erosion near Curupichu, east of Parícutin cone.
Note massive interior between brecciated basal and upper zones. (*Photo by Carl Fries, Jr.*)

Fig. 64. Front of lava flow advancing over newly fallen ash, east side of Parícutin cone, February, 1944. (*Photo by Tad Nichols.*)

diminished and increasing quantities of liquid lava poured from the same vent system, moving slowly out from under the pile of cinders. Observers watched whole segments of the cone being rafted away, carried on top of the lava. In a few days or weeks the breech was filled by new accumulations.

Even before reworking by rainwater the ash was stratified, owing to fluctuations in the eruptive outbursts and wind direction plus the fact that fine ash settles more slowly through the air than does coarse. As in dune sand (page 198), the inclination of the layers of fragments in the cinder cone is parallel to the outer slopes and approximately the angle of repose for such material—about 32°.

The front of the lava flow usually moved only about ten feet per hour, guided by depressions in the topography, yet it reached distances of almost three miles from its source despite an average slope of only 3°. (Internal temperatures measured a mile or more from the vent were 1,050° to 1,100°C, as determined by instruments sensitive to the glow seen in cracks between cooler blocks on the surface.)

When seen in vertical section (Fig. 63) this lava consists of solid basalt between two irregular breccia zones. The upper breccia, as we have seen, was produced when the crust formed on the cooled surface was broken up by continued movement of the still viscous interior. The same kind of crust formed and broke at the front of the slowly advancing flow (Fig. 64), and it is the over-running of the resulting blocks that accounts in large part for the basal breccia. In some places the flow advanced over ash, as in Figure 64, giving rise to a very common volcanic accumulation, lava interbedded with breccia and tuff.

Central Eruptions II:
Mount Shasta

The upper scene here is a general view of Mt. Shasta from the northwest. This majestic cone rises 10,000 feet above its immediate surroundings to an elevation of 14,162 feet—second only to Mt. Rainier among the eight Cascade Range volcanic peaks that have summits more than 10,000 feet above sea level. Shasta's slopes steepen from 5° in the foreground to almost 35° near the summit.

The volcano is so new, geologically, that erosion has not yet cut very deeply into it. The lower view shows the east side and includes some of the largest canyons—most of them less than 1,000 feet deep. To judge from the cliffs above the talus along the canyon walls (especially those in the left half of the view) the cone is composed mostly of flows lying parallel to its outer slopes. These have an average thickness of about 50 feet and are only moderately vesicular. Blocky lava, like that in the Parícutin flows, is uncommon, and dikes are rare. If these limited outcrops are representative, ash seems to be definitely subordinate to lava in making up the estimated 80 cubic miles of the cone. (The total volume erupted by Parícutin during its nine years of activity was about half a cubic mile, of which about one-third was lava and two-thirds ash.)

From these observations it is evident that the eruptions took place from a central vent and were only mildly explosive. The cone must have been more substantial than the pile of cinders at Parícutin, for it was strong enough to contain the upwelling lava until most of it erupted from the top instead of oozing out around the base. The lava must have been more viscous, too, for it is unlikely that the Parícutin flows would have stopped moving, as some of these did, on slopes of more than 20° within a few thousand feet of the vent.

These differences are fundamentally explained by chemical composition. Parícutin is composed of basalt, while Shasta is composed almost entirely of andesite, a more siliceous lava which is less fluid than basalt under similar conditions of temperature and volatile content. (A more extreme case of viscosity is discussed on the next page.)

Above the snowline on the west slope (right side of Fig. 65) is a subsidiary cone known as Shastina. Its almost uneroded shape and the fact that the lavas that issued from it flowed over deposits left by the melting of Shasta's once larger and more numerous glaciers, make it clear that Shastina was born after Shasta was fully grown. Its composition is almost identical to that of Shasta. We may speculate that a last bit of magma found the main conduit so effectively plugged by solidified andesite—rock that may have been there for thousands of years—that it detoured around the obstruction to an easier exit on the flank of the cone.

Just right of center in Figure 65 is a conspicuous group of lava flows, crossed by the railroad and ending in finger-like lobes. These have a composition more like the Parícutin flows and like them are blocky on the surface. They evidently issued from a point quite far down the northwest flank of Shasta, very late in its history—perhaps within the last few thousand years. Close study of the group will show that at least four different outpourings can be distinguished, and that the longest was probably the next to last.

The small steep-sided dome of light-colored lava in the left foreground of Figure 65 is an example of the kind of eruption described on the next page. The unusual fate of a stratovolcano similar to Shasta but farther north in the Cascades is explored on pages 314–317.

The lenticular cloud cap in the lower view deserves comment. There is a strong flow of air over the summit from left to right. The rising current cools, causing its moisture to condense at about 13,000 feet, and the process is reversed at the same level on the descending side. The base of the cloud thus marks the condensation level, below which the moisture occurs as invisible water vapor and above which it becomes visible in the form of tiny water droplets.

Fig. 65. Mt. Shasta, looking southeast up Whitney Creek. Haystack Dome in left foreground
is 500 feet high, 3,000 feet in diameter and 9.8 miles from the summit of Shasta
(which is almost 10,000 feet higher).

Fig. 66. Cloud-capped Mt. Shasta looking west southwest from 13,000 feet.
Wintun glacier and Ash Creek in center.

Central Eruptions III:
Mono "Craters," a Group of Rhyolite Domes

Stratovolcanoes are familiar to people all over the world because they are conspicuous. Cinder cones are almost equally familar because they are so numerous. Volcanic domes generally occur in intermediate sizes and are interesting because they represent a less well-known form of volcanic activity. Between central Mexico and the Aleutian Islands there are many more domes than stratovolcanoes; Lassen Peak in northern California, for example, is a composite dome and there are more than a dozen others within ten miles of it.

The Mono "Craters" are a group of 20 domes situated at the east base of the Sierra Nevada, south of Mono Lake. Several of them appear in the upper view here. The extruded rock has taken two forms. One is the true volcanic dome, exemplified by the small one surrounded by a ring of cinders at the right. The other is the steep-sided thick flow with irregular shape and rough upper surface. Three flows and six domes are evident without searching the background of this scene.

Domes and flows alike are composed of glassy rhyolite, a lava containing about 75% silica and similar to granite in bulk chemical composition. Rhyolite is the most siliceous of the common igneous rocks and on this account may be expected to be more viscous than the andesite of Shasta (65% silica) or the basalts of Parícutin (55% silica) under similar conditions. This viscosity is responsible for the great thickness of the flows (400 to more than 600 feet) and the fact that none have spread more than a few thousand feet. It is also partly responsible for the lava being more than 90% glass, for viscosity retards the growth of crystals (page 10). Not all rhyolites are so sluggish or so glassy however, which probably means that the temperature here was relatively low, perhaps around 800°C, and that chilling was rapid.

As may be seen in the outcrop shown in Figure 68, at the right, the glassy rhyolite of Mono Craters is light to dark gray. The variations depend primarily on the distribution of minute bubbles throughout the rock. The highly frothy variety is almost white (= rhyolite pumice) and the solid glass is almost black (= rhyolite obsidian). (See also Figures 11 and 12, and page 11.) Most of the rock is like the flow-banded outcrop shown here, in which pumiceous layers alternate with darker, less frothy rock. All of it is light in weight and has the characteristic squeak and tinkle of glassy clinker when pieces are disturbed by persons climbing over the lava.

Explosive eruptions of pumiceous ash evidently preceded the lava extrusions, for a blanket of such ash underlies the flows and some of the domes have clearly risen within earlier craters or explosion pits composed of similar fragmental material (as in the right foreground). After this release of gas and pumice the viscous lava ascended slowly—1 to 50 feet per day, to judge by other domes that have been observed in the making. In some the lava is almost solid and rises as a plug or spine, but here it was evidently sufficiently plastic to spread a little.

Domes and flows alike are thoroughly fractured and shattered (note cracks, Fig. 68). The sides of the flows are everywhere draped with talus, implying that the original margins were very steep. The tops are a jumble of rubble and tumbled blocks derived in part from the collapse of originally greater unevenness. Domes that have been observed in process of formation are often accompanied by explosions of gas. This is one reason for the broken rock; another is the strains that are produced in rapidly chilled glass.

One of the flows in this scene (Fig. 67) overrides deposits left by a tongue of glacial ice that issued from the Sierra Nevada (background) during the Ice Age, which proves that some of the eruptions took place after that time. Indirect evidence suggests that the domes and flow nearest the lake may be less than 10,000 years old.

Fig. 67. View south over Northern Mono Craters from above Mono Lake. The circular crater in the
right foreground (known as Panum) is about half a mile in diameter. Light snow.

Fig. 68. Vertical face of large block of banded rhyolitic glass near summit of Panum (see Fig. 67).
Width shown is about 6 feet; black rectangle at right is dark slide from 4 × 5 film holder.

Fissure Eruptions—Mauna Loa

The island of Hawaii is the southeasternmost of the Hawaiian Islands, a volcanic chain 1,600 miles long. If the water around the islands were removed they would be revealed as one of the world's major mountain ranges—a great volcanic ridge topped by volcanic peaks rising 20,000 to 30,000 feet above the ocean floor. Hawaii itself, the highest, largest, and youngest mountain in this range, is a composite volcanic edifice built from five principal vents, two of which, Mauna Loa and Kilauea, are still active. Both the rim of the summit crater of Mauna Loa (in the center of the view below) and the summit of Mauna Kea (the snow-covered peak beyond the clouds in the background) are more than 13,000 feet above sea level.

Even the small part of Mauna Loa shown here demonstrates the very gentle slopes characteristic of a shield volcano—4° to 11° instead of the 32° at Parícutin and an average of about 21° on the concave slopes of Mt. Shasta. These gentle slopes result, as we will see, from the nature of the eruptions and the low viscosity of the lava.

The summit crater of Mauna Loa is 1½ miles wide and is connected, by a narrow breach, with dark-floored South Pit, which is itself almost half a mile across. Notice the straight dark lines to the left of South Pit and the parallel lines running through the pit in the foreground; these mark some of the cracks or fissures through which lava rises during eruptions.

The other view was taken from a plane flying over the summit crater of Mauna Loa during an eruption. Note that all the glowing (light colored) lava is issuing from a long fissure which crosses the floor of the crater and extends up the wall near the breach into South Pit (upper left): it then joins the set of dark lines mentioned above. The lava fountains along the fissure are about 75 feet high.

Each fissure is evidently a nearly vertical crack which taps some kind of reservoir of molten magma at depth. The shape of the shield volcano depends on the distribution of these fissures and how they release lava. In Hawaii the pattern is roughly radial and in the early stage of an eruption it is common for a fissure to be active along several miles of its length, the activity becoming locallized at a few points in the waning stages. Activity may take place near the summit or far down the slope of the mountain. When magma hardens in the crack a dike results.

Fig. 69. Aerial view northward over the summit craters of Mauna Loa, following a light snowfall. Mauna Kea in background. (*Photo courtesy of U.S. Air Force.*)

The typical Hawaiian lava is highly fluid basalt. (The average silica content is less than 50%, well below the lavas of Parícutin.) Some Hawaiian flows are of the blocky type (known as *aa*), but many produce smooth or billowy lava surfaces (known as *pahoehoe*) resembling free-flowing tar. Some of the latter have been observed advancing several miles in a single day, some streams even at rates of more than 20 miles per hour. Temperatures range from 1,050° to more than 1,175°C, and there is relatively little associated ash.

The result, structurally, is a gently sloping dome or "shield" built up from a great many tongues and sheets of fluid lava and cut by many dikes which are often arranged in parallel groups.

Mauna Loa, which accounts for about half of the island, is enormous—far larger than Parícutin or Shasta. Its volume above sea level is about 1,700 cubic miles and if the part under water is included the figure would probably be around 10,000 cubic miles. Yet during the past 130 years Mauna Loa, erupting on the average once every three years, has added a total volume of less than one cubic mile. (Of course a cubic mile is a pretty respectable quantity; one cubic mile of lava spread as concrete would provide 6-inch paving for a four-lane highway long enough to reach around the earth about 35 times.)

Delicate instruments installed on parts of Mauna Loa have revealed that thousands of tiny earth tremors accompany the rise of magma before an eruption and that the surface of the ground around a vent may rise and fall as much as five feet because of swelling as the lava nears the surface and subsidence during its release. The volume of this "breathing" is roughly equal to the volume of lava erupted. The instruments show that the deepest tremors originate about 35 miles down, which may indicate the ultimate source of the lava. During the weeks preceding an eruption the tremors become progressively shallower, apparently as molten dikes work their way toward the surface, probably forming one or more supplementary reservoirs along the way. The actual eruption seems to come from such a puddle only two or three miles below the surface.

Thus we begin to understand *how* the eruption takes place, but we cannot yet explain *why* magma exists where it does or seeks escape when it does.

Fig. 70. Aerial view southwestward across summit crater of Mauna Loa during eruption of December 2, 1933. (*Official U.S. Navy photo.*)

7 ORIGINAL STRUCTURES OF SEDIMENTARY ROCKS

Stratification

The fluctuating nature of the processes of erosion, transportation, and deposition of clastic sediment inevitably leads to a stratified accumulation (pages 28,30). Consider, for example, the scene in the upper photograph at the right. It is only during occasional high-water stages that rivers deliver such large amounts of muddy water to the sea. Here a strong wind blowing toward us is drifting the muddy water to the left and close to shore, the fresh river water floating on the denser sea water and the fine clay particles it carries dispersing and sinking slowly, as indicated by the change in color toward the left. On another day the sediment might be carried in another direction, and wherever it settles there is always the likelihood that it may later be reworked by other currents in the ocean. The variations inherent in these possibilities, as well as in the conditions that obtain between peaks in the river flow, cause different amounts and kinds of sediment to be delivered to any one place at successive times. The result is a layered deposit.

Individual layers, commonly referred to as beds or strata (singular: stratum), are usually distinguishable from one another by differences in grain size, color, thickness, and cohesion. For example, in the marine deposits shown in the lower photograph, there is an alternation of light sandy beds and dark shaly beds of differing thickness and differing resistance to weathering. Note how some beds protrude and some are recessed, and the differences in susceptibility to gullying. A thicker sequence of well-stratified rocks may be seen in Figure 277.

One of the elusive problems connected with stratification is: How much time is represented by each layer? At present the only deposits for which this can be definitely answered are the laminated clays that form in quiet waters subject to strong seasonal influences—for example, a lake whose surface and feeding streams freeze during the winter. While the lake is thus protected from winds and river currents and cut off from fresh supplies of sediment, the finest particles have months in which to settle through undisturbed water and complete the cycle of sedimentation that will begin again the following spring or summer, when thawing of surface ice and swollen meltwater streams inaugurate a new cycle. Such annual laminations are known as *varves* (Swedish: *varv*, turn) and are illustrated in Figure 293.

For the common sediments, however, we have no accurate knowledge of how long individual beds took to accumulate or of how much time elapsed between the deposition of each. For instance, records for the past 70 years indicate that a new layer of the gravels discussed on pages 2–5 is not likely to be added in any one place oftener than every 10 to 25 years and that hundreds of years undoubtedly separate the times of deposition of many adjacent beds. In the deeper, quieter parts of the oceans sedimentation is steadier but very slow; fine clays and shells of microorganisms (e.g., *Globigerina* ooze) probably accumulate at rates of less than 1 inch per 1,000 years. Closer to the shore or in smaller bodies of water, the rate rises sharply and becomes more uneven owing to the influence of stronger currents and to mud irregularly supplied from the land; the overall average for the Black Sea is estimated to be about 20 inches per 1,000 years and for the Gulf of California about twice that amount.

70

Fig. 71. Muddy water pouring into the Pacific Ocean from the mouth of the Russian River, 60 miles northwest of San Francisco.

Fig. 72. Horizontal beds of sandstone exposed in road cut at Del Mar, California. Note imperfect parallelism of individual layers.

A Closer Look at Bedding

Since the delivery of sediment to its final resting place is usually intermittent, and can take place in different ways, we may expect some beds to contain built-in records of how they were deposited. If some kind of current sweeps the material over and along the sea floor, for example, the heavier and coarser particles will tend to be concentrated near the bottom, since the finer ones will remain in suspension longer. In turbid flows this fine material, like the dust cloud that hovers over an avalanche, may finally come to rest on top of the coarser material beneath, or it may be wafted away to another site by gentle currents too weak to disturb the material that has already settled. In any case, the particles are likely to be sorted to some degree, the larger being concentrated toward the lower part of the stratum and the finer toward the top. Such beds are said to be *graded*, and where the sorting is unmistakable and repeated it serves to distinguish the top of each bed from its bottom, a distinction that may be vitally important in the interpretation of later deformation.

The illustration at the right shows two tilted beds of sandstone. In each the gradation from coarse to fine sand shows that the original top of the accumulation is toward the upper left. This, in turn, proves that the 50° tilt of the beds must be the result of a counterclockwise rotation of that amount and not a clockwise turning of 130° from the original horizontal position. The texture of this sandstone, as seen under a microscope, was shown in the lower part of Figure 45; its angular grains of many sizes set in a muddy matrix are characteristic of turbidity current deposits. Also, the upper bed contains an indistinct slab of finer sandstone with included streaks of shale (labelled "inclusion" in the drawing), a common occurrence in turbidity current deposits and difficult to explain by any other origin. We may conclude that each of these graded sandstone beds was probably deposited by a single turbid flow, the material arriving in a matter of minutes or, at most, hours. The shale beds are accumulations of finer material that could have settled only in quiet water. The sharp contact surface between each shale and the overlying coarse sandstone must record the abrupt arrival of a new turbid flow.

Any current that can move sand can also move silt and clay—for these, once in suspension, are much more easily transported. If such a current comes to rest in quiet waters this finer material, which settles more slowly, should ultimately form the top of the deposit and we would expect the sandstone to grade upward into a shale. Indeed, in a typical deposit of this type every sandstone bed is sharply set off from the shale below it but grades into the shale above it. But here the top of the lower sandstone is also a sharp boundary, succeeded by shale less than an inch thick. This probably means that any hovering cloud of finer material that arrived with the first flow either drifted off to some other locality before it settled or was swept away by later currents that were not competent to move the sand. This in turn implies that the thin shale separating the sandstones is of quite independent origin; it must have been deposited later—perhaps much later—from suspended muds stirred up by storm waves (Fig. 176), introduced by swollen rivers (Fig. 71), or drifted over this sand layer from a turbidity current in a nearby area. Whatever its origin, the thin shale probably took longer to accumulate than did the thicker sandstone, since such fine material takes days or weeks to settle. There is no way to tell how much time elapsed between the accumulation of the beds. This interval, marked by the contact surface of no thickness between them, could easily represent more time than either of the beds—a time probably measurable in years if not centuries.

From this analysis one can understand how uneven the rate of sedimentation is. Some thick beds accumulate in a short time, some thin ones take much longer, and in all probability the periods of nondeposition that separate most layers represent far more time than is represented by the strata. As Charles Darwin pointed out over a hundred years ago, with far fewer facts to go on than we have today, from the standpoint of time the sedimentary record is very incomplete—just an entry now and then with long pauses between.

Fig. 73. Two 8-inch beds of graded marine sandstone separated by a thin parting of shale and offset
a few inches, near the top, by a small fault. For the texture of this sandstone under
the microscope, see Figure 45, *right*. North of Duarte, California.

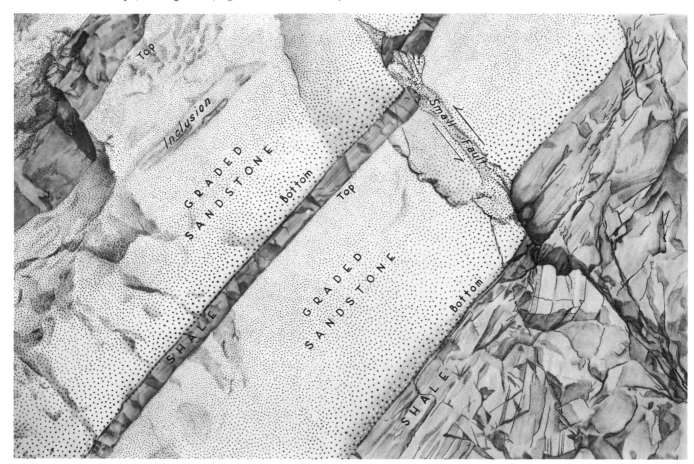

Fig. 74. Two beds of calcareous sandstone, the upper one distinctly cross-bedded.
Supai formation, Grand Canyon, Arizona.

Cross-Bedding

Currents of relatively clear water, in streams, lakes, or seas, are much more common than turbidity currents; in fact, they deliver most of the materials from which turbid flows develop. Water moving about one foot per second can roll coarse sand grains along the bottom, and if the channel is rough enough to induce strong cross currents the resulting turbulence helps to keep smaller particles in suspension, further increasing the effectiveness of the current as an agent of transportation. However, there are significant differences between the deposits of such bottom currents and those of a turbid flow.

A turbidity current is an occasional phenomenon. When it moves it transports a mixture of the available material and the resulting deposit, as we have seen, usually consists of a clay or mud matrix in which coarser particles are set in chaotic fashion or only the larger grains are sorted. Also, since the site of deposition is where the flow died, so to speak, its burden can be gently laid on even a soft bottom with remarkably little disturbance of pre-existing layers.

Water currents, on the other hand—such as those in a stream or river—often pass continuously over the same spot for long periods. The selection of grains these currents can move is largely determined by the velocity of the current; the greater the velocity, the larger and heavier the grains transported. At any given water velocity, there is a fairly definite maximum size of particle that can be moved; among grains equally accessible to the current, those smaller or lighter than this will be swept on while the larger or heavier ones are left behind. If a wide range of particle sizes is available, and the current is fluctuating, it can be a matter of delicate and shifting balance whether erosion or deposition will take place at any given location and moment. When the current subsides, the first deposit will be composed almost wholly of the larger particles, producing a relatively "clean" sand with little mud in the interstices. As the velocity decreases the grains deposited become progressively smaller. Thus, in further contrast to a turbid flow, a water current tends to sort all the particles, including the finest. The graded bedding thus made possible is usually on a small scale in nature, however, probably because most currents are too irregular in behavior to deposit thick or extensive sheets of sand in this manner.

Two original structures, commonly observed in beds of clastic sedimentary rock, are closely related to these sensitive relations between water velocity and sediment transport: *cross-bedding* and *ripple marks*. Cross-bedding is a general term for thin laminations within a stratum that are inclined to the bedding planes between which they occur. Note in the example shown at the left (Fig. 74) that the laminae run obliquely through the bed, their upper ends cut off by its top or truncated by other laminae. The lower ends may curve into tangency with the base of the bed, or be truncated there too; both conditions can be seen in this example.

Figure 75 is a close-up of cross-bedding in a thick bed of coarse sandstone; the true bedding, not shown, is horizontal. Note the thin layers of coarser and finer grains, most of them sloping about 15° down toward the left. This rock contains shallow water marine fossils which, with other features (pages 267–268), indicate that its cross-bedding is the work of nearshore currents perhaps associated with waves.

The dry stream bed in Figure 76 illustrates one origin of cross-bedding. A sediment-laden current flowed from the right into a shallow pool at the left, building a small steep-fronted delta whose natural slope is seen to the left of the hammer and whose structure is exposed on the cut face in the foreground. The inclined laminae, whose deposition build forward the steep front of the delta are called *foreset beds*, to distinguish them from the nearly horizontal layers deposited upstream (under the hammer) and downstream (on the floor of the pool) from them. Note how the foresets terminate abruptly against a finer layer near the bottom of the cut; this contact is a true bedding plane. The cross-bedding is basically a result of deposition in a depression on the floor of the stream.

In more general terms, most cross-bedding is a natural result of the alternating scour (erosion) and fill (deposition) that accompany normal fluctuations in most currents. When a stream changes course as it wanders over a flat valley floor (page 144), or consists of a shifting network of subchannels as in braided flow (Fig. 140), the sites of erosion and deposition are constantly shifting. Alternating scour and fill may also be brought about by changes in the velocity of the stream, by small shifts in the position of the main current within it, or by swirling vortices in the moving water. Any of these can produce erosion followed by deposition at the same place. The scour will cut into existing laminations, and the new fill, when the current wanes, will almost inevitably be introduced from a slightly different direction or with a slightly different velocity (and therefore size range) which will mean that its laminations will be unlike those of the old fill, and discordant to them. Cross-bedding is the result.

Because older laminae can be cut into only by younger ones, this relation is often useful in distinguishing top from bottom in deformed beds. If Figure 74 is held so that the bedding is vertical, one can still tell which stratum is the younger. Also, since most cross laminae are deposited with their slopes facing downcurrent, like the fronts of small deltas, this feature can be used to indicate current directions.

Cross-bedding is most common in sandstones, but may sometimes be seen in sandy conglomerate and occasionally in sandy limestone.

Fig. 75. Cross-bedding in coarse-grained Tapeats sandstone, Grand Canyon, Arizona.

Fig. 76. Natural (background) and artificial (foreground) exposures of foreset beds in front of small delta built into pool along a desert stream. West of Salton Sea, California.

Ripple Marks

Ripple marks are another original structure that is characteristic of sandy deposits accumulated under moving water. Although they are probably more conspicuous and familiar than cross-bedding to most people, ripple marks are actually less common in sedimentary rocks because more specialized conditions are required to form and preserve them.

Current ripples on the bottom of a swift clear stream about ten inches deep are shown at the right (Fig. 78). The water is moving from left to right, sweeping grains of sand and shining flakes of mica up the long gentle upcurrent slope of each little ridge. The crest produces an eddy which helps trap the grains on the downcurrent side, keep the small cliff steep, and hold dark leaves and twigs in the trough along its base. (Note also the scoured depressions at top and bottom of this scene; when a change in current fills these hollows with sand, the deposit will be cross-bedded.)

Ripple marks are not always so asymmetrical. The vertical section through recent Colorado River mud in Figure 77 shows some in which the downstream slopes are only slightly steeper

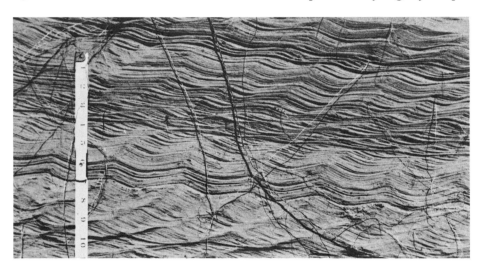

Fig. 77.
Vertical cross section of ripple marks in recently deposited silts at Dubendorff Rapids on the Colorado River. (*Photo by E. D. McKee.*)

than those facing upstream. Note here that the crests migrated slowly to the right as the deposit was built up through the addition of thin layers; both this and the difference in steepness prove that the current came from the left.

Figure 79 shows ripples exposed on a tidal flat. Some, especially in a zone just in front of the shovel, are almost perfectly symmetrical; these are commonly produced by the to-and-fro motion of water on shallow bottoms beneath waves and are therefore known as *oscillation ripples* to distinguish them from asymmetrical *current ripples*. Others, in the foreground, suggest that flow from the right dominated their final shaping, which probably took place under shallow, localized sheets of water draining toward the sea during the receding tide.

The similar ripple marks in the lower scene (Fig. 80) occur on bedding planes of sandstone that was deposited about 100 million years ago and was later tilted by the uplift of the Rocky Mountains. Their symmetry implies oscillating or very weak currents, and numerous dinosaur footprints in some of the layers prove that the water was shallow; perhaps it dried up entirely from time to time. Yet there are marine fossils in strata nearby that are in sequence with, and less than 200 feet above, these beds. The area must have been a mud flat near sea level.

No doubt the height, spacing, shape, and pattern of ripple ridges depend partly on the coarseness of the sediment and velocity of the current, but the exact nature of these and other possible relations have not been worked out. Experiments indicate that ripples are produced only within certain narrow ranges of current velocity and that some of those associated with

Fig. 78. (*left*) Actively moving asymmetric ripples about 8 inches apart on the sandy bottom of Clear Creek, west of Golden, Colorado. Current and lighting are from left to right.

Fig. 79. (*right*) Oscillation and current ripples exposed on tidal flat at Cholla Bay, upper Gulf of California (*Photo by E. D. McKee.*)

rapid flow may not survive the waning current to become part of the sedimentary record. This would help to explain their relative scarcity.

Current ripples, like cross-bedding, can be used to establish ancient drainage directions if enough are available. With care, ripples can sometimes be used to determine the top sides of deformed strata, but neither their shape nor the distribution of coarse and fine grains is a wholly reliable guide to this. At one time ripples were considered proof of shallow water, but recent photographs of the ocean floor have disclosed their existence at depths of more than 9,000 feet. They do, however, always indicate a current of some kind.

Fig. 80. Ancient ripple marks exposed on bedding planes of tilted sandstones at the east base of the Rocky Mountains, north of Morrison, Colorado. View is toward the south.

Mud Cracks and Surface Imprints

One of the most interesting and useful properties of the submicroscopic mineral flakes that constitute clay is their ability to attract and hold water molecules at their surfaces and along cleavage planes. The water thus adsorbed in a body of wet clay often accounts for more than half its weight, and of course increases the volume of the mixture: if lumps of certain kinds of clay are placed in a glass of water they will begin to swell almost immediately. By the same token, when clays dry out they contract, and the resulting shrinkage produces the characteristic cracks evident in many clay-rich soils during a dry season.

Most mud is a mixture of clay and silt, the latter being composed of particles larger than clay but smaller than very fine sand. Consequently most muds shrink when they dry, and the resulting *mud cracks* are often preserved in the sedimentary sequence.

The photograph directly below (Figure 81) shows part of the floor of a recently emptied reservoir. (For a general view of the site, see Figure 85.) This picture was taken in a slack-water area where the bottom was muddy rather than sandy. These contraction cracks have developed in the course of a few weeks. The cracks are widest at the top, where the deposit has dried out the most, and narrow downward into the progressively damper mud. Here and there are small gas vents topped by tiny mud craters resembling miniature volcanoes; they are produced by the escape of gases believed to come chiefly from decaying organic matter trapped in the mud. By now the surface is so solid that a man's weight makes no impression, but earlier one or two coyotes evidently traversed the spot. If there had been a scattering of hail or rain drops when the surface was still soft, they too would have left their record.

Fig. 81. Mudcracks and footprints on the muddy bottom of San Gabriel Reservoir a few weeks
after it had been drained for repairs to the dam. Note the many small gas vents.
The scale is 7 inches long.

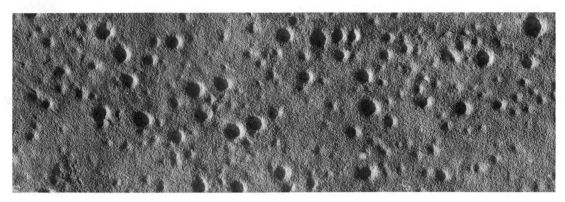

Fig. 82. Imprints left by a scattering of raindrops or hailstones on a bedding plane in the Coconino sandstone. The largest pits are about one-half inch across. (*Collected by Raymond Alf from near the south rim of the Grand Canyon.*)

These and other surface markings may be preserved if the succeeding stratum settles without destroying them, and we are most likely to see them later if the surface on which they are imprinted becomes a plane of easy separation between beds.

The pictures on this page show some of these features as preserved in sedimentary rocks that are at least 200 million years old. Figure 82 is a bedding plane in sandstone pockmarked by hail or raindrop prints—evidence of a storm that passed more than 200 million years ago. Figure 83 shows the underside of a layer that was deposited on a mud-cracked surface, the network of ridges consisting of material that filled the open cracks. At the upper left is the cast of an imprint somewhat resembling a large human hand; this was made by the right rear foot of *Chirotherium*, an extinct dinosaur-like reptile with a maximum stride of almost four feet. The "thumb" at the left is actually the fifth digit, on the outside of the foot. Note the preservation of the "slurp" pulled up in the center of the heel-mark (seen here in reverse, of course) as the reptile raised its foot out of the mud. Our meager knowledge of this animal is based solely on its footprints; no bones have yet been found.

In addition to adding colorful details to a record written long ago, these features are as useful as graded bedding and ripple marks in providing an indication of "which way was up" at the time they were deposited, and therefore in identifying the top sides of strata that have subsequently been turned on edge.

Fig. 83. Underside of layer of sandstone deposited on an ancient mud-cracked surface and bearing, at the left, a footprint of *Chirotherium*, an extinct dinosaur-like animal. Length of slab about 4 feet. (*Specimen collected from Moenkopi formation in Arizona by Raymond Alf.*)

Fig. 84. Thin-bedded sands deposited on steeply inclined schist. (Tailings deposits at an old mine.)

The Insight of Nicolaus Steno

It is self-evident that in any undisturbed sequence of strata each layer is younger than the one below it and older than the one above it. This basic geologic principle, that strata as originally deposited are successively younger upward, is known as the *Law of Superposition;* it was first clearly stated by Nicolaus Steno (1638–1686), a Danish physician who spent much of his life in Florence, Italy. In 1669 he wrote with compelling logic:

At the time when any given stratum was being formed, all the matter resting upon it was fluid, and, therefore, at the time when the lowest stratum was being formed, none of the upper strata existed.

Steno went on to offer two further observations. The first of these was that the top surfaces of all strata are originally approximately horizontal whereas their bottoms and sides may conform to the shape of the floor and confines of the basin of accumulation. In the same sentence he pointed out that this would mean that all but the lowest layers would consist of nearly horizontal sheets, and thus that strata now found in vertical or inclined positions were not originally so. This conclusion of Steno's is now known as the *Law of Original Horizontality* and means that undisturbed true bedding planes (not cross-bedding) are nearly horizontal. Actually they usually slope less than 3°.

Steno's third observation was perhaps the most penetrating. Somewhat freely recast in modern terms, it is: As originally deposited, the edges of strata are never visible; the layers either lap out against the sides of the basin or spread laterally until they thin to the vanishing point.

Hence it follows that in whatever place the bared sides [edges] of the strata are seen, either a continuation of the same strata must be sought, or another solid substance [i.e., the confines of the basin] must be found which kept the matter of the strata from dispersion.

In other words, wherever the edges of strata are visible, something has happened; not only must the disappearance of the water be accounted for in most cases, but the layers must either have been cut into by agents of erosion or have been broken by dislocations of the crust. Let

Fig. 85. Deposits of sand and mud on the floor of San Gabriel Reservoir
in Southern California. Looking upstream.

us call this Steno's *Law of Concealed Stratification*. It means that every outcrop in which the edges of strata are exposed demands an explanation. Whether it be the work of a highway crew or of the Colorado River, of a landslide or of faulting (page 94), some later event must have occurred if the edges are visible.

The two accompanying photographs demonstrate Steno's second and third observations with the help of sediments that were deposited under known conditions. In the one at the left sediment accumulated in a pond behind an earthen dam which later broke, allowing most of the sediment to be flushed out, but leaving this exposure. Note that the thin-bedded sediments display marked horizontal stratification despite the uneven surface of the underlying schist on which they were deposited.

The view above is upstream along a steep-walled canyon blocked by a dam built about 1930 and located out of sight at the left. The maximum height of the resulting lake is clearly shown by the levels at which waves have stripped the vegetation and terraced the canyon sides. When this photograph was taken, more than 25 years after the dam was built, the reservoir had been emptied and the stream was cutting into its own deposits. In the banks of the winding watercourse we can see that the strata are horizontal right up to the canyon walls and that except where the stream was cut into them there is no place where their edges show. As originally deposited, only the top surface of the youngest layer is visible. The canyon was being filled this way:

not this way:

The astute observations of Steno on some of the original structures of sedimentary rocks form a basis for interpreting their subsequent deformation—a key to the problems of tilted, folded, inverted, and repeated strata, and axioms that focus attention on the events responsible for exposing them. An example of their use is given in Part IV.

8 DEFORMATION OF SEDIMENTARY ROCKS BY FLEXURE

Introduction

Sedimentary rocks are of unique importance in geology for two basic reasons.

First, they exhibit greater variation in composition and texture than other rocks, and the origin and significance of these differences is more fully understood than it is in most other rocks. Their materials may be derived from any pre-existing rocks—igneous, metamorphic, or another, older sedimentary rock—or from several varieties of each in any combination. This derivation can be highly selective, resulting in sediment composed entirely of one mineral, as in rock salt, gypsum, pure limestone, or pure quartz sandstone. Or it may produce mixtures of great complexity that include igneous minerals, organic remains, and water-soluble salts, both as discrete grains and as fragments of other rocks. These materials may be transported by a variety of agents, such as wind, glaciers, rivers, ocean currents, and landslides. They may accumulate in many different environments, some of the most obvious being bogs, lakes, river valleys, desert basins, the land beneath melting ice, and the floors of seas and oceans.

Each of the three factors in the origin of a clastic sediment—its source, the transporting agent, and the environment of accumulation—is likely to leave some imprint on the resulting sedimentary rock. The important thing is that these factors are accessible for study, so that despite the complexities involved we usually can do a better job of narrowing down the possible conditions of origin of a given sedimentary rock than we can for rocks belonging to the other two classes. Add to this the special value of fossils, which occur almost exclusively in sedimentary rocks, and these rocks are indeed without equal as records of prehistoric conditions at the earth's surface.

(This is not to belittle the value of igneous and metamorphic rocks. They bear evidence regarding a different part of the crust, and we have already pointed out some of the methods used in their interpretation.)

Second, whether on land or under water, sedimentary rocks always form on the earth's surface, in layers that are approximately horizontal, and often with a built-in record of which side of each stratum was the original top. These geometrical attributes, shared with the much less common lavas and ashfalls, are uniquely important. The fact that these rocks were originally nearly flat and level means that when they are found tilted, folded, or broken they afford a graphic picture of what has happened. Sedimentary strata are unexcelled as recorders of deformation in the outer crust.

Sedimentary rocks are thus unusually rich in the two kinds of information the geologist looks for in a rock—a record of the conditions under which it formed and a record of what has happened to it since. For these reasons, the use of sedimentary rocks in reading the autobiography of the outer crust will be emphasized throughout the rest of this book.

We will begin by examining some relatively simple examples of deformation, first those involving bending of strata and then those involving breaking and dislocation.

Fig. 86. The Colorado Plateaus east of Hurricane, Utah. Looking northeast.

Monocline

The opening of the Far West in the second half of the nineteenth century was a stimulating epoch in American geology. The lure of gold in California and silver in what was later to become Nevada, combined with the promise of open land and growing commerce on the west coast, helped set in motion one of the world's great migrations. In order to learn more about this vast country west of the Rocky Mountains the government organized a number of expeditions to gather facts about its natural resources, the navigability of rivers, and possible routes for railroads, and to determine accurately the locations and sizes of mountain ranges and fertile valleys.

Grove Karl Gilbert was one of the ablest of the hardy company of geologists who, with biologists and engineers and protective military escorts, participated in these remarkably fruitful journeys. From this little-known western territory they brought back a wealth of scientific data as well as vivid accounts and sketches of primitive life and scenery. These records played an important role in the later establishment of some of our most-visited national parks and monuments.

Gilbert was both a keen observer and a lucid writer. In 1876, describing the vast area drained by the Colorado River and its major tributaries, such as the Green and San Juan rivers, he wrote:

From the western edge of the Plains to the Pacific Ocean the characteristic features are mountains. The strata are bent and broken and upturned at all angles. The typical structures are structures of displacement. Within this region of great disturbance is a restricted area of comparative calm. Dislocations of strata . . . are less frequent, less profound, and less complex than in the surrounding region . . . Its mountains are few and scattered, and its typical topographic form is the table or plateau. It is called the Colorado Plateau Province.

Figure 86, above, shows a typical sample of this province, which occupies more than 140,000 square miles (about the same area as Montana) centered near the lower San Juan River, about 150 miles northeast of the Grand Canyon. It is everywhere characterized by tablelands standing at different heights (hence the common name "Colorado Plateaus"), most of them more than a mile above sea level, and by deep and steep-sided canyons.

As visitors to such places as the Grand Canyon, Monument Valley, Zion and Bryce Canyons, or Mesa Verde well know, sparse vegetation and colorful rocks make it possible to see

Fig. 87. Looking south along the monocline two miles east of Mexican Hat on the San Juan River (foreground) in southeastern Utah. Note the horizontal strata at left and right, on either side of the flexure.

the geologic structure over great distances. The observer concludes almost at first glance that the plateaus are probably flat and level because they are composed of horizontal layers of resistant rock. (For other scenes in this province, see Figures 238, 259, 286.) Under these circumstances the few departures from this predominating horizontality are especially conspicuous. The three most notable structures that disturb the "comparative calm" of the plateaus are isolated volcanoes and related intrusions (e.g., Figs. 5, 19, and 20), small vertical breaks or faults, and monoclines (e.g., Figs. 87 and 88). The most distinctive are the monoclines.

As may be seen in these views of the northern Colorado Plateaus, a *monocline* is "a double flexure connecting strata at one level with the same strata at another level" (Gilbert's words). One of the longest and best known is the "Waterpocket Fold," part of which is seen in Figure 88. The extensive horizontal layers at the left are bent down along the monocline and pass beneath younger horizontal strata at the right. We must agree with Steno that the flexed layers did not accumulate with this bend in them and that the exposed edges call upon us to look for their continuation or account for the lack of it. In this case the missing portions have been eroded from their position on top of the horizontal beds to the left of the monocline. By measuring the thickness of the strata between the oldest flat-lying bed exposed at the left and the youngest flat-lying bed at the right we can determine the thickness of the strata thus removed—here, about 6,300 feet. Then if we imagine these beds restored to the left half of the scene, as in the drawing, and measure the elevation of any single *stratigraphic horizon* (such as the bedding surface between two distinctive layers) where it occurs among the horizontal strata at the left and again at the right, this difference in elevation is the amount of deformation—the amount by which the plateau at the left has been raised along the monocline relative to the plateau on the right. This displacement here is close to 7,000 feet. (If erosion had reduced the plateau on the left to the level of that on the right, the thickness of the strata eroded would also be 7,000 feet.) Judging by present-day measurable movements of the crust the deformation of the strata probably occurred in very small increments, totaling perhaps a few inches to possibly a few feet per century. Whatever the rate of deformation, erosion would start to work on the uplifted rocks long before they reached maximum elevation and we can be sure the scene never looked like that suggested in the background of the drawing. But putting the eroded strata all back in this way does dramatize the magnitude of the total job done by erosion.

Here then is compelling testimony, in earth's own handwriting, that the crust has been active. Two adjacent parts of it have moved vertically more than a mile relative to each other and a similar thickness of rock has been eroded from the uplifted block. There are dozens of such flexures within the Colorado Plateaus Province, with lengths up to 150 miles and displacements up to 14,000 feet—yet Gilbert termed this an area of "comparative calm"!

Fig. 88. Aerial view, looking northwest along the Waterpocket monocline in southern Utah. Dirt road in left foreground gives scale. In the drawing, an imaginary cut has been introduced to show the subsurface structure, and the eroded beds have been restored in the background.

Anticline

Strata are almost never absolutely horizontal. If undeformed, they generally slope very gently downstream or toward deep water: if deformed, they may slope any amount in any direction.

To be at all precise in discussing sloping strata one must indicate how steeply they are inclined and in what direction. The steepness is referred to as the *dip* and is measured in degrees down from horizontal. The direction is expressed in terms of the *strike*, which is the direction of the imaginary hingeline along which the inclined layers may be visualized as having been bent down; the strike is always exactly horizontal and at right angles to the direction of dip. For example, the steepness and direction of the dip would be traced out by a drop of water allowed to trickle down the inclined bed, and the strike would be a line drawn on the bed at right angles to this trace.

The view of the monocline in Figure 88 on the previous page is along its strike, which is somewhat sinuous but approximately northwest, and the maximum dip is about 45° toward the northeast (right). At any one good exposure of the bedding planes the direction of the strike and amount of dip could be measured within a few degrees, and a large number of readings intelligently spaced throughout the area and plotted on a map would provide an accurate and quantitative representation of the structure.

Because they involve accurate plotting of direction in the horizontal plane (*azimuth*), dip-and-strike symbols can properly be used only on maps; however, a few have been added to the drawing at the right as though they were painted on level ground, in order to demonstrate their relation to structure. (Note that they are independent of the slope of the ground.) In the foreground corners of this scene the strata on the right dip to the right and those on the left dip to the left. Comparing these two areas in greater detail, it is evident that there is a symmetry in the sequence of strata; from the central ridge outward in either direction one encounters similar groups of weak and resistant layers. Traced toward the middle distance these belts of *strike ridges* join in a succession of loops, like nested hairpins, establishing beyond doubt that they involve but one sequence of strata.

A local up-arching of layered rocks such as that shown here is the most common form of *anticline* (from Greek roots meaning opposed inclinations or dips). The name alludes to the fact that the two *flanks* or *limbs* of the fold dip away from each other; this is clearly shown in the wall of the trench in the drawing at the right.

The *axis* of an anticline is the *direction* of an imaginary line drawn on the surface of a single stratum, parallel to the length of the fold. It is most easily visualized in a position along the crest of the fold on a given bed. In looking along the axis one sees the fold from the end, i.e., in cross section. The axes of the folds in a curtain are vertical; those of the upfolds in a wrinkled rug are horizontal, or gently bent down at the ends where the wrinkles die out. The axes of most geological folds are neither horizontal nor vertical but, as at the right, inclined. The angle of descent, or dip, of its axis is the *plunge* of a fold.

The slightest amount of plunge causes the strikes of the two flanks of an anticline to converge toward the plunging end. In the example at the right, we are looking along the axis and in the direction of the gentle plunge of the near part. Erosion has removed the crest of the arch in the foreground and since the ground is fairly level, the outcrop pattern shows the convergence of the strike ridges along each flank and the loops where successive strata plunge underground along the central part of the fold.

Anticlines are most spectacular when, like this one, they have had their anatomy exposed by erosion. Here it is easy to see that the oldest rocks must be at the core and the youngest on the outer flanks; in fact this is another way of defining an anticline, and a better way of recognizing them if they have been turned on their sides, on end, or upside down.

Fig. 89. Aerial view northward along the axis of the Virgin anticline in southwestern Utah. Note highway and buildings at left for scale. In the drawing, an imaginary trench has been cut to show the subsurface structure.

Structural Domes

The oval pattern in the upper scene at the right is the result of a local uplift that produced outward dips on all sides—a *structural dome*. Erosion has removed its top, affording a better view of its anatomy than we would have if it were a smooth grassy hill. Each of the concentric strike ridges is formed by one or more relatively resistant beds; the troughs between them follow weaker rocks. To this extent the topography reflects the structure; but note that the center of the dome, where we know the uplift was greatest, is lower than many of the flanking ridges. The same was true of the anticline on the previous page. In both cases erosion, in cutting into the structure, has found the rocks near the core less resistant than some of those stratigraphically higher, and so has hollowed out a valley in the central portion—a topographic low superimposed on the structural high. The erosion has modified the topographic expression of the fold, but not its internal structure; the dips and strikes of the rocks (which refer to the slope of the bedding planes and not the topography) have not been changed by this surface sculpture, and by carefully projecting them upward over the dome we can quite accurately reconstruct its whole shape.

Domes come in many shapes and sizes, although elongate ones are most common. Long and narrow ones are sometimes referred to as doubly-plunging anticlines, which indeed they are. Some are rather flat-topped, like the one in the right background of Figure 91; some have moderately pointed summits, like the one in the center of the same view. This last, however, is only the core of a much larger dome—note the outward dips at the edges of the picture. The fact that the broad lowland drained by the river in the foreground is uniformly bounded on both sides by ridges of outward-dipping resistant strata indicates that it is underlain by a sequence of weak layers that lie between the two sets of resistant ones whose exposed edges form the ridges near the center and margins of the view. We may therefore be quite sure that this lowland is underlain by strata whose dips conform with the rest, and that the domal structure extends the full width of the foreground of the photograph.

A useful term in dealing with geological shapes is *relief*, denoting the difference in elevation between the highest and lowest points in a specified area. California has a relief of 14,775 feet (Mt. Whitney to Death Valley); the scene in Figure 90 less than 500 feet. Steep, rugged mountains have high relief; low, rolling hills and plains have low relief. As thus used the term is understood to mean topographic relief. By extension of the same concept, *structural relief* may be used to denote the relief that would be produced by a given structure if no erosion had taken place. Structural relief is based upon the shape of the surface of a representative layer (i.e., a stratigraphic horizon) involved in the deformation. To know this, of course, it is necessary either to restore (hypothetically) the eroded part above ground or to deduce or probe (as by drilling) the concealed part underground, or both, with the objective of defining the exact shape produced by deformation alone. Obviously the structural relief of the anticline and domes illustrated here is considerably greater than their topographic relief. Where the eroded rocks have been restored in the background of the drawing in Figure 88, the structural relief is the relief of the imaginary land surface; in the foreground the structural relief is the same, but the topographic relief is much less.

In thinking about anticlines, domes, and other varieties of deformed strata it is important to visualize the structure in three dimensions and entirely apart from the effects of erosion. The terms used to describe folded strata, for example, must be valid regardless of how much the rocks have been cut into by erosion. Such sculpturing, after all, does not change the anatomy of deformed strata any more than carving a piece of wood alters its grain. In both cases careful examination of the surface pattern (which does change in accord with the sculpturing) gives insight into the internal structure.

Fig. 90. Dome in strata 6 miles east of Rawlins, Wyoming. Highway and railroad at right
give scale. View is toward the southwest.

Fig. 91. Domes northwest of Riverton, Wyoming. Roads and earth dams give scale. The snowy
peak in the left background (Grand Teton) is exactly 100 miles west of the center
of the dome in the middle of the photograph.

Syncline

A *syncline* is a downward fold, the opposite of an anticline. The word is concocted from the Greek and means "inclined together," in reference to the fact that the sides, or limbs, dip toward the central axis. Like other folds, synclines are most easily recognized and measured when they occur in sedimentary rocks that have been cut into by erosion; the best-exposed synclines are therefore found in mountains where there has been both wrinkling and uplift of the upper crust.

The view at the right is along the axis of a small syncline. The bottom, or trough, of the fold is evident in the cliff in the background. By noting, with the aid of the drawing, the dips shown in small outcrops in the foreground, this structural trough can be traced forward almost to the point from which the picture was taken, thus delineating a segment of the total length of the fold.

In simple synclines such as this, the youngest rocks are cradled along the center of the fold and successively older strata are found farther out—to the right and left in this view. This is the reverse of the comparable relationship in anticlines.

Anticlines and synclines are said to be symmetrical if opposite sides have approximately equal dips, asymmetrical if one limb is steeper than the other. The syncline in Figure 93 is symmetrical; the anticline in Figure 89 is asymmetrical, the left limb being steeper than the right—a fact shown by the figures (giving amount of dip, in degrees) beside the dip-and-strike symbols, and represented in the shape of the fold in the imaginary trench across the drawing.

Anticlines and synclines commonly occur together in elongate groups, like belts of wrinkles in the earth's outermost skin—which indeed they are. (Part of such a belt is shown in Figure 373). Although the examples presented here are gentle flexures in otherwise horizontal strata, the terms can be applied to folds of any size in almost any kind of rock and in any position. If it can be determined in which direction the rocks are progressively younger, anticlines can be distinguished from synclines regardless of their orientation—even if they are upside down (e.g., pages 396–398).

As might be expected, the development of anticlines and synclines is usually accompanied by a variety of mechanical side effects; there is tension (stretching) on the outside of the bend and compression (squeezing) on the inside; beds tend to slip past each other in the limbs, like cards within a bent pack; some layers may break, others may flow plastically. For such reasons, the details of even a simple fold are likely to be complex. Note, for example, the details of some of the layers in the photograph below.

Fig. 92.
Folded strata near Borah Peak in central Idaho; looking northwest. Pine trees in lower part give scale.

Fig. 93. Looking eastward along the axis of the Barstow syncline in the Mojave Desert of southern
California. The drawing helps to clarify the dips in the weaker rocks near the foreground.

Fig. 94. Looking northwest over joint pattern in massive sandstone on flank of gentle Salt Valley anticline north of Moab, Utah. Cracks are about 50 feet apart.

9 DEFORMATION OF SEDIMENTARY ROCKS BY FRACTURE

Joints

There are three fundamentally different ways in which rocks yield to deforming movements in the earth's crust. One is by *plastic flow*, which was discussed and illustrated in the section on metamorphic rocks (pages 52–53). Plastic flow occurs where the confining pressures are high enough and/or the strength of the rocks low enough to permit slow internal flowage, analogous to the behavior of toothpaste when being squeezed out of a tube. A second way is by *flexure*, which includes the relatively simple bends we have just been examining—monoclines, anticlines, and synclines, domes and some downwarped basins—in which the rocks are not so thoroughly distorted as in plastic flow and the strata retain approximately their original thicknesses and degree of parallelism. A third way is by *fracture*, with or without movement along the cracks. If there is no detectable slipping of the walls past each other, the crack is referred to as a *joint;* if there has been such displacement it is called a *fault.*

These different ways of accommodating movements in the crust often combine or overlap. In tightly folded strata, the steep limbs are often thinned and the crests and troughs thickened because rock has flowed from regions of high confining pressure toward those of lower pressure. Joints, too, usually accompany folding, especially if the folded rocks include some brittle types.

Joints are found in all kinds of rock and in nearly all outcrops. (Figure 59 shows them in metamorphic rock and Figure 114 shows them in plutonic rock.) Usually they occur in sets of approximately parallel cracks, and often two or more sets running in different directions divide the rock mass into irregular blocks. Such a combination of intersecting joint sets is a *joint system.* The remarkable regularity of these over large areas, sometimes many hundreds of square miles, implies some reason for their system-wide orientation, but in only a few areas has this been worked out with reasonable certainty. Probably most joint systems include sets

Fig. 95. Evolution of pinnacles and "monuments." Looking west on the north side of the Colorado River near its junction with the Green River. Dirt roads on the shelf topping the pinnacles give scale.

made at different times. In sedimentary rocks the earliest set may have developed while the rock was still being buried and compressed by the weight of overlying layers. The latest set may have been produced by accompanying unloading when erosion exposed the strata. Between these episodes, deformational tension and shearing or twisting may have produced still other sets. (Relatively simple systems, produced by contraction, were illustrated on pages 24–25.)

Since it is not easy to distinguish joint sets either by age or by origin, the examples shown in the accompanying photographs are hard to interpret. In the lefthand one the fractures in the massive light-colored sandstone apparently have little effect on the overlying dark shale (right foreground), probably because the shale is less brittle. Since the fractures are parallel to a large gentle anticline (central axis along valley at left), perhaps they reflect simple tension. But if so, why are there two sets? And why do the cracks curve? It is difficult to pick the right answer from many possibilities.

Whatever their origin, joints in rocks at or near the ground surface provide channels for the deeper penetration of air and water and thus contribute importantly to the weathering of the rock. The tumbled blocks below most cliffs and steep slopes are constant reminders of this process. A variation on this theme is illustrated in the photograph on this page. The white sandstone forming the prominent shelf is strongly jointed, evidently in at least two directions. Although the less resistant dark beds beneath may have fractured in the same pattern, it is more likely that they have been weakened in this pattern by water percolating downward through the cracks in the overlying white layer. When first carved out, each pinnacle wears a joint-trimmed white cap to whose shape and protection it owes its existence. Later this slab falls off and the spire slowly shrinks as individual grains and small pieces of the dark rock are dislodged by pelting rains.

Joints may also be important below the surface. Dikes sometimes occur in "swarms" whose patterns fit expectable cracks—like those around Ship Rock (Fig. 19). Other swarms comprising hundreds of individual dikes are known in Montana, Colorado, and California (Fig. 102). These strongly suggest that many dikes are formed when underground joints are expanded by injected igneous rock.

93

High-Angle Dip-Slip Fault

Earth fractures along which there has been demonstrable slipping of one side past the other are known as *faults*. Geologists often speak of the surface along which the rock masses rubbed against each other as the fault "plane," but it is usually a warped surface, curving gently as it is traced over the ground or followed downward in mines or drill holes. Furthermore, the fracture often has measurable thickness, being a zone of crushed rock from a fraction of an inch to many feet wide.

Some faults can be followed for hundreds of miles (e.g., page 96); others can be traced only a few feet along the surface. Less is known about the depths to which they extend. Faulting is found in the deepest mines and wells, and the fact that most earthquakes originate at depths of several tens of miles certainly means that sudden movements of some kind, such as slipping on faults or changes in shape or volume, also occur at these depths.

Faults of all sizes are found in all possible orientations from horizontal to vertical, and strike in every compass direction. Added to this variety are possibilities of relative movement in any direction on each fault and of movement in different directions at different times. The resulting number of possible combinations makes description cumbersome and some kind of classification essential. But in practice even a classification limited to geometric relations is difficult to use because, for most faults, it is impossible to determine all the variables accurately. Some of the reasons for this will emerge from the examples that follow.

To begin with, let us divide all faults according to whether the fault plane has a steep dip (more than 45°) or a gentle dip (less than 45°). These are often spoken of as high-angle and low-angle faults, respectively. Then let us further differentiate, on the basis of the displacement of rocks that can be matched across the fault, between movement dominantly up or down the dip and movement dominantly along the strike, or, more briefly, between dip-slip and strike-slip faults. It must be emphasized that these are arbitrary and oversimplified categories; for example, the movement on many faults has been obliquely up or down the fault surface.

The view below (Fig. 96) shows part of a cliff more than 100 miles long that owes its existence to movement on a steep fault along its base. (This cliff is an edge of the Colorado Plateaus,

Fig. 96. Looking north along Hurricane Ledge at the Arizona-Utah line. Along the Hurricane fault, which follows the base of the 1,000-foot cliff, the rocks on the left have moved down relative to those on the right by about 4,600 feet.

Fig. 97. Looking northeastward at the same scene from a lower altitude. The dark mesa at the upper right, marked 15 in the drawing, also appears at the upper right in Figure 96.

whose flat surface and horizontal strata on the right abruptly give way, across the fault, to a variety of tilted younger rocks on the left.) The sinuous trace of this fault almost precludes any horizontal component of movement—the two sides are keyed together and could only slip past each other approximately parallel to the dip. Similarly, on a smaller scale, in the few places where the nearly vertical fault surface is exposed, grooves and flutings run straight up and down it, showing that at least the latest movement has been in the direction of the dip. There can be little doubt that this is a high-angle dip-slip fault.

Part of the same cliff viewed from a lower altitude is shown in Figure 97. Here a measure of the amount of vertical displacement along the fault can be obtained by matching strata on the two sides. For example, the beds (marked 15) topping the dark mesa beyond the fault at the far right are identical to those on the sloping hilltop in the foreground at the far left, which is well on this side of the fault; this is substantiated by the fact that the sequence of lighter layers (14) beneath them is also identical in each place. On the far side of the fault this group of dark layers (15) is horizontal and stands about 5,600 feet above sea level. On the near side it dips gently toward the northeast (upper right) and is overlain by strata up to and including those marked 16, which are exposed in the left center of the figure and in the small hill which lies directly against the fault. (This hill also appears in the left foreground of Figure 86.) By measuring the thickness of the strata between 15 and 16 at the left we can predict the depth at which the dark layer (15) lies beneath the outcrop of 16 in this small hill: this is about 2,800 feet. Since the summit of the small hill is 3,800 feet above sea level, the dark layer must here be at an elevation of about 1,000 feet. Now, if the same layer lies at 1,000 feet on this side of the fault and at 5,600 feet on the other side, the *stratigraphic displacement*, or separation, of the parts of this layer by movement on the fault is 4,600 feet. The lavas (18) are offset less—only about 1,000 feet. They are younger, having flowed across the fault long after its inception, and have participated only in its later displacements.

Farther north the total displacement increases to more than twice 4,600 feet, while to the south it gradually diminishes to zero in a distance of about 100 miles.   *95*

Fig. 98. Air view looking southeastward along the trace of the San Andreas fault 12 miles west of Taft, California. Length of trace shown is 3 miles.

High-Angle Strike-Slip Fault

Probably the best known geological feature in California is the San Andreas fault. A small portion of this great fracture that cuts obliquely across more than 600 miles of the state is shown in these two views. No other fault shows such consistent and well-exposed evidence of geologically recent activity over so great a distance, and few other faults have been so intensively studied and so widely publicized. It was sudden slippage of at least 16 feet along part of the San Andreas fault which produced the earthquake that wrecked much of the city of San Francisco on April 18, 1906.

Throughout most of its exposed length, from north of San Francisco to near the Gulf of California, the trace of the San Andreas fault is clearly apparent—especially from the air. Almost everywhere it is marked by abnormal drainage lines and anomalous ridges and trenches. As may be seen in Figure 98, the remarkable alignment of these features produces a conspicuous crease-like line across otherwise smooth slopes and valley floors.

What has been the kind of movement along this fracture? How far have the sides moved relative to each other? How old is it? Studies along the fault, most of them made since 1906, have provided partial answers to these questions.

1. The shift at the time of the San Francisco earthquake was chiefly horizontal, the northeast side moving southeast relative to the southwest side. (This is conveniently referred to as *right lateral* displacement because when one looks across the fault from either side the opposite side has moved to the right.)

2. Repeated redeterminations of the positions of many surveyed points on either side of the San Andreas fault during the last 100 years seem increasingly to demonstrate that the

coastal part of California is slowly drifting northwest relative to the country northeast of the fault. This creeping distortion, which takes place between earthquakes, apparently amounts to at least an inch per year in some places. The release of this accumulated strain by sudden slippage along the fault at intervals of tens or hundreds of years is evidently the immediate cause of such earthquakes as the one that struck San Francisco in 1906.

3. In addition to the sudden release of strain at the time of earthquakes, there is also slow slippage on the fault itself in at least one place. Seven miles south of Hollister, 85 miles southeast of San Francisco, the concrete floor and walls of a winery built across the trace of the fault have ruptured and the parts are sliding past each other horizontally in a right lateral sense. The movement is irregular but seldom sudden; accumulated displacement was about 6 inches in the 12 years between 1948 and 1960 and an estimated 24 inches in about 50 years, or an average of about half an inch per year.

4. Many gullies and small drainage lines are offset to the right where they cross the fault; this is clearly shown by the pair in the view below. Since normal erosion in these would, with time, enlarge the gullies and erase the kinks, they prove geologically recent movement—perhaps at the time of the Fort Tejon earthquake of 1857 in this case. Just beyond them is an older gully which appears to have been completely separated from its headwaters (the drainage is from right to left) by accumulated horizontal shifts along the fault.

5. As is true of all faults, the rocks do not match across the San Andreas. Here, however, we find that instead of looking higher or lower for the displaced portions of a given rock body, we generally have to look to the right. In fact, the older the feature cut by the fault, the farther it seems to be offset to the right, a circumstance from which we infer that the many miles of displacement now seen have been accumulating a few feet at a time for millions of years. The total displacement is at least tens of miles and may exceed 100.

The relative straightness of the fault trace, which shows well in the larger view, is characteristic of strike-slip movement. (Contrast the Hurricane fault, Figure 96.) Note also the number of drainage lines that either do not cross it or show marked changes where they do.

No single historical earthquake has involved movement along the entire length of this fault; displacement dies out in both directions from a localized maximum. In 1906, about 200 miles of the surface trace was active. Assuming equal participation by both sides this means that approximately 8 feet of movement on each side was absorbed in a linear distance of about 100 miles—a demonstration of the compressibility and elasticity of rocks near the earth's surface.

Fig. 99.
Close-up of the two offset gullies just above the center of the previous photograph, looking in the opposite direction (northwest). Slope and drainage is from right to left.

Some Further Thoughts on Strike-Slip Faults

The trace of the Garlock fault, shown in the figure below, can be followed for at least 150 miles in a gentle arc, concave to the south, across southern California. This view is toward the west. In the foreground the fault separates rugged topography on the left from smooth alluvial slopes (page 134) descending toward a basin that is out of sight at the right. In the background the relations are reversed, the basin (white area) being on the left and mountains on the right. Several possible explanations for these relations come to mind.

1. Perhaps the rocks are all about equally resistant to erosion and the movement along this segment of the fault has been scissors-like—up on the south in the foreground, up on the north in the background.

2. Perhaps the rocks are of contrasting resistance to erosion, in which case there are a number of combinations of original rock distribution and subsequent displacement that might account for the present scene.

3. Perhaps horizontal shifting of already deformed rocks has created the illusion of vertical movements. Thus at A and C in Figure 101 the movement along the fault *seems* to have been down on the near side, and at B down on the far side, although the actual movement has been strictly horizontal. Note that if rocks, rather than topography, are compared across the fault, there will be evidence of these apparent shifts as long as any strata remain, regardless of changes in the topography produced by erosion.

When from place to place along the length of a fault the displacement seems to have been in opposite directions, as in Figure 101, strike-slip movement is a simpler and more plausible explanation than opposite vertical displacements along adjacent segments of the fracture. But this does not prove that only horizontal movement has taken place; oblique slip is also possible, as are combinations of different movements over a long period of time. For example, the white

Fig. 100. General view of the Garlock fault, looking westward in the Mojave Desert of southern California.

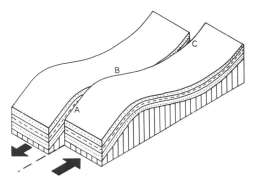

Fig. 101.
Apparent vertical displacement
produced by strictly horizontal
movement of gently deformed rocks.

floor of the valley in the background of Figure 100 represents the top of the sedimentary fill in a rock-bound basin that is probably two miles deep. Regardless of what is across from it, such a depression against one side of a fault probably means local sagging of the crust through downward displacement at some time in the past.

On the other hand, there is also good evidence of substantial horizontal movement. A 100-mile segment of the Garlock fault is shown on the simplified geologic map in Figure 102. Note, in addition to the alternation of mountains and valleys along either side of the trace, the two swarms of dikes. These groups, diagrammatically represented here, have much in common. Not only are they similar in trend, but the average width, length, and spacing of the dikes is nearly the same in both. In all the dikes the same two kinds of rock are dominant, one of them being especially abundant on the east side of each swarm, and the dikes are injected into similar plutonic host rocks. Unless we ascribe all these similarities to coincidence we can only conclude that these two groups represent a single swarm that has been cut and separated by movement on the Garlock fault. Such an explanation is supported by the fact that the western swarm does not continue south of the fault and the eastern one is not found north of

Fig. 102. Generalized geologic map of eastern 100 miles of Garlock fault. Note locations of views in Figures 100 and 104. (*From G. I. Smith, Bull. Amer. Assoc. Petroleum Geologists, vol. 46, 1962.*)

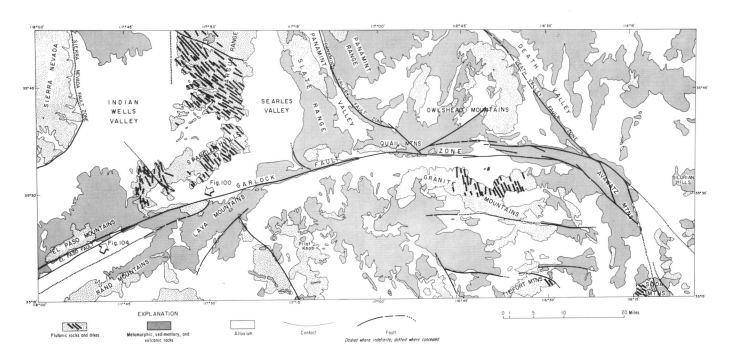

it even though some of the host rocks present in both places are similar to those intruded by the dikes. As exposed today the average dip in both groups is within 10° of vertical; if this was also true of their now-eroded upward continuations, no reasonable amount of vertical movement and accompanying erosion could account for so great a displacement of the disrupted parts of the original swarm. Such considerations lead to the speculation that, whatever the other motions, there has been left-lateral horizontal movement of about 40 miles on the Garlock fault since these dikes were emplaced.

With the possibility of some horizontal movement in mind, let us look more closely at the portion of the fault shown in Figure 104. The view is north (toward the right in the background of Figure 100) and the fault trace follows the near side of the rough topography in the foreground. Just beyond the fault trace drainage from the mountains in the background has cut the otherwise smooth slope into roughly parallel ridges and ravines.

Yet, obliquely crossing the ridges and ravines, a series of flat-bottomed shallow trenches extends from the right rear toward the left foreground. These trenches could hardly have been produced by running water; their direction is almost at right angles to the ravines, and their shape is obviously different from the shape of the ravines the streams are now producing in these materials. The shallow trenches must, then, have been produced by changes underground. Since they are belts of relative subsidence in the loose gravels being washed out of the mountains, we must look for something in the underlying bedrock into which they could have collapsed.

Noting the evidence from the dike swarms, we reason that if there has been any horizontal movement along this fault it was probably left lateral. As indicated in the drawing (Fig. 104), such relative movement in the area (large open arrows) would tend to distort any square patch of ground into a rhombus. This tendency is symbolized by the dashed square and solid rhombus at the right. The short diagonal of the rhombus indicates a direction of relative horizontal shortening or compression, and the long one, extended with small arrows, a direction of horizontal stretching.

At shallow depths crystalline rocks can withstand very high compression, a ton or more to the square inch, but their tensile strength is low—they are easily pulled apart. (Rocks make better foundation blocks than they do rake handles.) Therefore, very little distortion is likely to take place in the direction of compression (Fig. 104, *lower*), but stretching would tend to open cracks in the bedrock beneath the gravels. Such cracks should form at approximate right angles to the direction of tension (long diagonal, with small arrows). There is very little chance that the cracks will be open, however; the weight of the rocks under the vertical force of gravity usually results in sloping fault planes, and dip-slip movement down these accommodates the horizontal extension. The gravels have probably ridden down on such blocks as they subsided between some of these faults, as suggested diagrammatically in Figure 103, which is a hypothetical vertical section across three of the trenches.

Since the twisting effect of the regional movements represented by the large arrows (Fig. 104) involves both sides of the fault, why are similar patterns not seen on the near side too? Probably the thick accumulation of sediment here, on the south side of the fault, is not coherent enough to fracture and has yielded where necessary by plastic reshaping; any surface manifestation of these more subtle adjustments has almost surely been buried by the continuing sedimentation in this basin.

Fig. 103. Sketch showing, in vertical section, the probable relation of trenches in alluvial gravels to subsided blocks resulting from local tension on the north side of the Garlock fault.

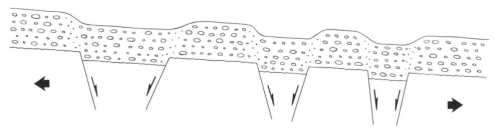

Fig. 104. *Upper:* Details along a 2-mile stretch of the Garlock fault where it truncates an alluvial fan 5 miles northeast of Saltdale, California. Highway and railroad give scale. Looking north.
Lower: Drawing of the same scene showing relation of shallow trenches to probable distorting forces. (Highway and railroad omitted.)

Low-Angle Thrust Faults

So far, we have considered two types of steep faults, those on which there has been dominantly vertical movement (e.g., the Hurricane fault) and those with dominantly strike-slip displacement (e.g., the San Andreas and Garlock faults). Among the many possible movements on gently dipping fault surfaces, one is notable for the geologically spectacular displacements it has produced. In this type the dominant movement of the rocks above the fault surface has been up the dip of the fault. Such faults are commonly referred to as *thrust* faults, because the rocks above the fault seem to have been pushed or thrust over those below, although it is only the relative movement of the two masses that can be proved, of course.

The best evidence of thrusting is the presence of rocks from deep in the crust faulted into a position on top of rocks produced near the surface. This usually means old rocks on top of young ones. In many thrust faults the total movement can only be estimated but the minimum for some is many thousands of feet vertically combined with tens of miles horizontally. In such faults the dip of the fault surface may be less than 10° for distances of many miles. Some of the best known examples occur in the Alps, the southern Appalachians, the central and northern Rocky Mountains, and southern Nevada. (One from the last area is described on pages 322–325.)

The thrust fault shown at the right is neither large nor famous, but it has the advantage for our purposes of having the rocks involved readily identifiable in the photographs. Those below the fault are massive pebbly sandstones with indistinct bedding that dips gently away from the mountain. In other nearby exposures this deposit shows the same dip, but the texture varies, becoming finer in the upper beds and with distance from the mountain. Wherever its base is exposed it may be seen to rest with depositional contact on deeply weathered gneiss and schist. It was evidently deposited by streams originating high in the range and is probably part of an old alluvial fan (cf. page 154). The rocks above the fault are schists and gneisses whose roughly horizontal banding can be seen in both views. Notice how these rocks are crushed and sheared in a zone two or three feet thick immediately above the fault plane (left foreground, lower view). The fault plane dips about 25° into the mountain—the only direction from which the metamorphic rocks could have come.

In short, here are rocks that can form only deep in the crust faulted into a position on top of rocks that can form only at the surface: rocks that make up the core of a mountain range faulted on top of sediments derived from the same mountains. The upper part of a crystalline mountain mass has been shoved sideways out over some of its own debris. This is an unequivocal example of a thrust fault.

There is not much evidence in this example to indicate how much of the total movement was vertical and how much horizontal. The pebbly sandstone is less than 1,000 feet thick, so the minimum vertical component of movement need be little greater than this; probably the total amount of slipping would be expressed in thousands of feet rather than in miles.

Fig. 105. Upslope view (looking south) of the Voorhies thrust fault as exposed in upper Blackhawk
Canyon on the north slope of the San Bernardino Mountains, California.

Fig. 106. Close-up of the same outcrop, looking west.

Recognition and Evaluation of Faults

Faults dislocate rocks. The recognition of faults, therefore, depends on our ability to distinguish slip surfaces from unbroken intrusive and depositional contacts. Along most faults, however, the actual slip surface is not exposed. In some, movement has occurred so recently that, as in the San Andreas fault (Figs. 98 and 99), the effect on the topography has not yet been erased by erosion. In others, faulting has placed resistant rocks next to weak ones, as along the Hurricane fault (Figs. 96 and 97), and erosion has carved the junction into a cliff. More often the existence of a fault is inferred from its disrupting effect on the bedrock geology; no amount of erosion can make dislocated rocks match again, so regardless of whether or not a fault is expressed in the landscape there should always be some evidence of it in the rocks.

Having recognized that a fault exists, the next problem is to determine in what direction, how far, and when, the rocks moved. The scenes of the last few pages demonstrate the special value of sedimentary rocks in providing this information. Because these have such built-in properties as superposition, original horizontality, and originally concealed stratification we can usually determine more about what has happened than we could if only crystalline rocks were involved: at least some components of the displacement can be measured by calculating the movement necessary to restore the continuity of strata across the fault zone. But despite these advantages there can be problems in evaluating what has really taken place.

In the upper example on the facing page, for example, we are looking along the trace of a vertical fault through strata dipping about 30° to the right and away from us. The dark beds in the rugged cliff high at the left are the continuation of those in the similar cliff lower on the right. We are tempted to jump to the conclusion that there has been a vertical offset, the left side having moved relatively up—or the right side down—several hundred feet. But if the left side had moved horizontally away from us—or the right side toward us—the effect, after a little erosion, would have been exactly the same (cf. Figs. 100 and 101). And why consider only vertical and horizontal movement? In short, unless we can find other evidence, we do not know what actual succession of shifts produced the present structure. We must speak then of the *apparent* offset, meaning that although we can measure the separation of the disrupted strata we leave open the question of the amounts and directions of the movements that actually took place. (Note that if the strata and movement were both horizontal, or both vertical, or in general parallel to each other, there could be movement without any offset of the layers.)

In the lower scene, as along the Hurricane fault (Fig. 97) and the San Andreas fault (Fig. 99), there is evidence that permits us to go one step farther. Here the drainage lines have been offset horizontally so that the canyons in the background do not line up with their mouths (between the dark mesas) in the foreground. Because we are dealing essentially with offset lines (the canyon bottoms) instead of layers, we can be sure here that the actual movement was very nearly horizontal (= strike slip). If we could be sure which mouth on the near side belonged with which canyon on the far side, the sense of relative movement (right or left) would be known and the amount of slip could be measured—at least that which has occurred since the canyons were cut.

In fixing the time of the movement of a fault, or "dating" the displacements, there is one basic axiom: A fault must always be younger than the youngest rocks it cuts, and older than the oldest undisturbed sediment or lava laid across it. By such reasoning, some faults can be shown to have been active at widely separated times; as already pointed out, the Hurricane fault moved about 3,600 feet (in the area of Figure 97) before some lavas flowed across it, and later moved another 1,000 feet. Another example, of two stages of movement on two faults, is given on page 324.

Fig. 107. View northwest along a vertical fault cutting sandstones in the south limb of the Barstow syncline, Mojave Desert.

Fig. 108. Looking north over offset channels and alluvial gravel deposits along the San Jacinto fault 20 miles southeast of Hemet, California. The fault crosses the scene diagonally from the lower right corner.

10 UNCONFORMITIES: RECORDS OF SHIFTING SITES OF SEDIMENTATION

The Concept

Up to this point, we have been dealing with uniform successions of strata that accumulated without interruptions greater than the short pauses that help create stratification. Such accumulations consist of practically parallel beds, and are said to constitute *conformable* sequences of strata. But such sequences are not infinitely thick, they do not extend completely around the earth, and they do not all represent the same time span. In terms of the earth's whole surface each one of them is a local, thin, and irregular patch, usually a few tens or hundreds of miles across and at most a few thousands or tens of thousands of feet thick, which differs somewhat in age from its neighbors.

What starts and what stops these accumulations of sediment? The answer lies in the broad generalization that most of the earth's surface is either undergoing erosion or receiving sediment, depending usually on whether it is above or under water. Strict neutrality between these states, for any significant length of time, is rare. With few exceptions sedimentation begins when an area subsides enough relative to its surroundings to form a basin or go below sea level, and it stops when the same area is elevated enough to be affected by downslope movements, rain, streams, waves, or glaciers.

It follows that a conformable sequence of sedimentary rocks almost always rests on an erosion surface—one converted to sedimentation by deposition of the first layer in the sequence. In the typical situation, in which land is slowly inundated by the sea, as might occur if the Gulf of Mexico or Hudson Bay were to spread slowly inland, this conversion takes place at or near the spreading shoreline. The bottom contact of any conformable sequence is therefore of special interest because it usually records a fundamental change, the start of sedimentation where there had been none in the immediate past. The ensuing history is recorded in the succeeding strata but the immediately preceding history must be read in the details of the surface on which the first layer rests. Because this contact always represents a break in the continuity of the record that is inherent in conformable strata, it is referred to as a surface of *unconformity*. It is usually a buried erosion surface.

The events which lead to an unconformable relationship between two bodies of sedimentary rock may be visualized with the aid of these scenes. The intricate outcrop patterns in the foreground of Figure 109 indicate deformed strata and record three events: (1) the deposition of sediments, (2) their deformation by earth movements, and (3) the erosion that has practically removed any hills formed by the deformation, thus exposing the structure it produced. In the background is the Salton Sea, a mere puddle in the upper end of the trough occupied by the Gulf of California (see page 376). If sea level should rise (or the land sink) about 200 feet, the entire area shown here would be submerged. In the resulting body of shallow water the nearly horizontal beds that would be laid,

Fig. 109.
Looking eastward across a portion of the San Felipe Hills to the Salton Sea in southeastern California.

Fig. 110. Detail of the angular unconformity exposed in the wall of a gully in background of Figure 109.

like blankets, over these contorted strata, would lie *unconformably* across their eroded edges. Indeed, to a limited extent this has already happened; the smoother landscape in the background is formed by such a blanket of new, horizontal sediments.

Where recent erosion has cut through this blanket we can find places like that shown in Figure 110, where, in the vertical wall of a gully, the new horizontal beds lie unconformably across the tilted and truncated older ones. Because the lower beds make an angle with the overlying ones, this relationship is called an *angular unconformity*. The unevenness of the surface of discordance separating the two series of strata reflects the relief of the buried erosion surface—an ancient landscape in profile.

The process is summarized in Figure 111. In the foreground are tilted and folded strata of an early sequence; at the left are younger horizontal beds that unconformably overlapped the truncated edges of the older strata and have since been partly removed by erosion. Here we see both "before" and "after" stages in one view. Note that the relative age of the two series is unequivocal; the deformed strata could not possibly be the younger of the two.

A more complicated situation, involving deformation and erosion of two series of strata and the unconformity that separates them, across all of which a third series is laid horizontally, is examined on page 326.

Fig. 111. Looking northward over unconformity about 10 miles west of Alcova, Wyoming.

Analysis of an Angular Unconformity

In the bottom scene at the right (Fig. 112, E) horizontal dark sandstones lie across the truncated ends of white shale beds that dip about 25° to the left. The surface separating the two series of beds is one of angular unconformity.

This same scene may be recognized within the frame near the center of the drawing (D) immediately above it, which has been prepared at a smaller scale in order to show (diagrammatically) the regional setting of the photograph. Like the other drawings, this includes a vertical cut in the front to show the relationships below the surface. In D the horizontal dark beds may be seen capping a number of mesas; they are obviously remnants of a once continuous sheet which erosion is carving into isolated flat-topped hills like the one photographed.

The drawing above this (C) represents the earlier time when these dark sandstone beds were being deposited—the stage before uplift and erosion had produced the isolated mesas. These sandstones contain great numbers of thick-shelled clams and coiled sea snails of the type found close to shore, so we postulate shallow water.

In the outcrop shown in the photograph (Fig. 112, E), and in many others like it, the surface of unconformity itself is conspicuously flat and even, despite the deformation of the underlying white beds. This means that the surface on which the first dark beds were deposited had been levelled by some kind of erosion. This process is shown in progress in drawing B, in which a river is widening its flat valley floor and waves are chewing away at the coastline. Together, these agents of erosion will gradually reduce the landscape to an even surface which, as the sea rises or the land subsides, will be flooded by the encroaching sea.

But we must still account for the beds of white shale. Before they could be deformed and then truncated by this erosion surface, they had to be deposited. Like the sandstone, they contain abundant shallow-water fossils, but these are types representing life in quiet, protected waters, and the rock is finer grained than the overlying sandstone. So we reconstruct this episode in the top drawing (A) with a shallow bay in which the white strata are accumulating. No rocks older than the white beds were exposed in the original hill, so we do not know on what kind of rocks they lie. Accordingly the shape and structure of the floor beneath the first white layer cannot be shown in detail; all we know is that there must be one.

Using evidence from the modern scene we have now reconstructed a scene that existed about 20 million years ago. Let us recapitulate, starting at the top of the page.

A. In a shallow and protected embayment the white strata are deposited.

B. These beds are deformed, uplifted, and subjected to erosion. The uplift and deformation may have occurred separately, but the likelihood is that they are related, the forces that produced one being also responsible for the other. As soon as uplift brings the sediments above sea level, erosion begins. In this scene we show the condition after most of the deformation has ceased; erosion is the dominant process modifying the landscape and the sea is again encroaching on the land.

C. Now, somewhat later, the hills of the preceding stage have been reduced by erosion to low residual humps and the sea has invaded a large part of the scene, probably because of subsidence of the land. The first layers of dark sandstone are being deposited in this sea, across the truncated folds of the earlier white series.

D. A second episode of uplift, this time without folding or faulting in this area, has elevated the land until the dark beds stand a few hundred feet above sea level. Two summits of older rocks protrude through the dark beds, marking hills that were never completely submerged in the preceding stage. Erosion, perhaps governed by the same main streams that survived the last stage, is again the dominant process and is carving the whole scene, leaving the dark deposits as isolated mesas. One of these is shown in closer view in the photograph at the bottom of the page (E).

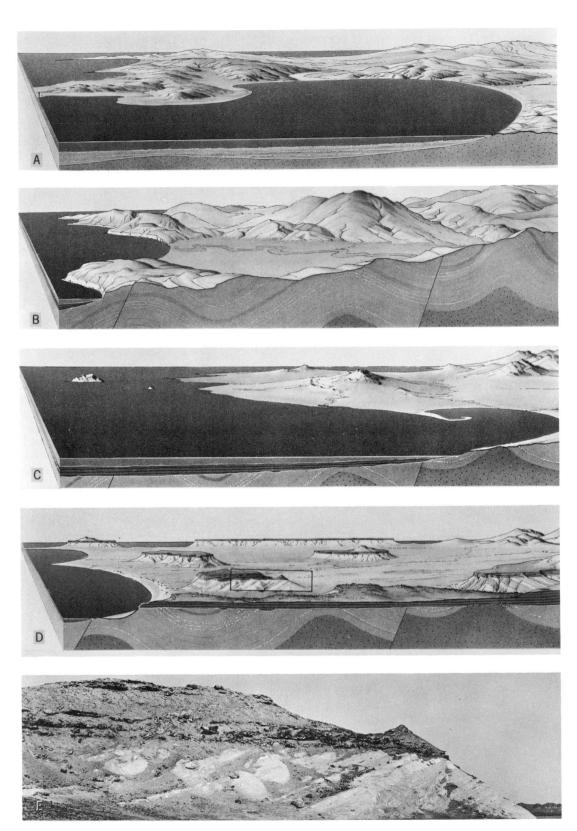

Fig. 112. A–D are drawings reconstructing the major steps leading up to the conditions in E, a photograph of dark sandstone beds lying unconformably on tilted white shale strata near Turtle Bay, Baja California. Both rocks are marine and highly fossiliferous. Box in D outlines area shown in E.

Nonconformity, a Variety of Unconformity

An unconformable relationship can exist only at the base of an unbroken sequence of surface deposits (sediments, lava, or ash). The chief differences between unconformities, therefore, arise not from the deposits laid down on the erosion surface, but from the nature of that surface itself and the rocks immediately below it. It is simply a question of how many different kinds of terrain there are on which sedimentation can begin. For descriptive purposes three such surfaces are commonly recognized.

Deformed older layered sediments (such as those shown in the last two examples). As pointed out on page 107, because of the angular discordance between the two sets of strata such a relationship is referred to as an angular unconformity.

Undeformed older layered sediments. In these the two sets of strata are parallel and identification of the unconformity depends on being able to distinguish an erosion surface from a bedding plane. This type of unconformity is the most difficult to recognize; it is called a *disconformity.* An example is described on page 275.

Plutonic or metamorphic rocks. When the erosion surface cuts across uplifted crystalline rocks of deep-seated origin it is sometimes distinguished as a *nonconformity;* an example is pictured at the right.

In this view there is a profound difference between the rocks in the upper and lower parts of the cliff. Above are nearly horizontal beds of silty limestone forming cliffs and ledges, and, especially near the top, some interbedded zones of sandy shale that produce gentle slopes. Below is a mass of dark schist with prominent vertical foliation and horizontal jointing, penetrated by a complex of light-colored granitic sills and dikes. (A closer look at this rock was shown in Figures 59 and 60; the origin of the granite was discussed on page 52.

At the base of the lowest limestone layer is the surface of nonconformity, the erosion surface on which sedimentation began at the time and place here recorded. By imagining all the sedimentary rocks removed, we can go back in time and picture this surface just before sedimentation began—probably a low-lying plain near shore, where a small rise in sea level or a slight subsidence of the land could easily inundate it and bring about sedimentation. Then we realize that to obtain such a plain on this kind of rocks it is necessary to bring them up from depths measured in tens of thousands of feet; and further, that to expose them to view these same tens of thousands of feet of overlying rock must be removed by erosion.

The overwhelming fact embodied in a nonconformity is this record of vertical miles of uplift and vertical miles of erosion. As pointed out on page 20, the same thing is true of any present outcrop of granite or gneiss, the difference being that the nonconformity is a record of such an outcrop in the past, perhaps hundreds of millions of years ago.

Fig. 113. Sediments of the lower Grand Canyon series resting nonconformably on Vishnu schist, as seen from the Kaibab trail in the Grand Canyon. Total height of view approximately 700 feet.

III Sculpture

*The ways the exposed parts of
Earth's rocky structures have been carved
and modified by surface processes to
give the landscape the shapes we see.*

"*The result, therefore, of our present
enquiry is, that we find no vestige
of a beginning,—no prospect of an end.*"
JAMES HUTTON, 1788

11 SOME PERSPECTIVE ON THE LEVELING PROCESSES

In Parts I and II we examined some of the rocky materials of the earth's crust and their geologic structure—the substance and anatomy of the lands we live on. But the shape of the surface of those lands is only indirectly determined by rocks and structure. The immediate and often dominant factor in the origin of the landscape is an entirely independent group of processes that operate on, rather than within, the crust. These are weathering, erosion (including downslope movements), and deposition. Their relation to the accumulation of sediment has already been touched upon: the following pages contain an analysis of their effects on the landscape.

Weathering gradually converts fresh hard rock at and near the earth's surface into soft crumbly material offering little resistance to erosion. Whether they are scattered by raindrops, agitated by waves, or transported by running water or glacial ice, the particles loosened by weathering cannot escape the influence of gravity; in the long run they can move only from higher to lower places. The persistent influence of the various processes of weathering, erosion, and deposition is therefore toward leveling the landscape—lowering high ground through removal of rock debris and filling in low places by deposition. For this reason, when their effect on topography is under consideration, they are referred to as *leveling processes.*

The giant among leveling agents is running water. Unlike waves, wind, ice, and groundwater, each of which is effective only in limited areas, running water (from rain or melting snow) is found in virtually all parts of the world. Rivulets, streams, and rivers, fed by downslope movements, carry away most of the weathered rock produced on land—even in deserts. Because it is thus the pattern and character of the streams that dominates most familiar landscapes, *fluvial* erosion (Latin: *fluvius,* river) will receive the greatest emphasis here. But before examining the leveling processes in detail it will be helpful to pause long enough to consider some general questions.

Why, since the leveling processes have been at work as long as there has been rainfall on the land—probably at least three billion years—have they not succeeded in removing all mountains and filling all depressions? Is it possible to estimate, either from conditions today or in the past, the degree to which a given landscape has been leveled? Can we say that this area has been leveled more, or less, than that one?

In present-day landscapes there is evidence that the leveling processes have virtually flattened the land in some places and have made relatively little impression on it in others. In North America, for example, the well-leveled areas include such flat and low-lying regions as the piedmont belt between the Appalachians and the Atlantic seaboard, some of the lowlands along the Mississippi Valley, the Great Lakes region, much of the Great Plains, and large parts of central and eastern Canada. In some of these the dominant leveling process has been erosion, in others deposition; in most places a combination of the two has been active. In sharp contrast to these are the areas that are poorly leveled: hills, high plateaus, and mountains, such as the Appalachians and the Allegheny Plateau in the east; the Ouachita, Boston and St. Francois Mountains, in the midwest; and more than half of the area from the Rocky Mountains to the Pacific Coast. In these the leveling processes are vigorously at work, but they have not yet subdued the local relief.

In the geologic record we find evidence of similar contrasts in the past. A widespread unconformity, such as that discussed on page 110, marks a well-leveled area; folded, faulted, intruded, and metamorphosed strata, by analogy with similar regions today, were probably mountainous at one time.

The mere existence of well-leveled areas that have been cut across crystalline rocks of deep seated origin at different times in the geologic past is proof that there has been time enough for the leveling processes to have flattened all the continents more than once. That they have not done so—that the continents are ribbed with mountains today—is proof of crustal unrest. Except for volcanoes, the hills and mountains of today exist *only* because of deformation of the crust; entirely apart from their structure, which often records folding and faulting, the world's mountains prove crustal deformation simply by being there.

From these observations it follows that any given landscape is a product of the relative achievements, at that place and time, of crustal unrest and the leveling processes. This is a fundamental concept: in the long view of geologic time the landscape is forever changing. Each time a section of the crust begins to rise above its surroundings it invites its own destruction through acceleration of the leveling processes, and the renewed erosion adds its mark to the shapes left from previous conditions or produced in the course of the uplift. Each time an area subsides below its surroundings it invites its own burial.

In this multibillion-year history of change there is, in most regions, no recognizable point at which the interplay of these upbuilding and downtearing processes can be said to have begun. Rather we must interpret each landscape in transit, so to speak, evaluating its present state and recent history by determining the relative effectiveness of the opposing influences.

In order to do this we need to know something about (1) the probable result if deformation of the crust were allowed to go unmodified by erosion, (2) the probable result if leveling were allowed to go uninterrupted by deformation, and especially (3) the ways in which each of the processes involved tends to assert itself—the characteristic imprint each leaves on the landscape.

In Part II we considered the more important forms produced by deformation. If these forms were to develop free from attack by erosion the result can readily be imagined: There would be smooth ungullied monoclines (like the restoration at the back of Figure 88) and perfectly rounded anticlines, domes, and synclines. The steep faces of tilted fault blocks would have exactly the dip and strike of the fault surface itself and would retain every groove, scratch, and bit of polish. Instead of being cut into so that their anatomy is exposed, as are the examples we examined, these shapes would be topped by the surface that existed when they began to form. Many of the upfaulted and upfolded areas would reach staggering heights that would dwarf the eroded stumps that we now see. With no sediment to fill them, subsiding areas such as Death Valley and the Dead Sea depression would be thousands of feet below sea level. Topographic and structural relief would be identical.

On the other hand, if the crust were to hold still long enough for the leveling processes to complete their work, the land would be reduced to low plains rising only slightly above sea level. All basins on land would be filled, and ultimately all drainage, and all sediment, would move slowly toward the oceans. The sands, muds, and clays thus added to the seas would probably displace enough water to bring about a gradual rise in sea level and a consequent flooding of the continental margins.

There is no evidence that either of these hypothetical extremes has ever been reached on a world-wide scale, and only the second has even been approached over areas of continental size. Since, then, we must study landscape evolution while it is in progress, our most useful tool will be a knowledge of how each of the separate processes operates. We have already examined the geometry of crustal deformation: the rates at which this deformation takes place, and some of its possible causes, are discussed at the end of this book. It is the nature of the leveling processes and the shapes they tend to impose on the landforms that crustal deformation tends to produce that constitute the theme of Part III.

12 WEATHERING

An Example of Weathering in Plutonic Rocks

Why do old tombstones become difficult to read? Why, even in the relatively dry climate of Egypt, have the sharp corners of the blocks in the great pyramids become rounded? Why do most cliffs have piles of broken rock heaped at their bases? How can plants and trees grow in crevices in seemingly solid rock near the timberline in high mountain ranges?

The answer to each of these questions is *weathering*, the slow but inevitable disintegration and decomposition of rock (and, indeed, nearly all materials known to man) under the influence of air and moisture. The effects of weathering are about us on every hand. We try to halt them with paint, with chromium plating, by using brass screws instead of iron, but it is a losing battle—paints oxidize, woods decay, metals rust and corrode, limestone and marble dissolve, granitic rocks crumble.

Consider the example at the right, which is part of a highway-cut in plutonic rock—coarse-grained, uniform in texture, close to granite in composition, a typical granitic rock. The rather well-developed joint system includes three prominent sets, one dipping gently to the right (lower part of cut) and two that are vertical, one trending into the cliff and one almost parallel to it. It is probably no accident that the white dikelets of quartz and feldspar tend to parallel these directions.

Parts of the land surface in the background look as though they were strewn with giant boulders; but a closer look discloses that the "boulders" are all composed of the same kind of rock—a circumstance that would be highly improbable if they were a sedimentary accumulation. Furthermore, some of them can be seen to be attached to the underlying plutonic rock, as in the upper right corner and in several other places along the top of the cut. Just right of center there is a "boulder" embedded in the pluton at least 20 feet below the present surface.

Considering the whole face of the cut, note that there are fairly large areas where the rock is angular and blocky—the shapes being largely determined by the joints—and there are others where the rock is rounded and subdued. Most of the angular rock lies toward the lower part of the view and most of the rounded rock near the top, but the pattern is irregular. The difference between the two results from the degree to which the rock has been decomposed by weathering penetrating from the top; the angular blocky rock is relatively fresh, the other is decomposed enough so that edges and corners cannot retain their shape. Some of the irregularities arise from subtle variations in the composition or texture of the rock; slight variations in the amount of mica (whose easy cleavage and susceptibility to alteration weaken a rock) probably account for some of the difference. However, the fact that in general the weathered rock is near the surface and fresher rock occurs with increasing depth, points to weathering, penetrating downward from the original land surface, as the root cause of the changes.

Notice the joints: throughout the exposure they have become grooves. Evidently the rock decomposes most rapidly along them, least rapidly at the centers of the solid blocks. It is apparent that weathering is not only progressing downward from the surface of the ground, but is also penetrating the rock along every joint, from which decomposition evidently spreads into the blocks. The attack on the face of a block comes from one direction only, on an edge from two directions, on a corner from three. The result is that in homogeneous rock, such as this, the corners succumb first, the edges next, and the faces last. Rounding of the blocks is the inevitable outcome; hence the "boulders."

Fig. 114. Steep cut in granitic rock, Viejas Mountain, San Diego County, California. The exposure is about 40 feet high.

Weathering and Soils

To the farmer or gardener, *soil* is the stuff in which plants grow. To the geologist, soils are also indicators of weathering processes and climate and, in the "fossil" state, records of buried land surfaces.

The upper scene here is unspectacular because it is common—and it is this very commonness that makes it important. It is a sample of the evidence, visible in road cuts, excavations, and quarry faces all over the world, that solid rock grades into soil at the land surface wherever the decomposed rock produced by weathering is not eroded away or otherwise removed. Here the bedrock is gneiss, dipping about 30° to the right. The rock grades upward into two or three feet of dark soil that supports grass and small bushes. The soil is unstratified, has an irregular and indefinite lower limit, and contains many small fragments of the underlying gneiss in its lower part. This shows it to be a *residual soil*, formed in place, as distinct from a *transported soil* that might have been washed in from nearby uplands.

Although they are not stratified in the sedimentary sense, soils normally change character with depth. For purposes of comparison it is useful to distinguish the changing processes that cause this. At the top is a zone from which soluble constituents have been removed by the leaching action of downward percolating waters—the *A horizon* of soil scientists. Below this is the *B horizon*, the zone in which this leached material is deposited, along with a few particles of rock and organic matter physically washed down from above. A's loss is B's gain, and together the two constitute the "soil" of common usage. The *C horizon* comprises the weathered rock below the zone of accumulation, and grades downward into fresh bedrock. In this scene, the A horizon is dark and only a few inches thick; the B horizon is lighter and one to two feet thick, grading irregularly into C, which is still lighter and extends below the base of the exposure.

Soils exhibit variety primarily because of differences in the parent rock, in the climate in which they developed, in the vegetation under which they formed, and in age or maturity. In the long run climate is probably the dominant influence, both directly and through its control of vegetation; mature soils in different parts of the world seem more to reflect the climate under which they have formed than the kind of bedrock from which they were derived. Thus, in the northeastern United States, which receives 40 or more inches of rainfall annually and is covered by forest vegetation, percolating waters carry much of the decomposed organic material (humus) from the A to the B horizon and leaching depletes both in the more soluble products of weathering, such as salts of sodium, calcium, and magnesium. Such soils tend to be acid, with light A and dark B horizons. In the western plains and the Southwest, on the other hand, where the annual rainfall is less than 25 inches and there is a long dry season, humus accumulates near the surface and the soluble salts get only as far as the B horizon, where they accumulate and often form concretionary lumps, hardpan, and limy layers known as *caliche* (see page 240). Such soils tend to be alkaline, with dark A and light B horizons.

Under special conditions a soil may be preserved under sediments deposited on top of it. Such "fossil" soils are found most frequently in wind-laid and glacial deposits. Their presence proves that there were interruptions in the accumulation—interruptions that were long enough for a weathered zone to develop at the surface.

The dark zone in the lower scene at the right is one of many buried soils in Alaska. The roots are those of trees in a dominantly spruce forest that grew here more than 2,000 years ago. The overlying gray gravels are part of a deposit over which glaciers later advanced, breaking off the spruce stumps (far left) flush with the surface of the deposit and leaving the splinters aligned in the direction of ice movement. Since the gravels were undoubtedly deposited by torrents of glacial meltwater, this outcrop records a complete cycle in the extent of the glaciers, and carries corresponding implications regarding changes in the climate.

Fig. 115. Residual soil on gneiss in highway cut, eastern foothills of Rocky Mountains.

Fig. 116. Buried soil, east side of Muir Inlet, Glacier Bay, Alaska. (*Photo by W. S. Cooper.*)

13 DOWNSLOPE MOVEMENTS

Talus and Rock Glaciers

Weathering is one of those inconspicuous natural processes, like evaporation, that go on around us all the time, but whose true magnitude and importance are not immediately obvious. One way to evaluate such subtle processes is to study their products—so let us turn to the movement and accumulation of weathered rock for any light these may shed on the nature of weathering.

One of the simplest products of weathering is loose broken rock, or rubble. This can be observed in places where disintegration (breaking apart) of rock predominates over chemical decay. Such conditions are found chiefly on deserts and near or above timberline in high mountains, as illustrated in the two scenes at the right. In both places angular shapes of the fragments show that although the rock is being split, the pieces are sufficiently fresh and coherent to retain their sharp edges and corners (cf. Fig. 114).

The processes that break rock in this way are not yet fully understood. The cracks must form from stresses developed within the rock, for nature has no analogue to the stonemason's hammer and chisel. These stresses can arise from changes in the surrounding pressure and from the expansion and contraction that accompany changes in temperature. Any rock that formed within the crust or has been deeply buried and is then uncovered by uplift and erosion has been subjected to such changes; at the surface the temperature is much lower and, probably more important, a confining pressure of thousands of pounds per square inch has been removed. These changes, often accompanied by stresses arising from crustal deformation, are probably the basic causes of joints (page 92): under certain circumstances, they seem able actually to shatter rock. (Among the hazards of deep mining are rock bursts—the spontaneous and sometimes violent expulsion of rock from mine walls when support has been removed from one side of rock that is under pressure.)

Once cracks of any kind are initiated, other factors help to extend them. Chemical decomposition goes to work in all cracks, plant roots wedge themselves into some of them, and in high mountains or cold climates the expansive force of freezing water confined in the cracks can widen them.

All these factors are considered phases of weathering, and all of them—deformation, unloading, cooling, chemical decomposition, plant root invasion, and frost wedging—help to produce broken rock at the surface. The simplest accumulation of this rubble is the *talus pile*—angular rock fragments stacked against the cliff or steep slope from which they have fallen. The upper view at the right shows a typical example involving volcanic rock in an arid climate. The surface of such a pile usually has a slope of 30° to 35°, depending on the size and shape of the fragments; this is known as the *angle of repose* because, inasmuch as every piece is introduced at the top, the slope is automatically the steepest at which the debris will stand.

In the lower scene, although small talus piles can be seen at the base of the shadowed cliff at the upper left, most of the rubble has moved on down the slope and acquired the pattern of lobes and surface wrinkles that is characteristic of viscous flow. The similarity in form and occurrence of such accumulations and some mountain glaciers has led naturally to the term *rock glacier*. Rock glaciers are most abundant near or above timberline in high mountains; the example shown here starts at about 12,000 feet. Rock glaciers almost certainly differ from talus piles in only one essential feature—the slow mobility conferred by seasonal freezing and thawing of abundant rain and snow water. The creeping movement thus produced reduces their average slopes far below 30°.

Talus piles and rock glaciers are products of the movement of weathered material under the influence of gravity. Since the weathered rock moves down mountain slopes, hillsides, and valley walls, the changes of position are called *downslope movements*. (See also page 32.)

Fig. 117. Talus piles below cliffs of volcanic rock in the Alvord Mountains, Mojave Desert, California.

Fig. 118. Rock glacier descending from a 12,000-foot ridge crest in the San Juan Mountains, Colorado.

Historic Landslides

The white scar on the forested mountain slope shown in the upper view at the right marks the place where an estimated 50 million cubic yards of sandstone slid into the valley of the Gros Ventre River in western Wyoming on the afternoon of June 23, 1925. There are no published eyewitness accounts of the slide in motion, but the story can be pieced together from the record of ensuing events and a study of the scene shown here. Like the 1959 Madison River slide (see page 172) this one must have moved rapidly, for some of the rubble was carried 350 feet up the opposite side of the valley while the main mass formed a dam 225 to 250 feet high. Probably the whole event took only a minute or two.

The lake that immediately began to form behind the dam rose 60 feet in less than 18 hours and was 200 feet deep by the end of three weeks. However, owing to seepage and relatively dry seasons, no water topped the dam for almost two years. Then on May 18, 1927, heavy spring runoff brought the level to the top of the dam and between 10 A.M. and 4 P.M. the lake discharged over 43,000 acre-feet (more than 14 billion gallons). This overflow cut a channel about 300 feet wide and 100 feet deep and lowered the lake accordingly. The resulting wall of water raced down the valley, did much damage and, despite warnings, cost the lives of several householders who were caught trying to save some of their belongings. Driftwood lodged in trees 10 to 14 feet above the valley floor. As the channel through the dam deepened it also became longer, which retarded the downward cutting. At the same time the lake was shrinking and seepage continuing, so that a balance was soon reached. There has been little change since that time, though of course from the geological point of view the lake is temporary because of the inevitable infilling by sediments and slow lowering of the outlet by erosion.

This is only one of many landslides in this valley, and there is a good geological explanation for each. Here the forested slope at the right is underlain by interbedded sandstone, limestone, and shale, and—as can be seen in the outcrop near the center of the view—the dip of these strata (about 20°) is approximately parallel to the surface of the ground where the slide occurred. During the winter and spring of 1925, water from heavy rains and melting snow penetrated the rock, reducing cohesion and adding to its weight. Some of the shales became quite slippery, and with these as skids the mass began to move. In descending 2,100 feet in a little over a mile the different rock types became thoroughly mixed, indicating that the material ultimately flowed rather than slipping as an intact mass. Slides of this type are known as *debris flows*.

The lower view shows a modern slide of a very different kind. It first became active in January, 1929, and the photograph shows the accumulated displacement 30 years later. A small escarpment outlines the inner limit of the approximately five acres of land affected. Other scarps and fissures and the fragments of the curved street near the bluff show that different segments have moved different amounts and that none has moved more than about 200 feet. Note that the original flat surface of the land is preserved on most of the blocks, despite some tilting. So far as is known, the movement has occurred chiefly at times of unusually heavy rainfall and has never exceeded an inch a day. Borings show that, like most large slides, this one has a definite lower surface or *sole* on which it is moving, and that this coincides with the dip of the strata. As in the Gros Ventre slide the movement here is down the dip of sandstones and shales (dipping 10° to 15° to the left), but the material is shifting in large intact blocks which move along a sole without much disruption.

Both slides, like all downslope movements, were prepared by weathering and powered by gravity.

Fig. 119. Looking east over the 1925 debris flow and resulting lake on the Gros Ventre River,
4 miles above Kelly, Wyoming.

Fig. 120. The Point Fermin slide near San Pedro, California.

Prehistoric Landslides

Recognizing ancient landslides is a geological pastime that can be practiced in all but the flattest country. The skill thus acquired is not without practical value; indeed, lack of it may lead to disaster. Shortly after World War II, many persons began to build homes in the Palos Verdes Hills near Los Angeles. In the next few years 150 houses were built in an area from which there is a beautiful view of the sea, despite the fact that the land was clearly shown as an old landslide on a United States Geological Survey map published in 1946. By 1956 the water from lawns, cesspools, and heavy rains—and possibly a redistribution of weight resulting from extensive grading—had reactivated the slide. Creeping along at 10 to 30 mm a day, the variation being seasonal and apparently related to rainfall, 350 acres of ground began to crack and heave, tilting and breaking houses, rupturing roads and pipelines, and rendering the entire area uninhabitable. In three years the total horizontal movement exceeded 60 feet and the property damage was estimated at more than 10 million dollars. Legal wrangling over the responsibility was bitter and lasted for years. Such costly mistakes are less likely to be made by the homeowner familiar with the character of unstable ground and the conditions that favor it.

The upper scene at the right includes a prehistoric debris flow of unmistakable outline. From the lobate lower end in the foreground the flow can be readily traced back up the canyon to an equally large steep-walled source area. A hummocky and wrinkled surface characterizes the ground that has moved. Some of its hollows are occupied by ponds and small lakes and normal stream drainage is not yet established over most of it, showing that the surface is little modified and that geologically it is a very recent feature. The smooth margin of the toe, where the mass moved out over the terraced valley floor, and the concentric repetition of this pattern in the hummocky surface of the thin, spreading lobe are evidences of viscous or plastic flow of fine debris. If there were a road-cut in the slide material, we would expect it to reveal a chaotic accumulation of fragments of the same rock that is exposed in the source area, probably in a matrix of fine-grained weathering products that are somewhat slippery when wet.

The characteristics just noted can be used to identify a more extensive but less-obvious area of sliding in the lower view. A broad, low, smoothly rounded ridge of dark volcanic rock extends from the left side of the scene to the upper right corner. Behind and in front of the ridge are flat-floored valleys. The whole center of the view is occupied by the uneven surface of a slide extending from a doubly arcuate steep headwall at the crest of the ridge to a broad lobate toe along the irrigation ditch in the right foreground. Beyond and to the right are the headwall escarpments of several other slides, or, more properly, the uneven upper edge of a composite slide involving the whole near side of the ridge to and beyond the limits of this view.

Many glacial moraines have the hummocky surface and chaotic internal structure found in slides, but slides have other characteristics that distinguish them. They are always limited in extent, the source area can usually be found close by, and the rock types of the slide match those of the source. Glacial deposits, on the other hand, (e.g., Figs. 218 and 320) may cover very large areas and usually contain rocks carried in from great distances; those that, like slides, are localized, are almost always associated with other unmistakable evidence of glaciation (see Fig. 208).

Other prehistoric landslides are explored on pages 328–331.

Fig. 121. Old landslide of the debris-flow type in altered volcanic rocks west of the
    Pahsimeroi River in south central Idaho.

Fig. 122. Old landslides in lavas on the north slope of the Saddle Mountains west of Othello, Washington.

Creep and Some Related Considerations

Landslides such as those just described are the most spectacular manifestations of downslope movement. Using the term broadly, they vary from a few square feet to many square miles in area. They may move a few inches per year or more than 100 miles per hour, and may contain material that is almost dry (as in the *rockfalls* caused by the collapse of steep cliffs) or material so saturated with water that it moves like thick soup (*mudflow*). In 1903 the mining village of Frank, at Turtle Mountain, Alberta, was wiped out by the sudden collapse of a steep mountain face. Rock avalanches have caused even greater disasters in Switzerland. The nearly simultaneous breaking of seven transatlantic cables at the edge of the Grand Banks off Nova Scotia in November 1929 is believed to have been caused by a submarine slump triggered by an earthquake in the same area. Associated more-fluid turbidity currents moved several hundred miles out to sea and reached a width of at least 100 miles. The times of successive breaks in other cables show that these currents apparently reached a top speed of 58 miles per hour and were still moving 14 miles per hour when they broke the last cable, about 295 miles from their apparent source.

It is events like these that attract notice, but when it comes to total influence on the landscape, whether measured in square miles affected or in tons of material moved, landslides are completely overshadowed by *creep*—the imperceptibly slow downslope flow of surface soil and weathered rock. Rates of creep seldom exceed a few inches per year and usually vary with the seasons. In fact, creep occurs essentially because the weathered zone swells a little with each wetting or freezing and shrinks with each drying or thawing, and when this occurs on a slope gravity imparts a net shift downslope. Creep does not produce the well-defined areas of slippage with a scar at the top and a bulbous deposit at the bottom that accompany most landslides, but it often causes fence posts and telephone poles to lean downslope and trees to develop curved trunks.

The upper scene at the right shows the effect of creep on steeply dipping beds of shale and sandstone. The ground surface slopes only about 10° to the left, yet movement in the weathered zone has dragged the upper edges of the strata into a distinct bend in the downslope direction.

Creep is most likely to occur where there are weak rocks, deep weathering, moderately steep slopes, and active disturbance of the soil by seasonal changes, animal hooves and burrows, and roots. The higher the proportion of clay in the weathered zone the greater will be the amount of swelling and shrinking with changes in moisture content. Under arctic conditions the ground is likely to be saturated the year round and the dominant activating agents are freezing and thawing.

Creep as a process is, of course, gradational with landsliding. As soon as the moving mass is outlined by breaks at the surface and develops a definite sole, or slip plane, it is classified as a slide. In the lower scene at the right a patch of soil several acres in area has pulled away from the crest of the ridge and moved a few tens of feet downslope, crowding itself into the head of a small canyon. Only a thin skin is involved, as shown by the very shallow scar at the top and the wrinkles on the surface of the sliding mass. These hills are carved out of soft shale; creep is very active on all of their slopes, and occasionally, as here, a small area breaks loose and becomes a short-traveled landslide. There are at least four other older slides in this scene.

Downslope movements are the dominant mechanism by which weathered material is removed from hills and mountains and delivered to the streams that carry it away. As one of the most ubiquitous and important varieties of downslope movement, creep is thus responsible, over long periods of time, for extensive modification of the outdoor scene.

Fig. 123. Deformation by creep in the upper part of a road cut in sandy shales.
Topanga Canyon, Santa Monica Mountains, California.

Fig. 124. Shallow landslide on shaly slope in the Puente Hills of southern California.

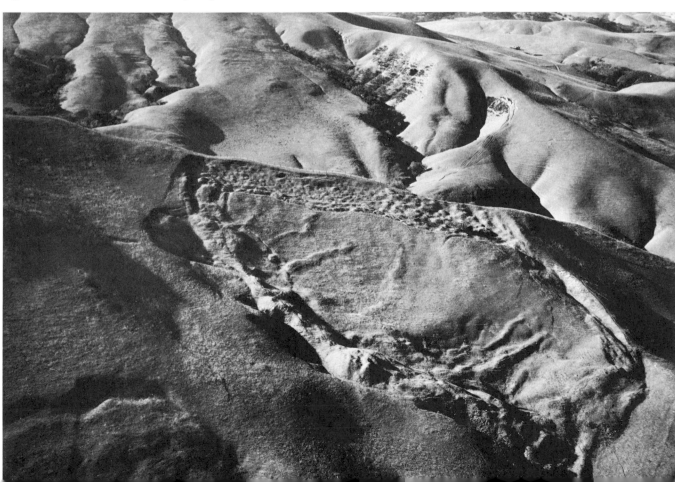

14 THE WORK OF STREAMS AND RIVERS

Introduction

Approximately 30,000 cubic miles of water falls as rain and snow on the land areas of the earth each year. Yet throughout most of recorded history man has believed that precipitation was insufficient to account for the amount of water carried by rivers. Perhaps this is partly explained by the influence on early philosophers of such important Middle East streams as the Nile, whose source was remote from and little known to the more-literate people living in the drier climate of Mediterranean shores. No doubt, too, men of all times and climes have been more impressed by the quantity of water in the swirling rivers they could not cross than in the raindrops of the passing storm.

During the Renaissance this imagined discrepancy was explained by postulating vast subterranean springs which helped to feed the rivers and which in turn were supplied either from hidden sources or from the oceans, whose water was made fresh by some vague de-salting process involving volcanic heat. Then, in the seventeenth century, Pierre Perrault *measured* the discharge of the Seine and also the precipitation on its upstream drainage basin. According to his data, the runoff actually amounted to only about one-sixth of the precipitation. This unexpected result introduced a new problem; instead of seeking the source of the water that supplied the rivers, it became necessary to account for the five-sixths of the rain and snow that fell on the land but did not appear in its streams.

Today many thousands of rain gages and many hundreds of stream-gaging stations (such as the one in the Grand Canyon; see page 30) provide enough information for a general explanation. The average annual precipitation on the United States, for example, is about 30 inches, but the runoff carried by its rivers is only the equivalent of about 9 inches a year. Precipitation on this area is thus more than three times runoff. This average includes individual drainage systems in which runoff ranges from about one-half to less than one-twentieth of the associated precipitation. Perrault's measurements now seem very reasonable.

What becomes of the rest of the precipitation on this land, the 21 inches a year that does not appear as runoff? Most of it evaporates directly from the land surface or is returned to the atmosphere through the transpiration of plants. Together, these processes return to the atmosphere more than twice as much water as drains off the land in rivers. A very small but geologically important amount enters into chemical reactions involved in weathering. Some fills the accessible pore spaces in soil and rock, where it is detained as groundwater, but since most groundwater is ultimately taken up by plant roots, enters into weathering reactions, or reappears in springs and seeps from which it joins the runoff, its final disposition follows the patterns already mentioned. Likewise, the water locked up in glaciers and ice caps is only temporarily out of circulation, for ice both evaporates and melts.

The whole complex process comprising evaporation, rainfall, and runoff is called the *hydrologic cycle*. The energy involved in this cycle—raising 30,000 cubic miles of water into the air and dumping it on the lands of the world each year—is almost beyond comprehension. Not only does most of this water pelt the land with an astronomical number of drops, each of which is capable of displacing a sand grain or soil particle many inches, but the 9,000 cubic miles of it that then runs off the continents has an enormous potential for erosion. Since the lands have an average elevation of about one-half mile above sea level the energy released by the runoff alone is equivalent to that of a waterfall half a mile high over which the equivalent of about 90 Mississippi Rivers pours continuously. This energy comes from the sun, whose heat keeps returning water to the atmosphere for continuous repetition of the cycle.

Let us examine some of the work accomplished by this falling and running water.

Fig. 125. Summer rainstorm in the midwest.

The Geological Importance of Seasonal Peaks in River Flow

In the discussion of the Grand Canyon segment of the Colorado River (pages 30–31) it was pointed out that the wide variations in the amount of water flowing under the suspension bridge are accompanied by even greater fluctuations in the load of suspended sediment. Some idea of the importance of these fluctuations can be gained by comparing discharge and load figures. Using, as before, the Colorado River water-year ending September 30, 1957, (prior to the completion of Glen Canyon dam, next upstream) the average daily discharge of river water was a little over 24,000 cubic feet per second and the average daily load a little over 425,000 tons. But these averages conceal the most impressive features of the record. One-sixth of the year's load of over 155 million tons was moved in just 11 days—the 11 of highest discharge, on each of which the flow exceeded 100,000 cubic feet per second. Or, taking the 61 days of highest discharge, 54 of which were in June and July, we find that more than 84½ million tons were carried; i.e., well over half of the total load for the year was transported during the two months of highest discharge.

These data illustrate a generalization that is valid for most streams: the average flow of water carries less than the average suspended load and the high-water and flood stages accomplished most of the transportation of sediment.

Studies of many rivers and streams of different sizes have shown that increases in the suspended load are usually proportional to something between the square and the cube of the discharge. Thus, when the flow of a stream triples, the load might be expected to increase approximately 9 to 27 times. However, whether this increase is attained, or maintained, depends largely on the amount and kind of sediment that is available. Not enough sediment may have accumulated since the last peak discharge to supply the load that this one is capable of moving; vegetation may now inhibit erosion of bars and islands formerly barren and more mobile; the coarseness or coherence of the debris may reduce its accessibility to the current. For such reasons, many peak discharges and floods are not loaded to capacity.

The cross-sectional shape of a river's channel and the nature of its banks also affect the movement of sediment during fluctuations in discharge. This can be illustrated by comparing the changes in the bed of the Colorado River in the Grand Canyon with those in the bed of the Rillito River near Tucson as each is subjected to increasing discharge.

During periods of accelerating discharge in the Grand Canyon the water frequently deepens faster than its surface rises because accumulated sediment is flushed from its bed. The walls of the inner gorge of the Canyon are steep and far more resistant than the sediments on its floor; although contributing very little to the load, they constrain the swollen river to flow in a narrow rock-bound channel. This results in higher velocity and, its carrying capacity thus increased, the current erodes sediment from its bed. It is quite likely that when this has been washed away the proportion of sediment load to water would decrease even if the discharge of water continued to increase.

On the other hand, the intermittent Rillito River, shown in both views at the right, flows between low sandy banks which cannot contain the maximum seasonal flow; the stream is obviously about ten times wider in February than in May. The low and easily erodible banks allow the stream to spread out at times of high discharge, thus adding to the supply of accessible sediment. Furthermore, a wide, shallow stream encounters more frictional resistance from its channel than a deep one. The stream velocity is thus lower than it would be in a confined channel, and this further inhibits the stream's capacity to transport sediment. As a result the bed is not scoured at high water; not even the bush in midstream is removed by the current, and despite the fact that a great deal of sediment is moved the channel looks about the same after a flood as before.

Fig. 126. *Upper:* View up the Rillito River, northeast of Tucson, in February, 1960. *Lower:* Same scene three months later, in May, 1960. (*Photos taken for the author by Tad Nichols.*)

Fig. 127. Portion of a steep artificial bank in loose sandy sediment showing erosion and deposition by recent rains. Width of view is about 10 feet.

The Tendency of Streams to Widen and Deepen Their Valleys

The scene above (Fig. 127) shows the effects of a single rainstorm on a steep bank artificially cut in loose sandy ground. The portion shown is about five feet high and ten feet wide. Such sights are familiar to all who have watched the runoff from a heavy rain convert ruts into ditches, cut grooves in fresh embankments, or furrow a freshly plowed field.

It is less obvious that similar changes are taking place, much more slowly, on steep mountain slopes. Yet if we compare the scene above with the upper one at the right (Fig. 128), in which the narrow parallel gullies and canyons also trench a steep slope and also give rise to low cones of sediment at their lower ends, the similarity is apparent. The mountain slope, which is carved from hard crystalline rocks, is over a mile high and 2½ miles wide—more than 1,000 times the dimensions of the eroded bank. Of course there is an even greater difference in the time represented; the first scene was produced in a day or two, the second probably required tens of thousands of years.

This comparison suggests that the slow progress of erosion on hills and mountains is similar to the much more rapid and observable changes seen in miniature all about us. Experiments with laboratory models duplicating the behaviour of the Mississippi and other rivers have demonstrated that this is indeed true of many aspects of stream erosion: channel cutting and the changing patterns produced by meandering (page 144) and braided flow (page 146), which take place over years or decades on large rivers, can be duplicated in hours or days on a small scale.

The processes by which valleys are widened have already been touched upon in discussing the Grand Canyon, and details of a small example were examined on page 32. In the lower scene at the right (Fig. 129) the whole valley-widening process is represented on an intermediate scale. Here the upper Rio Grande is flowing away from us in a valley 600 feet deep and over half a mile wide. Horizontal ledges of basalt in the uppermost 200 feet of the canyon walls indicate that the surrounding flat lands are a lava plain. Below these (and some talus) is a jumbled mass of tilted blocks, hummocky debris, closed depressions, and what seem to be

Fig. 128. Portion of the steep east wall of Death Valley between Bridge Canyon and Badwater. Skyline is about 5,700 feet above foreground, width of view about 2.5 miles.

remnants of the grassy soil from the surface of the plain. The jumble is the product of inter-mittent downslope movements ranging from creep to small landslides with associated slumped blocks and talus, facilitated in this case by the presence of weak sediments beneath the lavas. It is an illustration of the general axiom that under most conditions valley sides are not able to stand as vertical walls; weakened by weathering, and constantly pulled by gravity, they creep, slump, and slide into the stream. Valley widening is the result.

Figure 130 includes a stretch of the Colorado River, flowing away from us into the Marble Gorge, along which valley-widening processes are at present making little headway. The result is a channel that owes its cross-sectional shape almost entirely to deepening alone. This relatively uncommon condition is the product of special circumstances. Nourished by rains

Fig. 129. Looking south (downstream) along the Rio Grande near Taos, New Mexico.
(*Photo by Robert C. Frampton and John S. Shelton.*)

133

and snows in the Rocky Mountains, more than 400 miles away, the river flows through this semi-arid plateau with much greater volume than it could if it were dependent on local runoff. At the same time, the massive limestones of the plateau are highly resistant to the weathering processes that operate in a dry climate. This combination of a river that is abnormally vigorous and rocks that are highly resistant to downslope movements accounts for the chasm, which is more than 3,000 feet deep in the distance.

The process of valley deepening is essentially a combination of (1) rock decay beneath wet river-bottom sediment during periods of low water and (2) the flushing out of this material during floods. Direct abrasion of fresh bedrock in the channel floor is virtually limited to waterfalls and those uncommon segments of streams that flow on bare rock. Most channels are floored with *alluvium* (loose, stream-transported sediment) and to deepen such channels the stream must occasionally be able to set in motion all of the deposits along its bottom. Marble Gorge proves that the Colorado River has been able to do this. But as we have seen, some streams, and segments of many others, cannot do it because the available supply of loose material from banks and tributaries equals or exceeds the load-carrying capacity even of flood discharge.

If, over the years, the supply of load is greater than the capacity of a given part of a stream, there will be net deposition; the stream *aggrades* its bed in that part by depositing alluvium. Aggradation is common near the lower ends of most rivers. If, on the other hand, the average available load is less than the average capacity, erosion will take place; the stream *degrades*, or cuts into its channel. Degradation is characteristic of the upper reaches of streams.

In view of these facts it is helpful to distinguish three segments of a stream course, an *upper course* along which erosion prevails most of the time, a *lower course* where deposition prevails most of the time, and a *middle course* in which degradation and aggradation alternate. These distinctions are most appropriate in dealing with mountain torrents or with trunk streams that drain mountainous areas but have lower courses lying principally beyond the foothills. In Figures 127 and 128 the upper courses reach to the base of the steep slopes, and the lower courses are confined to the deposits beyond; middle courses are virtually absent, but will develop if the upper channels deepen and the lower deposits build higher. Figures 129 and 130 show upper-course segments of larger streams.

Obviously, true erosional deepening and widening of their valleys by streams and rivers can take place only along their upper and middle courses. Here they deepen their beds directly by erosion, but most of the widening is indirect, being accomplished primarily by downslope movements, which increase as downcutting steepens the valley walls. It follows that the overall steepness of the valley walls is determined by the ratio between the rate of downcutting by the stream and that of slope recession by downslope movements. These are not wholly independent of each other, for although downcutting encourages more colluvium to creep, slump, and slide into the river, the delivery of this material adds to the work of removal the river must do and so retards or postpones further downcutting.

The scene in Figure 131 is a reminder that in more-humid climates the products of weathering accumulate in a thicker blanket of soil whose creeping mobility smooths the shapes of ridges and valley sides. Under such conditions a vertical-walled canyon like that shown in Figure 130 is virtually impossible and the slump topography shown in Figure 129 would be difficult to recognize under its cloak of soil and vegetation. For this reason most of the examples used here are from semi-arid regions. Where there is very little soil and almost no vegetation the effects of erosion and deposition are easier to see, even though in humid climates the processes may be equally, or even more, active.

Fig. 130. Looking southwest (downstream) along the Marble Gorge section of the Colorado River in northern Arizona. Echo Cliffs at the left, Paria Plateau at the right, Kaibab Plateau and upper end of the Grand Canyon in the distance.

Fig. 131. Looking east over the Green River near Mammoth Cave, Kentucky.

The Tendency of Streams to Extend Their Valleys Headward

The third way a stream channel can grow, as it consumes the land, is in length. Again using examples from the less-disguised terrains of semi-arid regions, let us examine the process.

The photograph on this page shows an almost flat surface about a mile wide that slopes very gently from the cliffs at the upper left to the dry river bed at the right. The drainage on this surface is just beginning to cut definite channels; these form branching gully systems that begin at many points along the river bank and reach various distances across the sloping surface. Because the lower end of each system is fixed at the river, and the downstream end of each branch is fixed at the point where it joins the next larger one, the systems can grow in length only by extending existing gullies at their upstream ends or by sprouting new ones. Both constitute *headward growth*, which takes place because the water flowing from the flat into a gully drops over the rim with a steeper *gradient* (slope in the downstream direction), runs faster, and therefore has greater erosive capacity, than the slower-moving sheets and puddles on the ungullied surface. In addition, the gully systems collect runoff from a large area and concentrate it in a few channels, which, because they carry more water at higher velocities, tend to be eroded more rapidly than their surroundings. It is thus inevitable that the gully systems will grow at the expense of the flat surface. If soil and rock resistance are uniform, the upper end of each lengthening gully seeks out low places in the ungullied surface, for the incision must grow not only uphill but also toward the gathering grounds of the waters that feed its stream.

Although headward erosion is a universal tendency accompanying the expansion of a drainage system, there are many situations in which it does not take place. Wherever the upper ends of streams flowing in opposite directions meet at the crest of a dividing ridge, further headward growth is virtually stalled; free upstream expansion in length is no longer possible.

Fig. 132. Dendritic gully system expanding eastward from the right bank of the Chaco River (at far right), west of Farmington, New Mexico. View is southward; dirt roads indicate scale.

In the dry climates of the Southwest most of the rain falls in isolated high mountains, and the resulting short, intermittent streams, many of which are fed only at their upper ends, commonly disappear through percolation and evaporation before reaching the lowlands: these main streams do not grow headward from their mouths, although their tributaries may do so. It is also unlikely that the Mississippi-Missouri ever worked its way headward from the Gulf Coast to the northern Rocky Mountains. Such large trunk rivers are probably descended from drainage lines that are as old or older than the mountains; their ancestors may actually have grown mouthward as the seas withdrew from the land many tens of millions of years ago. Their present courses were produced, during this long period, by the complex interplay between a heaving earth and the tendencies we are examining.

The view at the bottom of this page presents a summary exhibit of the three growth tendencies—toward increased width, depth, and length. The well-defined light-colored bench at about mid-height on the mountain front was cut by wave action when the surface of a large prehistoric lake stood at this level. Below it, half way to the railroad at the bottom of the view, is a second, less prominent shoreline cut during a brief later and lower stage of the lake. These two shorelines divide the mountain front into three zones of contrasting topography. Above the upper shoreline, stream erosion has continued without interruption since before the lake appeared. When the lake came, these streams dropped their sand and gravel close to shore, where it was added to that produced by the waves in cutting the bench, and the resulting deposits helped smooth the lower slopes. The middle zone, between the shorelines, shows the effects of stream erosion that began when the lake level dropped to the elevation of the lower shoreline. The zone below the lower shoreline has been exposed to erosion only since this later stage of the lake disappeared.

Thus, from bottom to top, we can compare the results produced by stream erosion during progressively longer time intervals. The increase in depth and width that accompanies increasing age as well shown by the canyons in the left half of the scene, while at the right new gullies are growing headward into the zone between the shorelines. (See also Fig. 162.)

Fig. 133. Shorelines left by prehistoric Lake Bonneville as it shrank during and after the waning stages of the last Ice Age. Looking southwestward across Traverse Mountains; Utah Lake in background.

Fig. 134. Looking northward over site of imminent stream capture between competing south-flowing drainage lines in the Alvord Mountains, Mojave Desert, California.

The Tendency toward Stream Piracy

When, in the course of their normal tendency toward headward growth, the upper ends of two drainage channels begin to eat into the same upland surface, the stage is set for each to interfere in the affairs of the other.

The view shown here is upstream along the divide between two dry water courses in the Mojave Desert of California. The broad bush-speckled wash draining toward the lower left is part of the flat sandy middle course of a prominent drainage line that carries occasional runoff out of the hills in the background. At the right is the steep head of another drainage system whose badland topography indicates that it is working in easily eroded incoherent rocks. In the center the white bank of the main badlands stream is just about to tap the sandy wash; the divide between them is less than three feet high. It is clear that during one of the next few heavy cloudbursts either the badlands stream will eat through this divide or the sandy wash will cut into it and spill over. Once this has happened, the steeper gradient and increased discharge of the connecting link will cause it to grow rapidly headward into the broad wash to complete its capture of the latter's drainage above this point. The downstream part of the broad wash will be left without its source area, and will be fed only by what runs into it below this point. In short, the sandy wash will have been beheaded by a flank attack: the badlands stream will have captured its headwaters. Stream piracy will have taken place, the upper course of one stream having intersected the middle course of another.

Obviously, stream piracy is likely to occur only when one of two competing streams has a distinct advantage over the other. Such an advantage usually arises from the possession of a lower bed, and therefore a steeper potential stream gradient close to the point of capture. The reason for the lower bed may be a lower mouth, a more direct route to a mouth at about the same level, a course in more easily eroded rocks, or a combination of these. Weak rocks are usually a factor in the more spectacular examples of stream piracy, of which there are many in the Appalachians.

Fig. 135. Topography resulting from stream capture a few miles north of San Juan Capistrano, California. U.S. Highway 101 in foreground.

Now take a look at the scene above. The view is seaward over some coastal hills. The shore is 5 miles distant from the double highway in the foreground, and between the two there is a well-developed system of valleys draining away from us. In tracing these upstream (away from the coast) we would expect to find them ending in a number of successively smaller branches separated by a sinuous divide from other similar stream systems flowing in other directions. But instead, each of these coastal valleys remains approximately the same size until it ends rather abruptly at a saddle or low divide (for example, A and B), opposite very short straight gullies that drain into the valley through which the highway runs.

Two features of this pattern are strange: First, why should the divides or saddles be so close to the large valley containing the highway? The latter's size alone suggests that it has been here longer or is favored by weak rocks so that its tributaries should be the longer ones. Second, why are the seaward-draining valleys so large and straight and unbranched right up to their heads (at A, B, etc.)? They do not now carry enough runoff to account for this shape, which is typical of middle courses rather than headwaters.

The answer to both of these questions lies in stream piracy. The valley followed by the highway has, in working its way headward (toward the lower right), captured the normal headwaters of the streams that once flowed through the other valleys, so that now we see only their lower portions, fashioned in earlier times by a runoff they no longer receive. Subsequent to the capture —again because there is no large drainage area to feed them—only short steep valleys of insignificant proportions have developed on the near sides of the divides. This segment of the pirating highway stream, of course, receives most of its runoff from its left side—including captured drainage from the upper parts of the beheaded streams.

In the example in Figure 134 capture was imminent; in Figure 135 it is an event of the past. We can be sure that there was some good reason for the advantage possessed by the capturing stream, even though it is not apparent in this view what that reason was. It is quite clear, however, that the stream valley followed by the highway has cut its way obliquely across all the drainage lines in the upper and righthand parts of this view, and that a succession of stream captures has played a major part in the evolution of this drainage pattern.

139

Base Level, the Long Profile, and the Graded Stream

Examination of fluvial processes thus far has demonstrated the tendencies of stream valleys to become deeper and wider, to grow headward and sprout tributaries, and sometimes to cut into one another—in short, to extend themselves into, and to consume, the uplands that generate them. Barring serious interruptions, this expansion inevitably leads to a situation in which drainage systems must compete with each other for the dwindling high ground that separates them.

As already noted, however, a stream cannot erode its channel significantly below the level of the body of water into which it empties. The level of the surface of that water is the *base level* for all streams flowing into it. Ultimate base level for all rivers is, of course, sea level. Other base levels, such as those created by lakes or resistant rock barriers along the path of a stream, are local and temporary, inhibiting or preventing downcutting only during the time required for their own destruction.

For practical purposes, then, sea level—extended beneath the land, following the curvature of the earth—constitutes the absolute limit to which stream erosion can go. Actually, except near the mouth of a river, downcutting will never reach this base level, because as erosion lowers the channel the gradient decreases, which lessens the velocity and the energy available for transporting load. We may imagine the current utltimately moving so slowly that it can neither pick up nor hold in suspension any particles of sediment. In such a condition, clear water will flow lazily to the sea and all erosion except by chemical solution will have ceased. This is a theoretical state, for no whole stream system is known to have reached it; the penultimate stages take so long that crustal movements interrupt the process before it is achieved. The concept of base level is thus one of a limit, and streams, or segments of them, may profitably be compared according to how closely they approach this limit.

The *long profile* of a stream is the form of its surface as viewed from the side. It is usually represented by a graph on which elevations are plotted for closely spaced points distributed from source to mouth. The long profile of the Colorado River, from its source a few miles north of Fall River Pass in the Rocky Mountains to its mouth at the upper end of the Gulf of California, is shown in Figure 136. If the information were available, one might plot the profile from elevations on the channel deposits or on the bedrock instead of the water surface; each would have its special significance. But since the complete profile of most rivers is tens or hundreds of miles long, and the depth of water plus channel deposits seldom reaches even a few hundred feet, the differences between such pofiles are usually too small to show on the graph. They may be studied by enlarging a small part of the whole profile, as was done in Figure 38. Also, because many rivers wander within their valleys and the long profile follows all the meandering crookedness of the stream, the long profile of the stream is generally longer and of gentler slope than the corresponding profile of its valley.

The graphic representation of the long profile of a stream is an intentionally exaggerated picture of its gradient in every part. The average gradient of the Colorado River, for example, ranges from one foot per mile between Yuma and its mouth to over 400 feet per mile near its source. In round numbers its total length is 1,440 miles, in which distance it descends two miles. Drawn to true scale, therefore, the height of the graph would be 1/720 of its length or, on the scale reproduced here, about 0.2 mm, or 8/1000 of an inch.

As the long profile of a river like the Colorado is traced upstream, the influence of sea level becomes progressively weaker. In mid- and upper-course, especially, irregularities are likely to be determined by local base levels, differences in rock resistance, and the effects of tributaries. But it is difficult to know what these mean without some understanding of the normal profile toward which the river is tending. Let us examine some of the factors involved.

Erosion of the streambed lowers the channel and deposition raises it. But there are limitations on both processes; ignoring the usually insignificant depth of the water, no point can be cut below the level of the stream's mouth or raised above the level of its source. Within this range the stream erodes here and deposits there in response to a number of factors which, taken together, cause all natural streams to approach (usually, like the Colorado, imperfectly) a long profile that is concave to the sky and progressively steeper upstream. This is true even of canyons cut in the straight slopes of new volcanoes and on the flanks of recently bulged domes and anticlines where the land surface is actually convex to the sky. We can only conclude that this upwardly concave profile must represent a relatively stable condition, a preferred distribution of erosion and deposition along the length of a river, and that it is determined by the running water itself rather than the shape of the surrounding land. It is the normal profile.

Velocity is the single variable that most directly affects the load a stream can move and its ability to erode its channel. The faster the current flows the more and coarser the sediment it can keep in suspension and drag along its bed. But changes in velocity at a given point occur only in response to changes in other factors, chief among which is *discharge*—the total amount of water passing any point in a unit of time. Thus, during highwater and flood stages both the proportion of sediment in the water (assuming suitable grain sizes are available) and the total quantity of water rise sharply: no wonder the total load in motion increases in more than direct ratio to either velocity or discharge.

Velocity is also directly affected by the smoothness, straightness, and cross-sectional shape of the channel—a group of variables often referred to as *channel characteristics*. The same amount of water will flow faster in a narrow deep channel than in a wide shallow one because of differences in the total surface friction against bed and banks. Straightness and smoothness also favor increased velocity; crooked and rough channels tend to retard the flow.

Finally, steep gradients tend to produce high velocities, other things being equal, and gentle ones low velocities.

But the relation between velocity and each one of these factors is a two-way affair. Although increased velocity makes possible much larger suspended loads, the larger loads tax the energy of the current and reduce its rate of flow; in an extreme example enough fine material may be available to convert the current to a sluggish flow of soupy mud. While straight and smooth

Fig. 136. Long profile of Colorado River. Note coincidence of many irregularities
with changes in rock resistance or the junctions of tributaries.

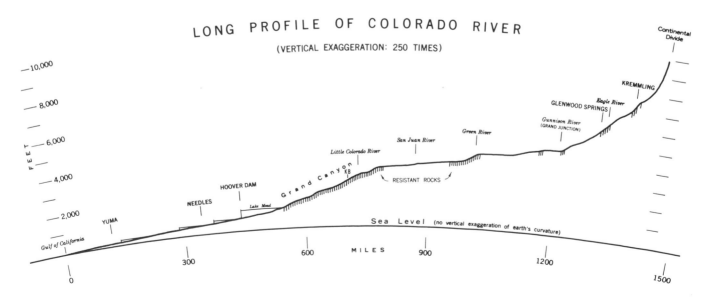

channels favor faster currents, the latter are more erosive and where thrown against an erodible bank will cut into it, converting the straight course into a crooked one. And although steep gradients favor high velocities, if the stream bed is erodible a faster current will pick up more sediment, thus lowering the channel, which will reduce the gradient and slow the current. In each case the interplay tends to restore conditions that existed before the balance was disturbed rather than to produce extremes.

Among these interrelated factors we can distinguish between relatively independent variables and dependent ones. In the long run only the available supply of water, and the available supply of sediment (potential load) are truly independent; these are imposed on the system from the outside and can be altered by changes not directly related to what is happening in the stream—as when a change in climate or stream piracy increases or decreases the supply of water, or a landslide abruptly changes the amount of available debris. Ultimately, channel gradient and, to a large extent, channel characteristics, are dependent variables; they are produced by the stream itself and change in delicate response to such variables as velocity and load.

Consider, for example, the long profile of the alluvium-floored Colorado River. The amounts of water and sediment supplied to it change at each point where it is joined by a tributary, every time a bank caves in, every time a storm passes over the drainage basin, with the seasons, and even following forest fires, to say nothing of possible gains and losses from stream piracy. These are all independent variables and the different segments of the river must be almost continuously adjusting slope, channel, velocity, and amount of load in motion to accommodate their fluctuations. Yet over the years the average supplies of water and sediment along the river are constant enough to be considered characteristic of this particular drainage system. These average conditions would be distinctly different if the stream were located in a different region; the Columbia and Potomac are not as muddy in mid-course as the Colorado.

How are these differences expressed in the river? *In the long run, the gradient and channel characteristics tend to adjust themselves in such a way as to provide, with the available supply of water, just the velocity required for the transportation of the load supplied from the drainage basin.*

This means that in any part of a stream that is not running over hard rock, but instead has plenty of available alluvium along its banks and bed, the amount of load in motion and the velocity are sensitively related and their effect on the gradient is self-correcting. If the potential load of such an adjusted stream is increased, as by a landslide or the sudden deposition of a bouldery delta at the mouth of a steep tributary, the excess cannot be moved, unless it is finer grained or more easily moved than material previously available. The accumulation raises the bed, increasing the gradient downstream from the disturbance; this accelerates the flow and increases the competence of the current to handle the new load. (This is the cause of most of the rapids in the Grand Canyon at points where the Colorado is joined by steep tributaries.) If, on the other hand, the load is decreased (as it would be if the stream passed through a lake, which would serve as a settling basin), the now underloaded or "hungry" current will erode its bed, thus lowering the gradient, which in turn reduces the velocity, and will continue to do so until the velocity is so low that erosion is no longer possible—i.e., when the current is again just able to handle the load. This explains why the Colorado River has removed much sediment from its channel just below Hoover Dam since the latter was completed in 1935.

Changes in discharge are accommodated in the same way. If discharge increases, as below the confluence of a lightly loaded tributary, the increased capacity for load results in erosion of the bed, which lowers the gradient and slows the velocity, the adjustment automatically ceasing when the rate of flow is again just sufficient to move the load.

Without going into the complexities of associated adjustments in channel characteristics such as width, depth, crookedness, and roughness of the bottom, it is probably safe to conclude that gradient is the most consistently and completely dependent variable. It is for this reason that the long profile of a stream has basic significance. A stream, or segment of a stream, which has reached the condition in which slope and channel characteristics are adjusted to the average supply of water and load is said to be *graded* (compare aggrading and degrading, page 134). The graded stream represents a kind of equilibrium between external influences and the processes within the stream. This equilibrium is imperfect at any given moment, but because it contains a built-in feed-back mechanism that responds to disturbances in any of several interdependent factors, it results in constant changes in the direction of a norm that is prescribed by the average supply of water and load and is expressed in a smooth, headward-steepening, normal profile.

How does one recognize a graded stream? This is admittedly difficult because it is average conditions over a period of years or decades that must be evaluated. However, the graded stream deepens its valley extremely slowly and may not do so at all for long periods; the net deepening in several thousand years could be less than the fluctuations of one season. Accordingly, any evidence favoring stability of the profile would imply that a stream was graded. The long profile of a graded stream or stream segment is always smooth and progressively steeper toward the source. Under most conditions the banks and bed of its channel will be alluvium, not bedrock, and the greater part of this will be moved only during floods. A graded stream or stream segment should not show evidence of active downcutting, such as a steep narrow canyon hardly wider than the stream, with walls and bottom of exposed bedrock, or rapids, or waterfalls over resistant ledges. Graded conditions are usually confined to the lower and middle courses of a stream.

Applying these criteria to the Colorado River (Fig. 136), we can be quite sure that the Grand Canyon segment is not graded, that most of the section below Hoover Dam probably was graded before the lower dams were built, and that, at most, probably only short stretches of the upper river (the ones in weak rocks) are graded.

Why should we try to distinguish graded from ungraded streams? Because each means something quite different when it comes to interpreting the geologic histories in which they have participated. In general, ungraded streams, being out of adjustment with the supply of water and load they are handling, must be looked upon as still recovering from one or more major upsets, among which the two most likely are crustal deformation and the climatic changes during and since the last Ice Age. Graded streams, on the other hand, imply that the controlling conditions have remained constant long enough for equilibrium to be established. The time required for this varies greatly. The fact that graded segments are usually found only in those parts of a river underlain by relatively weak rocks may be taken as one more bit of evidence that the earth's crust is restless—too frequently so during recent geologic time to allow most streams to reach grade where they flow through resistant rocks. Each irregularity in the long profile is evidence for some kind of interference with the smooth and characteristic curve every stream strives to establish.

The Tendency of Stream Channels to be Crooked

Natural stream channels are almost never truly straight as seen from above. The closest approximations to straightness are found in short segments, seldom longer than about ten times the channel width. Even where the banks are straight, the axis (deepest part) of the channel wanders from side to side as the main current swings to right and left. Any irregularity in the channel capable of deflecting the current against a bank can set a whole series of swings in motion. The current will erode the bank, forming a curving embayment. The more deeply the bank is embayed, the greater the change in direction of the current and the more vigorously this curve sends the stream back against the opposite bank. There, the process is repeated and the current is returned to the first bank, farther downstream. The process tends to be self-perpetuating and accounts for such sinuous courses and crescentic *ox-bow* lakes as may be seen in the views at the right.

The tendency toward crookedness is apparently present in natural streams of all sizes. It is not very effective in steep channels between banks of bedrock, but as any stream segment approaches the graded condition, downcutting becomes very slow and bank erosion more conspicuous: every curve tends to enlarge because in rounding the bend the current is thrown against the outside, where it deepens the channel and undercuts the bank. At the same time deposition takes place in the slack water on the inside of the curve, which is the familiar location of sand and gravel bars and flats (white crescents in Fig. 137). In this way each loop is enlarged as it slowly creeps downstream. Such a river is described as a *meandering* stream.

Note that despite this tendency to expand, the meander belt has a definitely limited width—about half that of the valley floor in the upper scene. As successive loops on the same side of the valley are enlarged the neck between them narrows and finally is broken through. Thus each ox-bow lake is an abandoned meander loop. The width of the meander belt is determined by the size of the individual loops before they are cut off—which in turn is related to the size of the stream. In general the "wave length" of the meanders ranges from about 7 to 15 times the width of the stream, the ratio being greatest for large rivers.

In both scenes there are ox-bow lakes and meander scars in many places on the valley floor; these are not all close to the present course of the river, proving that the meander belt as a whole shifts its position. Also, the relatively straight valley walls have been cut into only here and there, now and then, as the meander belt swung against them; note, in Figure 138, the curving swipes thus left along their lower slopes.

Because of the continuous slow change in position of the channel, the entire valley floor of a meandering river consists of alluvium—stream deposits that are constantly being reworked as the stream reexcavates and redeposits the sediment. In so doing it gradually moves the alluvium down-valley. Since the accumulation is a composite of bars and deposits on the insides of bends, put down by a current that may have flowed in any direction, the alluvium is thoroughly cross-bedded. Such a valley floor is perforce both flat and very nearly at the high-water level of the stream. Being thus susceptible to widespread shallow flooding it is known as a *floodplain*.

Floodplains are normally produced in lower and middle course by graded or aggrading streams. If tectonic or climatic changes convert such a river to a degrading one it will deepen its channel and become *entrenched* at a lower level, from which it can no longer inundate the valley floor. If a new floodplain is developed at this lower level, remnants of the old floodplain may remain along its margins as *stream terraces* (page 162). In the lower view, all of the cultivated area (rectangular fields) on the far side of the river is situated on terraces above the present floodplain; at least four different levels can be distinguished.

Fig. 137. Meander belt and floodplain of the Animas River a few miles above Durango, Colorado.
Looking northwest; river flows toward lower left. Farm buildings and roads give scale.

Fig. 138. Meanders, floodplain (with haystacks), and terraces along the upper Colorado River
at the junction of Troublesome Creek (background), about 5 miles east of Kremmling,
Colorado. Looking north-northeast; river flows toward lower left.

The Braided Stream

One of the fascinations of running water is its almost infinite complexity. The interplay of such variables as discharge, velocity, turbulence, shape and roughness of channel, quantity of load and the sizes and shapes and proportions of the different kinds of particles, the ways in which these change downstream, and even the effect of the rotation of the earth, is so intricate and difficult to study that so far it has defied complete analysis by man. Progress is being made through the study of thousands of natural streams and rivers, controlled irrigation ditches, and small scale models in laboratories, but there is much yet to be done. No way has been found to sample the bed load dragged along the bottom of a natural stream without disturbing the currents. Rigorous mathematical treatment of turbulent flow is so complex as to be impractical, even when the role of the suspended particles is neglected. Yet it is just such particle movements that are important to engineers and geologists.

Humbled by this complexity, let us select a single manifestation, the *braided* stream, illustrated in these views, and try to develop a plausible qualitative explanation. Both streams are flowing toward us. Careful examination of either one shows that the stream pattern consists of channels that divide and join again and again. The unusual feature is that the current should divide at all in the downstream direction. Normally, when two streams join, their combined capacity for moving solid particles is greater than the sum of their individual capacities, for the proportionate frictional loss against bottom and banks is less below the junction than above (because the same amount of water that flowed between four banks above the junction is now constrained by only two). The single stream should, therefore, be able to scour out a deeper channel below the junction. To a limited extent this is true in these scenes; note that, in the upper one especially, the deepest stretches of the many abandoned channels tend to occur below confluences. But the effect does not persist; a little farther downstream each deepened channel broadens, becomes shallow, and divides again.

From these examples it would appear that perhaps a basic condition for braided flow is an abundance of available load of such coarseness that the average discharge and gradient of the stream can move it only with difficulty, or can move only the finer particles. Probably, under these conditions the slight advantage gained by the junction of two currents results in erosion of the channel floor just below their confluence, and this introduces a flat place in the long profile. The water is now faced with transporting a slightly larger than average load along a slightly gentler than average gradient; relatively speaking it must climb out of the hollow it dug. To do this it must leave behind some of its load. The deposit thus formed makes the channel shallower, forcing the stream to widen and erode its banks. This both slows the current and adds to its available load, thus destroying its advantage until it is joined by another current and repeats the process downstream.

Exactly why the deposits sometimes form a central bar aound which the current is deflected or divided is not known. It may be that fluctuations in discharge are a factor; local overflowing and piracy produce rapid changes in the sizes of individual channels and the amount of water they carry, and the whole system is sensitive to local weather—especially if glacial meltwater is a major source (as it is for the stream shown in Figure 139).

The two examples given here typify the conditions under which braided flow generally occurs. The upper one has a steep gradient, is moving coarse gravel, and is probably aggrading (note the trees being buried in the lower right corner and the absence of high banks). The lower one has a gentle gradient, is moving sand, and although it was presumably aggrading here at the close of the Ice Age, is now probably a graded stream. Both streams have an ample supply of available load, which is certainly characteristic of braided flow, if not essential to it. Actively degrading streams are "hungry" and would seem to have less chance to achieve the sensitive balance between erosion and deposition that apparently characterizes braided flow.

Fig. 139. Braided flow on gravelly bed of Muddy River a short distance above its junction
        with the McKinley River, Alaska. Looking upstream. (*Photo by Bradford Washburn.*)

Fig. 140. Braided pattern of sandy Platte River, 13 miles northeast of Grand Island,
        Nebraska. Looking upstream.

Leveling *vs.* Crustal Uplift I:
Dominantly Tectonic Landscapes

Every landscape is a geological battleground—the scene of a struggle between two complex sets of processes. On the one hand are the leveling processes of weathering, erosion, and deposition, which tend to subdue the relief of the earth's surface by removing rock debris from high ground and dumping it in low places. Landscapes dominated by these may be referred to as *fluvial* landscapes, *glacial* landscapes, and so on, according to the principal process involved. On the other hand are the more obscure processes which slowly elevate, depress, wrinkle, and break the bedrock of the outer crust from place to place and from time to time. Where shapes resulting directly from these are dominant, we will call the landscapes *tectonic* (Greek: *tekton*, builder). Most landscapes, of course, exhibit the effects of both influences.

In evaluating the interplay of these two sets of processes, let us begin by examining three types of landscapes involving local uplift, each illustrating a different balance between the opposing influences; first, in this section, dominantly tectonic landscapes, in which leveling processes are just beginning to assert themselves against the shapes produced by crustal deformation; second, (page 150), landscapes that bear approximately equal imprints of tectonic and leveling processes; and third, (page 152), landscapes in which leveling processes have almost obliterated the tectonic forms.

The steep mountain face in the upper view here, over 4,000 feet high, is composed chiefly of crystalline rocks and has an average slope of more than 25° in the lower part. This is several times steeper than most hills and mountains, yet there are only small accumulations of alluvium along its base. Furthermore, at some places the alluvium is broken into small steps by faults that parallel the base of the mountain (e.g., Figs. 147, 150, 340), an indication that there has been geologically recent uplift of the mountain relative to the valley floor. Erosion has made little progress in cutting into this local uplift; the gullies and canyons are narrow, shallow, extraordinarily steep, devoid of channel deposits, and almost unbranching.

This mountain front is the edge of a rising block that is little modified by erosion—a dominantly tectonic landform. Crustal deformation has here gotten far ahead of leveling processes, probably because of the resistant rocks, geologically rapid uplift, and arid climate.

The lower scene represents an analogous situation in nonresistant rocks. The little stream valleys are obviously just getting started. The smooth slopes, covered with grass, are underlain by relatively weak silty and shaly sediments. Remarkably little of the total area is occupied by drainage lines and their associated side slopes; none of the canyons has a bottom any wider than its stream and very few of the ridges between streams are sharp-crested.

Why should streams have made so little progress in an area of weak rocks, where we would expect just the reverse? Either the streams are extraordinarily feeble, or they have been at work only a short time, or both. That their feebleness is not the whole explanation is proved by the fact that nearby hills, including those visible in the background, are deeply cut by canyons, despite their more-resistant rocks. It follows that this grassy surface must have been available to the streams for only a short time. The reason for this geological newness, only faintly suggested by what can be seen in this view, is that the grassy slope is essentially a blanket of creep and landslide debris derived from the weak rocks. These downslope movements are a superficial effect of thrust faulting and folding near the base of the higher hills in the background. The grassy hills thus owe their height to tectonic processes; they would be even higher and steeper if the rocks were not so weak that they slumped. This landscape, born of crustal deformation and slightly modified by creep and sliding, is now in the very early stages of attack by streams.

Fig. 141. East wall of Death Valley, California, just south of Mormon Point. Looking northeast; highway along base of mountain gives scale.

Fig. 142. Looking southward over the Pleito Hills at the south end of San Joaquin Valley, California. Pipelines and roads give scale.

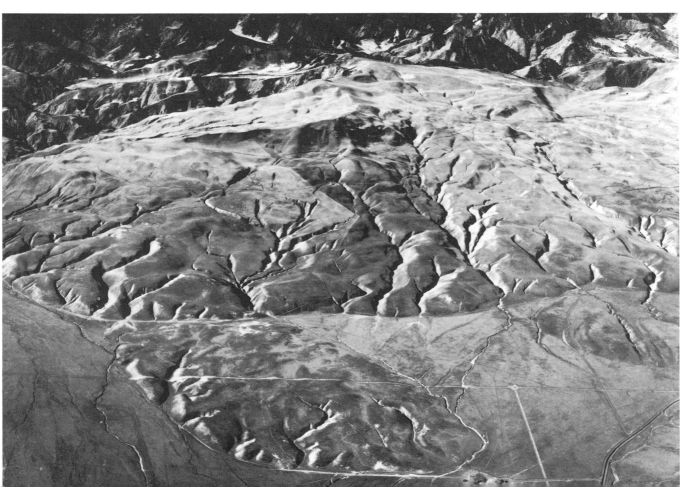

Leveling *vs.* Crustal Uplift II:
Landscapes Bearing Approximately Equal Imprints of Tectonic and Leveling Processes

These scenes illustrate the second of our arbitrarily chosen stages in the struggle between crustal deformation and leveling—landscapes in which neither process is clearly dominant.

The landscape in the upper one is being carved from weakly cemented sands and gravels in a dry climate; the area is less than 30 miles from that shown in the last scene (Fig. 142). It is apparent that here the normal increase in length, width, and number of members of the drainage system has consumed nearly all of any pre-existing surface. Except in the far background, the only obvious remnant is the smooth light patch at the right in the middle distance; the rest has been replaced by gully walls and stream bottoms. The divides between stream courses are now sharp-crested ridges. One of the streams (draining toward the lower left corner) has begun to develop into a main stream, probably partly by piracy, and is collecting drainage from a large area. Note that the cross-profiles of the larger stream courses in this scene are not sharply V-shaped; their bottoms are just beginning to widen and to develop the flat alluviated floors that work upstream as the middle course develops.

The progress of erosion is thus clearly evident, but the part played by crustal deformation must not be lost sight of. Although the sands and gravels of which this upland is made are stream deposits, the present streams could not have built it in its present position; they are now destroying it. These sediments must have been laid down under different conditions—in a basin or on an alluvial slope at the foot of mountains out of which vigorous streams carried them. The conditions probably resembled those now prevailing in the area shown in Figure 148. What can change a basin or an alluvial slope, a site of deposition that lies below its surroundings, into an upland, a site of erosion that rises above its surroundings? Because older deposits now stand at higher elevations nearby, the answer here is crustal deformation that has uplifted this and other margins of the filled basin toward which these streams are headed.

The lower scene shows an area of plutonic and metamorphic rocks; it resembles the previous one but is on a much larger scale, tall trees look smaller than bushes do in the first scene. But the stage of the fluvial processes is indicated by the same criteria: narrow ridge crests, most of the area occupied by slopes between ridge crests and drainage lines, the main streams just beginning to broaden their valley floors.

How do we know there has been uplift here? The chief evidence is that the mountains are composed entirely of deep-seated rocks that can have formed only at depths of several miles below the surface. As emphasized earlier, such rocks are visible at the surface only where these miles of overlying rocks have been removed, and erosion cannot accomplish the task unless the area is raised well above the base level of the streams that are going to do the work; at the very least this means substantially above sea level.

The time required for the leveling processes to achieve any given degree of dominance over crustal deformation depends, of course, on many variables. Fluvial erosion, as such, will progress more rapidly in weak rocks like shale than in resistant ones like granite, and more rapidly in steep terrain where downslope movements are vigorous and current velocities high, than on flat lowlands where streams wander lazily about. Climate is also a factor. The most rapid erosion takes place where there is little protective vegetation and where the rainfall, even though there is little of it, is concentrated in a few copious downpours—conditions typical of the margins of the world's desert belts. When to these variables is added the evidence (discussed on pages 416–418) that crustal deformation takes place at very different rates in different places and can be intermittent, it is easy to see how leveling might succeed in dominating one area before it had really started in another, even though the related crustal deformation began at the same time.

Fig. 143. Looking eastward over dissected ancient alluvial deposits at the east end
    of Cuyama Valley, California.

Fig. 144. View northward between the South Fork of the Boise River and the Sawtooth Mountains
    (in distance) in south central Idaho.

Leveling *vs.* Crustal Uplift III:
Dominantly Fluvial Landscapes

The third type of landscape we have chosen to examine, that in which leveling processes dominate, is illustrated by these scenes. It should be borne in mind that just as no large, perfect example of a purely tectonic landscape can exist because leveling processes will start to destroy it before it is fully grown, so no large, perfect example of a purely fluvial landscape can exist because crustal movements will deform it before it is completely fashioned by erosion and deposition. Every landscape therefore contains evidence of both processes and we must be content with examples of *relative* dominance.

The upper example at the right is a view downstream along a drainage system in relatively weak sedimentary rocks and a semi-arid climate. It has recently become entrenched a few feet and is sprouting new gullies, but the total effect of these on the landscape is as yet very small. Ignoring the gullies, or imagining them filled, we can picture the landscape before they were cut: a landscape dominated by wide, shallow, grassy valleys and broad, low, wooded hills. The total relief is not more than a few hundred feet.

There are two good reasons for ascribing this surface to erosional leveling. First, by comparing this area with its surroundings, it is possible to demonstrate that several thousand feet of overlying strata that are present nearby are missing here; erosion has lowered the land by this amount. Second, there are a number of gentle anticlines and synclines that trend across this drainage system without the slightest effect on the topography; any surface expression of these local wrinklings of the crust has been completely erased by erosion.

The scene thus qualifies as one in which leveling processes dominate tectonic processes. The fact that the stream has recently begun to deepen its channel simply proves again that conditions do not often remain stable long enough for streams, even those favored by weak rocks and sparse vegetation, to achieve complete ascendency. The cause of the entrenchment here is not surely known; it could have resulted from a change in climate, crustal movements, weakening of the grass cover by grazing cattle, or a combination of such factors.

The lower scene involves a very different climate and almost totally different kinds of rocks. The wooded hill is a remnant of resistant metamorphosed sandstone on a lowland that stretches as far as the eye can see. Beneath the well-developed soil in the foreground are belts and patches of schist, metamorphosed volcanics, slate, marble, sandstone, limestone, shale, and fresh volcanics—and several varieties of some of these. The structure of these rocks includes two major unconformities and several faults, but the many tectonic events thus recorded have virtually no effect on the present topography.

When fluvial erosion produces such an almost-flat lowland by working the landscape down almost to the theoretical limit (base level) the lowland is known as a *peneplain* (Latin: *paene* = almost). The peneplain is the nearest thing we know to complete victory for the leveling processes. Hills like the one in the background that rise above a peneplain because they are composed of especially resistant rock are often referred to as *monadnocks*—from Mt. Monadnock in New Hampshire, which was used as an example when the concept was being formulated.

Both the underlying structure and the monadnock imply that there probably used to be greater topographic relief in this area; restoring the eroded rock layers to their full thicknesses as measured in the bedrock would make mountains at least 15,000 feet high. Although erosion contemporaneous with the deformation undoubtedly prevented the attainment of anything like such heights, it is likely that there were times when erosion dominated the scene less than it does now. We may be sure that this placid stream system and its ancestors knew busier days, when there were steeper gradients and far greater loads to be carried away. With so much work behind it, perhaps it deserves to loaf in its bed for a time.

Fig. 145. Looking downstream along a 6-mile stretch of Dinnebito Wash
on Black Mesa, northeastern Arizona.

Fig. 146. Looking upstream along the Potomac River toward Sugar Loaf Mountain, Maryland.

Fluvial Landscapes in Dry Climates:
Alluvial Fan, Bajada, and Playa

In studying the opposing tendencies of crustal deformation and leveling processes there are several advantages to working with landscapes in a dry climate rather than those in a humid one. More bedrock is usually exposed, so that folds and faults are easily recognized. Downslope movements are less concealed and usually so intermittent that there is plenty of time to examine the effects of one change before the next occurs. The battle is fought in the open; camouflaging of weapons and removal of casualties are at a minimum.

The paucity of vegetation in semi-arid and arid regions leaves the scanty soil unusually susceptible to erosion. Consequently desert rains are highly effective, even though infrequent. It is difficult to conceive of the total impact energy of raindrops. The large ones fall with velocities of about 30 feet per second (= 20 miles per hour) in still air and strike even harder if wind driven. By the time one inch of rain has fallen on it, an acre of ground has stopped a total mass of 113 tons, equivalent to suddenly arresting the motion of 75 three-thousand pound automobiles traveling at 20 miles per hour. Small pebbles are jostled by the impact and finer particles are tossed into the air by the splash. (Mud and soil particles may be seen splashed up to heights of two or three feet on walls and fences after a good rain in any climate.) Such disturbance is enough in itself to coax the particles down the gentlest slopes, even if there were no subsequent runoff of the same water. Rainsplash helps to prevent the burying of ledges and cliffs on upper slopes of dry-climate ridges.

Another distinctive feature of fluvial processes in dry climates is the way alluvium accumulates on the lower slopes of mountains. The occasional hard rains produce turbulent torrents that charge down the canyons, heavily loaded because the loose and unprotected rock debris is easily moved by the current. Emerging from the confines of a canyon like that shown in the upper view at the right, the runoff spreads over the gentler slopes, losing velocity because of decreasing depth, and losing volume because of percolation into the porous deposits of previous outpourings. Usually the entire flow is absorbed in a few hundred yards, or at most a few miles beyond the canyon mouth, and most or all of its load is dumped within a short distance. The result is a land delta with its apex at the mouth of the canyon—an *alluvial fan*.

Note that there is no single well-defined channel down the fan; the water evidently shifts its course as successive channels become choked by alluviation. Braided patterns are common. The down-fan decrease in vigor of the current results in a down-fan decrease both in slope and in maximum size of fragments—often from boulders many feet across on slopes of 10° to 15° at the upper end to small pebbles and sand on slopes of less than 3° near the outer margin. The manner of accumulation precludes good sorting or well-defined bedding; a typical assemblage from about half way down a large fan was shown in Figures 1, 2, and 3.

Along the fronts of many desert mountain ranges growing fans from adjacent canyons have coalesced into a composite alluvial apron, known as a *bajada* (bah-hah′da) in the southwestern states. Several bajadas are illustrated in the lower view: their braided surfaces of gray gravel sweep down from the mountains, in some places surrounding outlying hills, and meet along the axis of the intermontane valleys. Here, low places along the axis are occupied by *playas*, or "dry lakes." Occasionally, in the infrequent years of heavy rain, the playas are flooded. The resulting lake is seldom more than a few inches deep and usually lasts only a few weeks, but the fine sediments and salts that accumulate in it produce a hard, flat, light-colored surface.

Alluvial fans and playas are built only by intermittent streams that seldom flow far beyond the mountains that generate them. The fan is a monument to the death of the stream that builds it, though some water usually continues toward the valley by slow underground percolation.

Fig. 147. Single alluvial fan at the mouth of a canyon in the steep east wall of Death Valley.
　　　　　Badwater at left; highway gives scale. Note light fault scarps across head of fan.

Fig. 148. Bajada slopes and playas along the south part of the Death Valley trough, California.
　　　　　Looking northwestward over Silurian Lake (playa) toward the Panamint Range on the skyline.

Leveling *vs.* Local Subsidence of the Crust

When a small area of the crust sinks below its surroundings, the leveling processes react with sedimentation; they begin to fill the depression while it is still sinking, just as erosion begins to cut down uplifts as they rise. This is probably almost as true beneath the oceans as it is on land. There are many subtle variations on this theme—depending, for example, on whether the subsidence is along the coast or inland, surrounded by plains or mountains, takes place rapidly or slowly. Some structural depressions are areas that have simply lagged behind in a general uplift.

We are interested here in recognizing depressions that result directly from movement of the crust and have not yet been entirely obscured by leveling, that is, *tectonic depressions*. Most areas that are still sinking show downward warping and faulting of recent sediments around their margins. Topographic basins that have no outlets and are notably deep despite active infilling must have subsided in the recent past. Well-known examples that illustrate one or both of these conditions include the Gulf of California and its northward extension the Salton Sink (Fig. 352), the Dead Sea (whose bottom reaches about 2,600 feet below sea level despite infilling from all sides), and the floors of such deep inland lakes as Tanganyika (more than 500 miles from the coast and more than 2,000 feet below sea level) and Baikal (more than 1,000 miles from the coast and reaching more than 3,400 feet below sea level).

The upper view here is northward along the lowest part of Death Valley, a sinuous 150-mile trough in southeastern California. Topographically it is a closed depression whose floor, about 285 feet below sea level, is 2,000 feet below the lowest point on its rim. Huge alluvial fans that poured into the Valley from the west have constructed a dark bajada, parts of which are seen in the foreground and at the left. Interbedded alluvium and lake deposits under the Valley floor are probably several thousand feet thick. Such conditions can exist only if the bedrock beneath the Valley has subsided relative to the surrounding mountains.

A closer look at certain features of the Valley throws some light on how this subsidence has taken place. Note that the fans along the east side are short and that many of them are isolated (e.g., Fig. 147) in contrast to the long bajada on the west. A small part of this asymmetry may be explained by the greater height of some of the mountains on the west side, but essentially it is the result of eastward tilting of the bedrock floor, which has encouraged the growth of the fans from the west while breaking and partly burying those from the east (note the light-colored fault scarp across the head of the fan in Figure 147). The shoreline of an ancient lake, believed to have existed about 2,000 years ago, is now 20 feet lower on the east side of the Valley than on the west. Indeed, the location of the present bottom of the Valley—not in the center of the trough but crowded close to the east side—proves the continuation of such tilting, and sensitive tiltmeters installed in 1958 indicate movements today which, if they persist for 1,000 years, will depress the end of a one-mile line by 10 feet or more.

Faults also cross many fans on the west side; the lower scene at the right includes an example. The dark bushes in the foreground mark the zone where the lower end of the sloping stream-deposited fan joins the flat lake deposits on the floor of the Valley. Up the slope is a sharply marked line of faulting along which the near (valley) side has been dropped about 15 feet. The older fan surface in the background is conspicuously cut into by drainage lines adjusted to the new lower base level on this side of the fault.

Even if, as seems likely, some of the displacement of some of these recent faults has been caused by settling associated with compaction of the valley fill, this only means that they are secondary instead of primary manifestations of the dominant fact: Death Valley has been and is an actively sinking basin.

Fig. 149. Looking northward into the central part of Death Valley. The lowest point in the valley, about 285 feet below sea level, is in the middle distance.

Fig. 150. Faulted alluvial fan on the west side of Death Valley near Tule Spring. Looking west, up the fan.

Evidence of Changed Conditions I:
Rejuvenation, or Accelerated Erosion

Normal fluctuations in the many conditions that influence the behavior of running water keep every stream and river in a continual state of change. Some adjustments are necessitated by movements of the crust. Others arise from stream piracy or the long slow climatic changes that attend the coming and going of an Ice Age. Even such ephemeral events as fires that burn off vegetation, or the introduction of grazing animals that eat it, can produce noticeable changes in the behavior of a stream.

It follows that most streams should show, somewhere along their lengths, evidence of changing conditions. This might include cutting into their own deposits (change from deposition to erosion), dropping sediments in the valleys they have excavated (change from erosion to deposition), or abandonment of old routes for new. The next six pages show three examples of evidence for recently changed conditions; the piracy illustrated in Figure 135 is a fourth.

In Figure 151 we are looking upstream along a valley whose entrenched stream now flows in a deep, tree-filled gully between flat remnants of its former floor, which now appear as broad benches or *stream terraces* on either side. Since the present condition is one of gully cutting and widening that will ultimately destroy the terraces, the old floor that the terraces represent must have been produced under different conditions. The stream has changed from one that was either graded or aggrading to one that is actively downcutting. There may have been an earlier change, too, for the terraces extend into the mouths of small side canyons in the manner one would expect if a body of alluvium had filled a pre-existing valley. If this is what happened, in the first stage the canyon was cut to a depth below anything now visible; it was then filled to the level of the terraces; and now, in the present stage, the fill is being trenched.

A stream such as this one, which can be shown by the presence of terraces to have recently incised its valley floor, is said to have been *rejuvenated*, in allusion to the renewed vigor which such accelerated erosion implies.

The lower scene also contains a two-stage landscape. Stretching from the mountains in the background almost to the river in the foreground is an old bajada on which the pattern of an older drainage system is still faintly visible (especially in the upper left). A new system of gullies and small canyons is working its way headward from the river, cutting into the bajada and actively destroying it. When the bajada was being built, by aggrading streams, the river could not have existed at its present level. Since then the river has either created or deepened its valley, thus lowering the base level of the drainage lines across the bajada. With their downstream ends thus steepened these have become degrading streams; they have been rejuvenated and are now engaged in reshaping the land to bring it into adjustment with the new base level.

For the sake of analysis, let us make the very plausible assumption that the river in this example was itself rejuvenated and passed the change along to these tributaries. What are the likely causes of such a rejuvenation? Basically, they include anything that will increase the river's capacity for load, either by causing an increase in discharge, or by causing a decrease in load from upstream. Increased discharge might arise from piracy or from climatic changes that increase the amount of water, or from lowered base level or downstream tilting, which would increase the velocity of the current. A decrease in load would result from the formation of a lake just upstream, since it would act as a settling basin.

Having such possibilities in mind is an important part of geological field work. Evidence of a former lake basin or of piracy upstream should not be difficult to find. For evidence of lowered base level we would study the mouth of the river, which, in this case, would involve us in the history of the larger one to which it is tributary. Regional tilting would produce different effects in different streams depending on whether they flowed down, across, or up the direction of tilt. In practice, a combination of several causes is often encountered.

Fig. 151. Looking upstream along the incised stream and terraces of lower
Hungry Valley south of Gorman, California.

Fig. 152. Looking northwestward over the Virgin River toward the Mormon Mountains
in southern Nevada.

Evidence of Changed Conditions II:
Alluviation and Underfit Streams

Just as any change which increases a stream's ability to move sediment will bring about rejuvenation, so any decrease in this ability will bring about aggradation, or (since the deposit along the stream will be alluvium) *alluviation*.

Alluviation can be induced by changes that lessen the stream's capacity or increase its load —the reverse of the changes just discussed. Possible causes include decreased velocity, either through a reduction in discharge (victim of piracy; change in climate) or in gradient (rise in base level; tilting), or a marked increase in load. Increase in load can result locally and temporarily from landslides and forest fires: increases derived from melting glaciers may be more widespread—the heavily loaded streams of meltwater issuing from the waning glaciers of the last Ice Age so choked many streams and rivers that the evidence is still apparent in many areas.

The upper scene here illustrates a common situation along some coasts—a stream valley whose lower end has been flooded or "drowned" by rising base level. As far as the stream is concerned it is not important that we cannot tell whether this means that sea level has come up or that the land has gone down, or both. The point is that this small stream has an unusual mouth; after cutting a deep gorge through hard rock on its way to the sea, it has apparently widened its valley and spread out to form acres of mud flats behind the white sandy barrier beach. Drilling has shown that the broad flat floor of this valley mouth (which includes the white salt marsh at the right) is an accumulation of lagoonal deposits—alluvium, sand washed in over the beach during storms, and plant and animal remains appropriate to the brackish (mixed fresh and sea) water in which it has accumulated. The depth of this infilling sediment is a measure of the relative rise of sea level—more than 200 feet. As the level rose, the current dumped its load in the ponded waters. This raised the river bed, lowering the gradient immediately upstream and causing deposition there also, until the marshy flat now extends more than a mile upstream and buries several hundred acres of the former valley.

Abrupt but enduring changes in the discharge of a stream can occur as a result of piracy; the flow of the pirating stream increases and that of the beheaded one decreases. Changes in climate can change the amount of discharge throughout a whole region. Evidence of a decrease in discharge is more easily recognized than evidence of an increase: it usually consists of a small stream in a disproportionately large valley—one it probably could not have produced even by extensive meandering. A striking example is shown in the lower view at the right. Draining toward us across the flat farmlands is a wide channel, tributary to a still larger one which it joins in the foreground. (A lake now occupies part of the larger channel.) The stream in the tributary is a mere thread, interrupted by lakes and ponds at several places upstream. There are no concave scars in the valley walls to indicate that this stream has impinged on them in the process of meandering across the full width of the floor. The floor is neither as flat nor as close to stream level as it would be if it were the floodplain of the present trickle. This is, therefore, an *underfit* stream, flowing in a valley inherited from a time when the discharge was much greater.

The reason for the reduced flow is not apparent in this view. This is a channel that carried water from melting ice during the late stages of the last Ice Age (pages 218–231); it may have been filled bank-to-bank as recently as 10,000 to 20,000 years ago.

Fig. 153. Looking onshore over the drowned mouth of the Santa Margarita River
north of Oceanside, California.

Fig. 154. Small underfit stream flowing in broad channel cut by glacial meltwater during late stages
of the last Ice Age. Looking northward up Cottonwood Slough from over the southwest
end of Lake Traverse, South Dakota.

Evidence of Changed Conditions III:
Stream Terraces

Stream terraces are the most common and conspicuous evidence of changed conditions along a fluvial valley. In the discussion of rejuvenation (page 158 and Fig. 151) we analyzed a clear and simple example of how a single pair of terraces may be produced when a stream becomes entrenched in its flat floodplain. But as can be seen in the more complex examples at the right, stream terraces may also occur as multiple benches of variable width running along the valley sides in steps at different levels above the stream. There are five in the upper view and three in the lower if only the conspicuous ones are counted. Furthermore, the levels on one side of a stream need not match those on the other.

By definition, stream terraces are remnants of former valley floors. The term refers to the shape only, the flat bench, regardless of the material from which it is fashioned. Since most valley floors are composed of alluvium, this is also the material underlying most terraces; they are simply remnants of former floodplains, composed of anything from silt and sand to cobbles and boulders according to the competence and load of the river that deposited them. These alluvial deposits are usually rather loose and unconsolidated because they have never been subjected to the burial, compaction, and slow cementation that make sediment into rock. Accordingly, most of the steep faces between terraces expose only alluvium, with stratification (if it can be seen at all) roughly parallel to the terrace above. But occasionally, near the valley sides where the alluvial fill is thin, or at a rocky promontory, the terrace may be cut into bedrock or only thinly veneered with river deposits.

Early stages in the formation of terraces were visible in Figures 145 and 151. The upper view here shows a more advanced stage in which the shifting course of the river has left at least seven terraces above the present narrow floodplain. As the steep banks between them show, each new terrace is cut into the margin of its predecessors and—unless the valley has had a very unusual history—the benches are successively younger downward. It is also evident here that a terrace may be completely removed, at least in some places, by the widening of lower ones; there are only three terraces on the far side of the river at the left because some of the intermediate ones, visible farther downstream, have been cut away.

The lower view is across the San Juan River, from the terraces on one side to those on the other. Their even, flat surfaces and step-like arrangement are especially clear on the far side. There are several places along the cliffs and steep slopes between terrace levels where bedrock strata are exposed, dipping gently to the left beneath a thin veneer of alluvium. On the near side, in the foreground, we see a sample of the well-rounded cobbles which, mixed with sand, compose the floodplain deposits of this vigorous river. The terrace below the one we are standing on has been notched by several gullies that drain out to the right, breaking up the once-continuous bench on this side of the valley.

The modern floodplain is over one-half mile wide (at the level of the telephone pole bases) and the river now flows over it in a braided pattern; should the stream deepen its channel in rejuvenation, this flat plain would become a new, still lower, pair of terraces.

There is considerable variety in the significance of terraces. In middle-latitude countries a great many of them top remnants of debris that was poured into river valleys by meltwater from the last Ice Age. In actively or recently growing mountains terraces often record successive uplifts which, by steepening the gradients of the streams, have caused repeated rejuvenations. Whatever their ultimate cause, well-preserved terraces are helpful in reconstructing segments of the former long profiles of the valley, and in some cases the stream, along which they are found. Sometimes these profiles converge or diverge or show warping in ways that record deformation. Terraces dipping upstream are particularly significant in showing deformation, since as originally formed all terraces must slope downstream.

Fig. 155. Stream terraces along the upper Pahsimeroi River, northeast of Borah Peak,
Idaho. Looking north (downstream).

Fig. 156. Stream terraces along the San Juan River 20 miles below Farmington, New Mexico.
Looking southeastward across the river.

Fig. 157. Dendritic channels developed along the northeast shore of the Salton Sea since the disappearance of Lake Cahuilla. Width of view at shoreline is 1.5 miles.

Drainage Patterns

The last three examples have been chiefly concerned with recognizable changes in the vertical position of a river. But what about the horizontal arrangement of streams—the *drainage pattern* of trunk streams and their tributaries as seen on a map? On a small scale, for example, the pattern on the flanks of a volcano is *radial* and that on a steep tectonic slope like the one shown in Figure 141 is *parallel.* As erosion cuts deeper will such patterns persist? If not, why and how will they change?

These questions lead to a more general one: What determines the overall pattern of streams and rivers on the land? The first thought might be that streams simply flow downhill from the mountains to the sea, with each segment controlled by the slope of the land. But this is countered by the realization that a great many details of the shape of the land are themselves, at least indirectly, fashioned by running water, and that the pattern can change through headward growth and piracy. The question now becomes: What really determines the direction taken during headward growth and channel deepening, and therefore controls the details of the resulting pattern? Three factors deserve attention.

1. *Slope.*—Slope clearly exerts the immediate influence on the direction of running water. The origin of the slope does not matter. Streams flow down the steep eastern and gentle western flanks of the Sierra Nevada (Fig. 360) or the east wall of Death Valley (Figs. 128 and 141) on tectonic slopes created primarily by crustal deformation. They also flow down the sides of the Grand Canyon (Fig. 34) and into many valleys on slopes directly related to erosion along main rivers. Still other streams flow down surfaces directly produced by their own deposits; Figures 147 through 150 include several examples.

But there is more to this than simply recognizing that water tends to run downhill. Figure 157 demonstrates what happens when a smooth gentle slope is newly exposed to precipitation. Less than 2,000 years ago this area was part of the floor of a large lake (see Lake Cahuilla, page 380). The surface seen here is chiefly loose sand and silt and descends 22 feet in the 1.1 miles from foreground to shore—a declivity of less than one-third of a degree. The branching *dendritic* drainage pattern (Greek: *dendron*, tree) which has developed is one that is found

Fig. 158. View upstream over upper Tule Wash, west of Salton Sea in southeastern California. Area, 6 square miles. The crescentic sand dune is discussed on pages 198 and 199.

under a variety of familiar circumstances, whether on a scale of inches, hundreds of feet, or miles. Formation of a dendritic pattern is evidently the natural way for a deepening initial stream on a gentle slope to expand the area from which it collects runoff.

Obviously, the steeper the slope the stronger its influence on the direction of flow. Steepness suppresses the growth of branches and tends to produce many subequal streams with short tributaries. If the steep slope is plane the streams are parallel, as in Figure 141; if it is curved in plan they tend to converge or diverge. Very steep slopes, however, cannot exist over areas larger than the flank of a mountain range, and even though a river system may have its source in high mountains, its overall slope is very gentle. Thus, although the headwaters of the Colorado River are more than 10,000 feet above sea level, its average gradient, from its source to the sea, is less than 1/10 of a degree, or about 7 feet per mile. From Pittsburgh, Pennsylvania, to Cairo, Illinois, the Ohio River descends less than 6 inches per mile. From observations such as these we may conclude: that development of dendritic patterns is the normal tendency on gentle slopes; that except near their headwaters most major drainage systems have very gentle gradients; and that therefore where such drainage systems exhibit other than dendritic patterns some special cause is to be sought. The dendritic pattern is thus a kind of norm toward which low-gradient streams may be expected to develop unless other factors interfere; in this sense it is analogous, in horizontal plan, to the smooth, upwardly concave normal profile which stream gradients tend to assume.

One of the most obvious causes of departures from the dendritic norm is the second of our three factors.

2. *Structure.*—It is almost universally true that as stream systems cut down into the earth's crust they encounter variations in the erodibility of the rocks. These variations may arise from gross contrasts in rock type, from differences between layers in a complex sedimentary sequence, or even from the presence of joints and crushed zones along faults in otherwise homogeneous rocks.

In Figure 158 we can examine an early stage in the influence of structure (as expressed by the pattern of weak and resistant rocks) upon a drainage pattern. Except for the hills in the distance this is another part of the old lake floor just examined. But here, in the hundreds of years since the water evaporated, all but a few patches of the veneer of lake deposits has been eroded away and a branching drainage system has developed with main channels up to 20 feet deep. Although the main gullies have a dendritic pattern, some of the newer ones (for example, in the left foreground) tend to follow weak layers in the contorted strata. The influence of contrasting rock resistance and structure is just beginning to modify the randomly branching dendritic pattern which originally developed on the uniform lake floor.

Other things being equal, tributaries working in weak rocks have an advantage over those in resistant ones; their headward growth is likely to be more rapid and to result in the capture of less-favored streams in the system. An example of a more advanced stage is shown in Figure 159, where the main stream is confined to a belt of weak rocks in a *strike valley* that follows the trend of the strata: in Figure 366 also, the ridges and valleys and principal streams are conspicuously controlled by the structure. An even later stage, in which only the most resistant rocks remain as ridges, may be seen in Figures 373 and 375, where most of the streams flow in broad weak-rock valleys on shale and limestone between wooded ridges of hard sandstone. These scenes demonstrate the giant etching process by which drainage patterns become adjusted to the structure, which is thereby expressed in the topography.

One might wonder how permanent these adjustments are. If a region remains stable long enough to reach the peneplain stage, in which even the most resistant rocks are subdued and deep soil covers most of the land, will the stream pattern inherited from earlier conditions be preserved? Or will the drainage system assume a dendritic pattern as structural controls are weathered into submission and the surface approaches flatness? The question seems almost unanswerable at present. Well-developed peneplains are very rare and their complete histories must necessarily be both long and complex; evidence of earlier drainage patterns on such a surface would probably be almost impossible to find.

Many drainage patterns, both small and large, cannot be explained simply in terms of slope and structure. Our third factor embraces the explanations for many of these.

Fig. 159. View upstream along Halls Creek, which follows the strike of weak rocks dipping to the right along the Waterpocket monocline, southern Utah. See also Figure 88.

3. *History.*—Many drainage patterns show lingering effects of special conditions or events in their past. In and north of the middle latitudes the most widespread modifications are those resulting from the last Ice Age. Figure 160 shows a simple example. When the ice melted it dropped an irregular blanket of deposits producing a hummocky surface strewn with giant boulders, which contrasts sharply with the otherwise smooth surface of this plateau at the right and in the distance. The ice and its deposits destroyed any preglacial drainage system that may have existed beneath them. Now, the most conspicuous drainage channel (center foreground) approximately follows the old glacial margin, probably because it once carried water from the melting ice (see also Fig. 320).

On a much larger scale, a section of the Upper Missouri River about 350 miles long and most of the Ohio River are also essentially ice-margin streams that were established along the border of the ice when, in spreading southward, it blocked a number of smaller rivers that flowed northward in preglacial times. A member of an extensive drainage system fashioned by Ice Age meltwater was shown in Figure 154, and a more spectacular and complicated drainage pattern inherited from the Ice Age is explored on pages 338–351.

The scenes on the following page illustrate another way in which events in the geologic history of an area can produce an anomalous drainage pattern. The view is downstream along a small river that flows through a gap in a rough ridge of granite. That the granite is relatively resistant to weathering and erosion is demonstrated by its steep slopes and the narrowness of the gorge. Had the stream detoured a few hundred yards to the left it could have avoided the hard rocks by going around the end of the granite ridge, now buried under unconsolidated alluvium.

Such a notch, or *water gap*, is a conspicuous violation of the tendency of streams to adjust to structure by seeking out courses in weak rocks. If this resistant ridge had been encountered in the course of the normal headward expansion of a drainage system the stream would have gone around it because any headwater-growing tributary that bypassed it would have had an

Fig. 160. View southeastward along curving outer margin of rough glacial moraine on Waterville Plateau west of Moses Coulee, Washington. Smooth unglaciated surface at the right; ice-margin stream channel in foreground.

Fig. 161. *Above:* View downstream over the Mojave River and its water gap at Victorville, California
The bajada in the left background of the photograph has been restored to its original
uneroded state in the background of the matching drawing. *Facing page:* A portion of the
Sweetwater River in Wyoming, printed in reverse to increase its similarity to an
earlier stage along the Mojave.

advantage over any that headed into it. To cut this gorge the full force of the whole river, not just the sensitive upper end of a tributary feeling its way like a tentacle, was evidently focused on this point. If this is so, the gorge must have been cut after this part of the river was well established, and the course of the stream must antedate the presence of the barrier: either the barrier rose across the river's path, slowly enough that the river could saw down through it; or the river's original course lay at a higher level on weaker rocks under which the ridge was then buried; or the history includes a combination of both.

If the river was there first and the ridge rose across its path the relation is described as *antecedence;* such events are often difficult to prove, but a remarkable example of an antecedent stream course appears in Figure 162 on the next page. If the ridge was buried and the river encountered it in the course of deepening its channel through younger rocks, the stream is said to have been superposed (let down) upon the ridge. The best evidence of *superposition* is the existence of recognizable remnants of the sedimentary cover which buried the ridge and on which the river earlier established its course.

The water gap shown at the left (Fig. 161, *upper*) is not very spectacular, but it is a valuable example because in the background, and still burying the nose of the ridge, can be seen an extensive alluvial surface which probably explains it. By restoring, in imagination, only the most recent erosion of this unconsolidated deposit, near the river, (as has been done in the lower view on the facing page) we can see that the granite in the area of the gorge would be buried. At this earlier stage the course of the river was probably strongly influenced by the alluvial fans which grew in upon it from both sides. The larger fans from the higher mountains to the left crowded it relatively close to the lower hills out of view at the right, increasing its chance of flowing over a buried spur from these hills. This condition is suggested in the background of the drawing and probably resembled the scene in the photo above, where a similar river is now crossing partly buried granite spurs (note the small outliers of granite to the left of the stream). Later, the growth of the large fans was halted when their upper ends were captured by the headward growth of other streams. It may be that this is when the river, thus deprived of a major source of load but with discharge remaining about the same, began to deepen its channel. As it did so, it encountered the buried ridge and proceeded to cut the gorge. *169*

Occasionally earth movements in the path of a stream take place so rapidly, or the stream is so feeble, that the drainage is directly affected by them. Two examples are shown at the right.

In the upper scene (Fig. 162) the view is obliquely downslope toward a low ridge with two gaps through it; beyond is the almost flat floor of the much-cultivated San Joaquin Valley of central California. Many natural drainage lines converge toward the deeply shadowed water gap near the center, through which the runoff flows to the valley beyond. The larger notch at the left is a low pass which drains in both directions from a high point near its middle. The ridge is a simple plunging anticline whose uparching must have taken place within the last million years or so because this is the approximate age of the youngest layers involved in the folding. The anticlinal structure is not broken at either notch, proving that neither was caused by crustal movements; they must have been cut by erosion.

This combination of facts is interpreted to mean that during an earlier stage of the rising anticline there were streams flowing through both gaps but as the fold developed the channel on the left was not deepened as fast as the ridge rose across its path. The defeated stream was diverted toward the right where it joined the flow through the central gap. Since then the greater flow across this lower point on the plunging anticline has been able to maintain its course. The center notch is thus a water gap resulting from antecedence, and the left one is a *wind gap*—the abandoned course of a stream through a ridge.

Note that the two gaps divide the ridge into three segments. From left to right these are successively lower and exhibit successively less dissection by gullies and canyons. This almost certainly means that they are younger toward the right, which suggests that the anticline has been growing in that direction and that the streams have had a tendency to go around its migrating nose. The distinct entrenchment of the small stream next to the highway (far right), where it crosses the anticlinal nose, implies that uplift has extended that far; perhaps some day the principal drainage will again shift one notch to the right.

The lower scene at the right (Fig. 163) records the direct effects of downfaulting on a drainage pattern. In the foreground is a bajada composed of coalescing alluvial fans that have been built out from the base of the mountains in the background. Across this from left to right is a band of subsidence almost a mile wide and at least 200 feet deep. The resulting broad trench is collecting runoff from all the mountain-front canyons in the left half of the view. Of the three main gullies in the foreground, the two at the left are abandoned, their headwaters having been captured by the trench; all of the runoff gathered by the depression now escapes along the freshly enlarged gully at the right.

In contemplating this scene it should be remembered that an alluvial fan is a depositional feature and the normal pattern on its growing surface is the braided one of shifting, aggrading streams (e.g., Figs. 147 and 148). But here the slope in the foreground is covered with small dendritic gullies that are cutting into the fan deposits and clearly belong to the present conditions, created when the trench cut off discharge from, and deposition by, streams from the mountains. Precipitation directly on the fan remnant is causing normal headward erosion by these small gullies. This does not explain, however, the two much larger gullies at the left, which could have been cut neither under present conditions nor during the growth of the fans. Using as a clue the present outflow from the trench through the steep-walled gully at the right we postulate that the other two are probably abandoned escape routes of the same sort. Indeed, the middle one looks less rounded by weathering and creep than the one of the left, implying that the outflow from the trench has occupied them successively from left to right—and hence that the trench has probably subsided progressively in that direction. That these drastic changes in drainage pattern have been accomplished by recent faulting is indicated not only by the geologic history but also by the many small linear fault scarps that parallel the trench on both sides and extend along its trend to the right.

Fig. 162. Looking northwest across the eastern end of Wheeler Ridge, a geologically recent plunging anticline at the south end of the San Joaquin Valley, California. Wind gap at left, water gap near center.

Fig. 163. Downfaulted trench cutting across alluvial fans built out from the west base of the Panamint Range, near Death Valley, California. Width of view at base of mountains is about 5 miles.

15 LAKES

Lake Basins Produced by Landslides

From the geological point of view almost all lakes are temporary. This generalization rests on the basic facts that for the vast majority of lakes the inflow must carry some sediment at least part of the time, and the outflow, because the lake acts as a settling basin, must carry almost no load and will be an actively eroding stream. As a result the lake bottom is raised by sedimentation at one end while the water level is lowered by erosion of the outlet at the other.

Of the many ways in which a lake basin may be created, the most common involve interference with a stream. One obvious and sometimes spectacular way a stream may be blocked is by a landslide. Just before midnight on August 17, 1959, a severe earthquake shook the area of the upper Madison River west of Yellowstone National Park. A day and half later the upper picture at the right was taken, looking downstream from a low-flying helicopter. A mass of approximately 35 million cubic yards of broken rock, triggered by the shock, had slid off the 1,300-foot ridge at the left with such velocity that it crossed the river and climbed 400 feet up the opposite side of the canyon.

Two conditions favoring instability were present—incoherent rock and an overly steep canyon wall. The rock was sheared and altered schist and gneiss except for a little white dolomite in the lower part, and the wall was the steep outer side of a bend in the canyon—undercut until its inclination ranged from 25° to as much as 45° in some places. (Natural slopes steeper than about 30°, other than cliffs, are rare and, being inclined at an angle close to the angle of repose for loose debris, are likely to be unstable.) It is interesting to note that the white dolomite, which started from a position low on the left slope, remained at the front of the sliding mass and climbed to the maximum height on the right slope, above rocks which possessed greater energy by virtue of having started from higher positions. It must be concluded that the dolomite was propelled by the schist and gneiss that followed it, and therefore that most of the slide moved all at once and not in sections or short tongues.

The foreground shows that in less than 40 hours, tall pine trees were disappearing under the rising waters. What would happen when the lake topped the dam? How long would this take? Remembering the disaster on the Gros Ventre River of Wyoming in 1927 (page 122) engineers knew that a major problem would be to control the overflow, and that the best way to do this would be to prevent rapid erosion of the spillway. The first foot of loose rubble is the easiest to erode and each additional foot by which the channel gets below lake level releases huge additional quantities of water. It was therefore important to prevent the vicious interaction in which an increasing flow of water accelerates the deepening of the channel and deepening of the channel increases the flow of water.

The calculation of the time available for taking preventive action involved the volume of the valley to the level of the lowest place in the dam and the rate at which water was flowing in. Emergencies like this place critical importance upon such existing and seemingly commonplace data as a contour map of the valley and discharge measurements on the stream. Those who perform these relatively routine tasks, often in remote places and perhaps with a limited grasp of the potential significance of their work, may take heart from the vital role played by the products of their labors on occasions such as this.

The lower photograph was taken 24 days later, looking upstream, after U.S. Army Engineers had prepared an artificial spillway channel and lined it with dolomite, the most durable of the available rocks. By making it 250 feet wide, flat-bottomed, and almost level for the first 300 feet, the overflow was made to spread as a shallow stream and its erosive power held to a minimum. This picture was taken a few hours after the lake began to use the spillway—while bulldozers and other heavy equipment were still at work.

Fig. 164. Looking downstream at the Madison Canyon landslide less than 40 hours after it took place. (*Photo by N. R. Farbman*, LIFE, *copyright 1959 by Time, Inc.*)

Fig. 165. Looking upstream at the Madison Canyon slide as the rising lake begins to overflow through the prepared spillway 24 days later.

Lake Basins Produced by Volcanic Activity

Volcanic eruptions may also obstruct drainage and produce lakes. In the scene below (Fig. 166) lavas issuing from the vicinity of the cinder cone have blocked a valley that drained toward the lower left side of the view. The resulting lake, a mile and half wide and in this view white with ice and snow, has no surface outlet; evidently losses from evaporation and by seepage through and beneath the lava are sufficient to balance the inflow.

Sometimes the crater of an extinct or dormant volcano becomes a lake. In the upper view at the right (Fig. 167) the original crater is 4 to 5 miles across and has been divided into two basins by later eruptions of basaltic ash near its center. These eruptions and the ones that produced the many small basaltic cones on the flanks of the old volcano (about 55 are visible here) are believed, on the basis of radiocarbon dating (page 309), to have occurred between about 2,000 and 9,000+ years ago. The lakes are supplied by precipitation and their outflow is through a creek behind the hill at the far left.

The lower scene shows the better-known Crater Lake in southern Oregon. This lake is more than 5 miles across and reaches a maximum depth of about 1,930 feet, which makes it the deepest known body of fresh water in the United States. Owing to the shape of its surroundings Crater Lake has neither inlet nor outlet in the conventional sense. Yet even without an overflow to control its level the water surface fluctuates less than three feet each year and during the past 60 years has remained within a range of 17 feet, responding only slightly to wet and dry decades. The average annual precipitation within the crater is about 60 inches and the evaporation about half this amount. The difference escapes by seepage and probably helps to account for the springs low on the outer slopes of the mountain. Increase in the surfaces of seepage and evaporation as the lake rises provides a self-correcting mechanism that helps to stabilize the lake level.

The origin of this crater is discussed on pages 314–317.

Fig. 166. Looking southeastward over Cinder Cone, snow-covered lavas, and Snag Lake in Lassen Volcanic National Park; in winter.

Fig. 167. Winter view northwestward over Paulina and East Lakes in the crater of Newberry Volcano in central Oregon.

Fig. 168. Looking southwest over Crater Lake in the stump of Mt. Mazama (page 315); note the many smaller volcanic peaks in the vicinity.

Lake Basins Produced by Faulting and by Glacial Scour

In view of the general restlessness of the earth's crust, lake basins that are the direct result of faulting are not as common as one might expect. Perhaps the reason is that they require a neat balance between climate, rock type, and crustal disturbance. If the climate is too wet, or the rocks too weak, formation of a lake in the inevitably shallow and slowly produced basin may be prevented by infilling sediments and erosion of the outlet; if it is too dry there may be a desert basin instead of a lake. The resistance of the rocks to erosion and the movements on the faults must therefore keep ahead of the changes wrought by surface waters, which, although needed to fill the lake, will probably also be the cause of its ultimate destruction. Tectonic basins are thus likely to be geologically young: i.e., to occur only in areas of relatively recent crustal disturbance.

The upper scene here includes a shallow lake, along the foot of the fault scarp at the left. The rocks are all lavas that have been broken by a series of faults, movements along one of which produced this escarpment. Other faults, outside this view to both right and left, mark similar lines of displacement and show that several thousand square miles of lava plain have been broken into gently tilted slabs. Accumulating sediment or strings of shallow lakes like that shown here, occupy the depressed sides of the slabs, each lake or sedimentary basin lying against the fault scarp that bounds the raised edge of the next block.

The lower view includes several lakes that lie above the timberline near the crest of the Sierra Nevada. Note that there is no evidence of dams, of whatever origin; the lakes occupy smooth shallow basins in solid granite, and there is unusually well-exposed fresh granite all around them. Hollows of this size in solid rock cannot be produced by stream erosion alone; streams are most effective in weathered material and cannot excavate basins of this type—they fill them.

The process responsible for these rock-bound basins is *glacial scouring*, the plucking and scraping that takes place under slowly moving masses of ice hundreds of feet thick (see page 216). In areas where joints are abundant and closely spaced, frost-wedging and the movement of ice can pry loose many blocks of rock. If these become frozen into the ice, even only inter-mittently, it can drag them out of a hollow because it is propelled by the weight of accumu-lating snow on the surrounding steep and higher slopes. As a result the ice excavates the bed-rock and scrapes it clean, producing shallow rock-bound basins.

Most present-day lakes fill basins produced by glacial erosion or deposition during the last Ice Age. Thousands of small ones like those shown here occur in mountain ranges from which the ice has largely melted. Thousands of others dot the lowlands of central and eastern Canada, some of them representing ice-scoured hollows, some ice-modified stream valleys, and some simply low places in the debris left by the melting of the ice. On a larger scale, the Great Lakes and the Finger Lakes of New York owe part of their origin to glacial scour: the ice partly reshaped pre-existing valleys and then partly blocked their rivers, at first with ice and later with debris dropped by the ice. It is only a few tens of thousands of years since the ice left the scene, and the streams have not yet fully recovered from this interference.

Fig. 169. Looking southwestward over Bluejoint Lake toward Hart Mountain and the
Warner Lakes, southern Oregon.

Fig. 170. Glaciated crest of the Sierra Nevada, looking southeast; Wallace Lake in foreground at left,
Wales Lake, and Mt. Whitney (nearer flat summit just below skyline at center).

16 WAVES AND SHORELINES

Sea Level

All over the world modern man expresses elevations in terms of sea level. The heights of Mt. Everest, Mt. Whitney, and Pikes Peak are given in feet above sea level; the depths of the Mindanao Deep, the Dead Sea, and the bottom of Death Valley are expressed in feet below sea level. At first glance this seems a simple and logical system; the connected oceans and seas of the world make up a great body of liquid whose surface, yielding readily to the influence of gravity, should everywhere be at the same level.

But suppose you were asked to go down to the shore and determine sea level within a fraction of an inch. The ceaseless rise and fall of the waves and tides pose an immediate problem. This has been solved by installing—usually on a pier that extends several hundred feet from the shore—permanent recording devices which, by means of a float and clockwork, make a continuous record on paper of the position of the water surface. In essence such a *tide gage* consists of a float within a vertical steel tube having small perforations below low tide level through which all water must come and go; this damps the rise and fall of the float from passing waves. A wire from the float passes over a pulley at the top, whose turns screw a pencil back and forth, tracing a line on a clock-driven drum. Examples of tide gage records are shown in Figure 175.

Study of these records from many stations has demonstrated that on the average it takes 24 hours and 50.4 minutes for one day's tides to complete their cycle, and that the pattern of tides changes slightly from month to month and from year to year. Experience has shown that the more important of the longer cycles are completed in about 18.6 years. Accordingly the U.S. Coast and Geodetic Survey considers that any continuous record 19 or more years long qualifies as a basis for a "primary," or first-class, independent determination of sea level. The average level determined from such a record is known as *mean sea level* at the station where the record was kept. There are more than 40 places where primary determinations of mean sea level have been made on the coasts of the United States, including Alaska and Hawaii.

By careful surveying methods these levels can be carried over the land for hundreds of miles with probable errors of less than an inch. In this way mean sea level at different locations can be compared. Doing this has led to the discovery that sea level is not the same at all points. On the shores of the conterminous United States, the east coast of Florida seems to have the lowest mean sea level. Using the level at St. Augustine, Florida, as a standard with which to compare others, sea level gradually rises along the Atlantic seaboard until it is about 15 inches higher at Portland, Maine. At Gulf Coast cities it is 8 to 10 inches higher than at St. Augustine, and along the Pacific Coast it rises from 23 inches (San Diego) to 34 inches (Oregon Coast) above this base. These figures indicate that Pacific Coast levels are higher than those on the Atlantic Coast by about 20 inches. There is even a 7-inch drop from west to east across the 120-mile-wide upper peninsula of Florida, similar to the 8 inch average difference from the Pacific to the Atlantic shores of the Isthmus of Panama. Many of these figures will undoubtedly be refined when more and better records are available, but the differences are far too large to be ascribed solely to uncertainties in measurement.

Wholly satisfactory explanations for all of these variations in sea level are difficult to find. Apparently the two dominant factors are barometric pressure and water temperature. Regions of low average atmospheric pressure tend to have higher sea levels because the water surface, being depressed under high pressure centers, rises beneath low pressure areas. Sea level is also affected by the fact that water expands when heated. The two effects are often combined. Other possible factors include prevailing winds, which tend to pile water up on windward coasts and blow it away from lee shores, variations in density caused by differences in salinity, the near-shore configuration of the bottom, and the effect of the rotation of the earth on ocean currents moving other than east or west.

All of these, however, are factors tending to maintain a difference in contemporaneous sea levels. Does sea level at any one place remain fixed over long periods? This question introduces another problem, for while it would seem obvious that the longer the record the more accurate the determination of sea level, this is strictly true only if the land supporting the pier on which the tide gage is installed has held still. The longer the record, the less likely it is that this condition has been met.

At this point the details of mean sea level become especially interesting to the geologist. Figures for the first half of this century at stations along the coasts of the United States show a general tendency for the sea level to rise (or the land level to sink, or some combination of the two) at rates of about 1 to 5 mm per year, the higher rates applying to the Atlantic Coast and to the more recent years. There are exceptions; at some stations sea level remained stable or lowered during decades when it rose at others.

This poses the question of how much of the change is to be ascribed to warping of the coastal land and how much to actual change in level of the sea. The latter, called a *eustatic* change, should be evident on the records from stations all over the world, and the part due to local warping should be recorded only at stations in the deformed region and, if the warped area is small, might produce systematic differences between the records of adjacent stations. By comparing records for the same period it is possible, within the limits of instrumental error, to assess these variables only approximately.

Significant world-wide, eustatic changes in sea level may be caused in three ways—or combinations of them: (1) by changes in the proportion of the earth's surface waters that is locked up on land in the form of ice, snow and lakes, (2) by changes in the capacity of the ocean basins through deformation, volcanism and introduction of sediment on their floors, and (3) by changes in the total amount of water on the earth's surface. The first produced a spectacular lowering of sea level in the course of the last Ice Age. As to the second, the interconnected oceans cover more than two-thirds of the earth's surface, and if the deformation of their floors is at all comparable in intensity and frequency to that observed on land it is remarkable that mean sea level is as stable as it is; perhaps over such huge areas the up and down movements approximately balance each other. Whether the third change is taking place is one of the unsolved problems of geology.

Nothing has been said yet about the shape of sea level. Since it is the surface of a fluid body, it is determined by the effective pull of gravity and faithfully reflects small differences in that pull from place to place. Near the equator, where the centrifugal effect of earth's rotation is greatest, the pull of gravity is slightly reduced; the result is the equatorial bulge. Close to the continents, especially along mountainous coasts, the gravitative attraction of the rock mass above sea level pulls the water up toward the land; an extreme case of this arises in some fjords where small arms of the sea are closely bordered by high masses of solid rock. These effects cannot be directly detected by ordinary surveying instruments because such instruments depend upon level-bubbles and plumb lines, which are subject to the same distorting influences. Calculations involving the volume, density, and proximity of the nearby rocks show that at the center of a fjord two miles wide the water surface would be warped upward about two inches above the level of the nearby open sea.

This actual shape of sea level (and its projected continuation under the continents), embodying its responses to small variations in the amount and direction of the pull of gravity, is known as the *geoid*, to distinguish it from the averaged, more uniform figure of the earth used for map projections, navigation, and calculations requiring a mathematically simpler shape. The known vertical differences between the two reach maxima of about 150 feet in only a few areas, all where there is mostly open ocean.

Wave Erosion

Everyone who has watched storm waves pound an exposed shore has sensed their power. It has been estimated that the average Atlantic wave strikes exposed windward shores with a force of several hundred pounds per square foot in summer and perhaps as much as 2,000 pounds in winter. The shifting and removal of huge solid blocks in breakwaters proves that storm waves sometimes exert forces of over three tons per square foot.

Waves are thus potent agents of erosion. Parts of the outer shore of Cape Cod shown in Figures 213 and 214 have retreated an average of about 3 feet a year in the past century. North of the mouth of the River Humber, on the east coast of Great Britain, comparisons of maps and descriptions covering a period of more than three centuries show that the shore has receded at an average rate of about 7 feet a year. Churches and villages have been swept away within historic times, and the positions of ruins show that waves have consumed one to three miles of the land since the Roman occupation. This unusually high rate of erosion is explained by the weak rocks (glacial deposits), the exposure to storm waves, and the fact that the debris is swept away to the south by longshore currents instead of accumulating offshore to interfere with incoming waves. A sand spit, growing at an even faster rate at the south end of this same shore, undoubtedly represents the redeposition of some of this material.

Heavy surf and collapsing cliffs must have been familiar sights to the eighteenth and nineteenth-century English scientists who contributed so richly to modern geology. Perhaps this explains why wave action was once considered a more effective means of flattening the land than weathering and fluvial processes; along coastlines it certainly is a more obvious one.

The effectiveness of waves in planing off rock stems partly from the concentration of most of their activity within the narrow vertical range between high and low tide. The photographs at the right were taken at very low tide in order to reveal as much as possible of the floor beneath the waves. A conspicuous feature in the upper one is an almost flat rock surface, the *wavecut bench*, here crossed by parallel ledges of lighter color. The lower view shows that these are the more resistant layers in a tilted sedimentary sequence. Such horizontal truncation proves that the bench has been cut (see Steno's third principle, page 81). At the landward edge of the bench is a beach deposit of well-rounded pebbles, cobbles, and small boulders. Most of these are durable volcanic rocks derived from a conglomerate that caps the higher parts of the cliff, but a few correspond to the resistant ledges on the bench itself and evidently were derived from them. Note the size sorting in the pebbly beach, the coarsest material being at the low seaward edge where the currents are the strongest.

From time to time, especially after storm waves at high tide have reached its base, a part of the cliff-face collapses onto the beach. The waves rapidly chew up the rubble, carrying the finer particles away and leaving behind a remarkably clean accumulation of cobbles. With each swash (= uprush) and backwash (= seaward return) of a wave on the beach, when the tide is in, the cobbles are rolled and tumbled against one another with a characteristic clattering sound; it is not surprising that they are worn smooth.

As long as the relative level of land and sea remains unchanged, the waves will slowly cut into the land, like a horizontal saw. The width of the bench will depend on the strength of the waves, the resistance of the rocks, and time. In any case, the retreat of the cliff will slow down as the bench widens because (1) the increasing breadth of the shallow water offshore will reduce the energy in the waves available at the cliff base, and (2) the increasing height of the cliff will mean that each successive foot of retreat produces a greater volume of debris to be worked over by the waves.

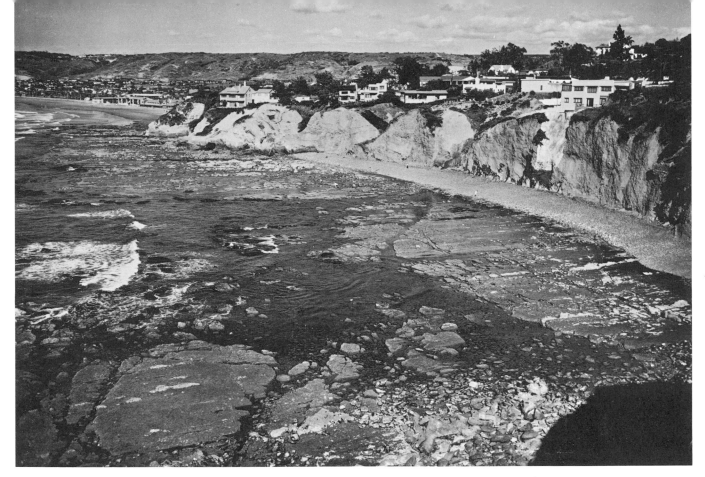

Fig. 171. General view of the wavecut bench at La Jolla, California, at low tide. The rocks are shales with lenses and beds of resistant sandstone.

Fig. 172. Looking seaward from the far end of the cobble beach in the scene above.

Wave Motion and Wave Refraction

The pictures on the preceding page demonstrate clearly that waves can cut into hard rock and that they can remove most of the debris. How, and why, do they do this?

The answers to these questions lie partly in the nature of wave motion itself. At the surface this can be observed by watching a floating object rise and move forward as the crest of a wave approaches and then sink and move back into the following trough. Viewed from the side, the object describes a nearly circular orbit whose diameter is the height of the wave crests above the troughs. This orbit advances very slightly with each cycle. Thus, although the wave form moves rapidly onward, the water itself makes almost imperceptible progress. (If this were not so there would have to be very strong currents away from all shores to return the water thus delivered by waves; see page 187.)

In open, deep water this motion within the wave dies out downward rather rapidly until at a depth equal to about half the wavelength it is negligible. By the same token, when a wave moves toward shore, the sea floor will not interfere much with the wave motion until the depth has decreased to about half the wavelength. From this point on, however, interference increases rapidly. First, the deeper of the nearly circular orbits are flattened into the to-and-fro motion familiar to underwater swimmers who have watched seaweed or sand movements just outside the breaker zone. At the same time the forward speed of the wave is reduced, which shortens the wavelength, and some of the energy of forward motion is converted into increased height. Ultimately the lower part of the wave will be slowed until it is moving toward the shore less rapidly than the water particles in its crest (at the tops of their orbits). This usually happens when a point is reached where the depth is about one and a half times the wave height. Now the wave must break, and most of its energy goes into the turbulent movements that characterize the water from the surf line through its uprush on the beach. The effectiveness of the wave in moving loose rock and sediment is greatest in this turbulent zone; here the wave reaches its peak as an agent of erosion. As the level at which this can take place is determined primarily by the tide and the height of the waves, both of which are usually restricted to a vertical range of a few feet, the bench is normally quite flat and is developed a few feet below mean sea level —as demonstrated in the scenes on the previous page.

Some of these considerations have a conspicuous influence on the pattern of waves as seen from above. In the upper scene here, for example, there is almost no wind and the wavelength of the gentle swell has been shortened close to shore by the retarding effect of shallow water. Similar retardation by the shallow water around the small island has bent the waves so that they cross on its near side. Indeed, wherever variations in depth cause different parts of a train of waves to be slowed by different amounts, the direction of advance must change. This change of direction, most readily seen in the curving of wave crestlines, is *wave refraction*. Numerous effects of refraction are visible in the lower view. The general progress of the waves is from right to left, but the crests coming toward us in the cove of the island in the foreground have been turned through 90°. Downwind from the island there is again a crossed pattern, and in the background the crests are swinging away from us. Refraction causes waves to converge on headlands and spread into bays; because of it straight ocean swells shape themselves to an irregular shoreline and yield nearshore waves that tend to roll onshore no matter what direction the beach faces.

Fig. 173. Wave refraction associated with small rocky island, one mile north of Prouts Neck, Maine.

Fig. 174. Wave refraction around Penikese (foreground) and Cuttyhunk (right background)
        Islands, Massachusetts. Looking south.

The Tsunami or Seismic Sea Wave

Wind is not the only means of generating water waves. Just as ripples in a pan of water can be created either by blowing on the surface or by disturbing the pan, so sea waves, which are normally generated by air movements, may also be produced by displacements associated with earthquakes, slides, and volcanic eruptions that affect the sea floor. The result of such displacements is sometimes a *tsunami* (Japanese) or *seismic sea wave*, often erroneously referred to as a "tidal wave" although it has nothing whatever to do with tides. Tsunamis may possess considerably more energy than wind waves and therefore do a great deal more damage when they reach shore; waves of this type well over 100 feet high have been observed at different times in Japan, Chile, Alaska, and the West Indies.

The tsunami that originated just south of the Aleutian Islands on April 1, 1946, must have been produced by a sudden displacement of the sea floor in that area; a major earthquake coincided in place and time with the source of the waves (map at right). The waves traveled the open ocean at speeds of more than 400 miles per hour, and the distance from crest to crest was about 122 miles. The crests passed any given point about 16 minutes apart and since they were only about 2 feet high (in the open ocean) ships were unaware of them.

The tide gage records from stations in the Hawaiian Islands and along the west coasts of North and South America as far south as Valparaiso, Chile, show that the waves arrived at different points with different amplitudes and different average speeds en route. Thus at Sitka, relatively close to the source, the waves could hardly be recognized (Fig. 175) because they had dissipated their energy in crossing shallow water dotted with islands before reaching the station. Yet at Valparaiso, over 8,000 miles away, the record is very clear (Fig. 175, *lower right*) and its similarity to other records implies astonishingly little loss of energy en route—a notable characteristic of most ocean waves. The seismic sea waves reached the Straits of Juan de Fuca at an average speed of 373 miles per hour and Honolulu at 490. These and other differences are evident in the pattern of advance shown on the map at the right, which was compiled in part from tide gage records like those reproduced here. Comparing speeds with water depths along different paths bears out the expected relation: the deeper the water the higher the velocity. Note how the lines of advance of the waves are bent by the retarding effect of the shallow water near the coast—an example of common wave refraction on a grand scale. Of course even the deepest parts of the oceans, which are 6 to 7 miles deep, are much shallower than half a wavelength of 122 miles and bottom topography should therefore be expected to have influenced the tsunami wherever it went.

Local extremes of this bottom effect were observed along the shores of the Hawaiian Islands. As with wind waves, the decreasing depth slowed and heightened the waves while the period (time interval between them) remained almost unchanged. In general the waves on the northern shores were larger than those elsewhere because of the energy lost by the latter in the course of refraction around the islands. Maximum high-water marks were left on shore opposite submarine ridges extending out toward the oncoming waves; these ridges slowed the wave advance to 15 to 20 miles per hour, causing convergence by refraction and consequent increased height. Some waves thus became over 30 feet high and washed debris up to 55 feet above sea level and half a mile inland in a few places.

Tsunamis come in all sizes and occur more frequently than most people think. In May, 1960, similarly destructive waves spread across the Pacific from earthquakes in Chile. Records covering more than 100 years in Hawaii indicate that severe tsunamis occur about four times a century and recognizable ones every four to five years.

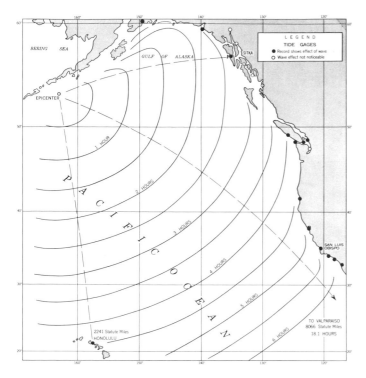

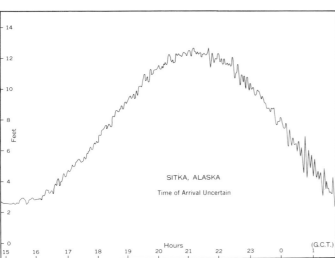

Fig. 175.

Map and selected tide gage records documenting the advance of the seismic sea wave of April 1, 1946. Note that the tsunami arrived at different places at different points in the tide cycle, and that in these examples the first sign of its approach was a small rise followed by a larger fall in water level; maximum amplitude was not reached until the third or fourth crest, at least half an hour later. These tide gage records represent some of the data used in preparing the map, which summarizes rates and directions of propagation. (*Modified from C. K. Green*, Trans. Amer. Geophys. Union, 1946.)

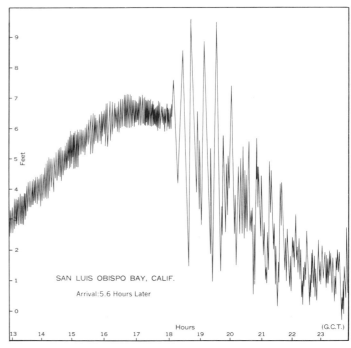

SAN LUIS OBISPO BAY, CALIF.

Arrival: 5.6 Hours Later

SITKA, ALASKA

Time of Arrival Uncertain

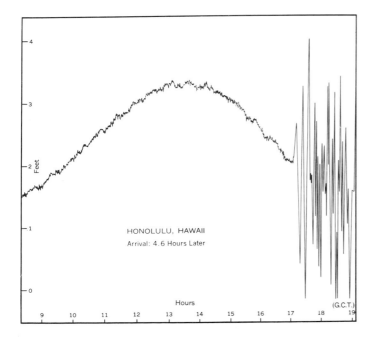

HONOLULU, HAWAII

Arrival: 4.6 Hours Later

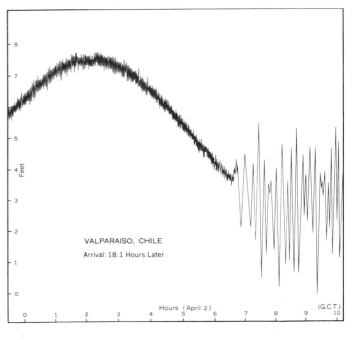

VALPARAISO, CHILE

Arrival: 18.1 Hours Later

Waves and Onshore-Offshore Movement of Sediment

The coming and going of sand on the beach is a familiar phenomenon to most people who have lived near a sandy seashore. Usually a beach loses sand during storms and regains it during calmer periods: thus, as in the example shown in Figure 176, many beaches are depleted in winter and restored in summer.

In a larger sense beaches are remarkably alike, whether moving slowly landward because the land is retreating under wave attack, as in Figure 177 (upper view) and along the outer shore of Cape Cod (Fig. 214), or seaward because of accumulating sediment, as along the Gulf Coast. Figure 189 shows a beach which has supplied the sand for several square miles of inland-moving dunes. From such observations it is apparent that neither long-continued gains nor losses of sand produce corresponding permanent changes in a given beach. We can only conclude that over the years the total amount of sand moved to, from, and along the world's beaches is enormous; that the currents associated with waves must therefore be effective movers of sand; and that the general uniformity of beaches under differing conditions implies that if they have access to enough material, waves are quite able to replenish losses and dispose of excesses in the process of maintaining a general equilibrium of their own design.

The scenes in Figure 177 demonstrate one of the few direct observations on this subject that can be made without getting wet. Both photographs were taken following stormy periods. The upper shows curling underwater "silt clouds" moving seaward, probably composed largely of muddy water from streams or generated when storm waves reached the debris at the foot of the shaly cliff. The streaks of localized offshore flow that carry these clouds are *rip currents*. The lower shows a lake shore along which storm waves chew at bluffs of weathered limestone and glacial deposits whose high clay content accounts for the delicate cirrus-like clouds of

Fig. 176. Part of Boomer Beach, a pocket beach between rocky promontories at La Jolla, California. *Below:* Summer conditions. *At right:* Winter conditions.

turbid water. Both photographs provide visible evidence that currents associated with waves can move fine sediment offshore beyond the breaking waves.

Rip currents represent the necessary seaward return of the water actually piled against the shore by waves. Some are located where refraction by low offshore ridges causes convergence of the waves at two or more points on the beach. The excess water delivered by the higher waves arriving at these points spreads into currents moving along the beach; where two of these currents meet, seaward flow occurs, and a rip current is the result. Others develop along smooth, straight beaches with oblique incoming waves; some of the water delivered by such waves moves parallel to the beach and the resulting longshore current builds up until, at intervals, it must break seaward in a rip current. Observed average rip-current velocities range up to at least two miles per hour, but their movement is normally pulsating, corresponding to the cyclic variation in the size of the arriving waves.

Waves moving over near-shore sandy bottoms tend to give the latter a complex seaward-sloping profile whose origin is but partly understood. In many places it seems to consist of three principal parts which, with tentative explanations, may be described thus (Fig. 178): (1) Farthest offshore, where the depth of water is equal to or a little greater than half the wavelength, the bottom has little influence on the shape of the waves and the to-and-fro orbital movements next to the bottom are small and symmetrical; if they are strong enough to disturb any grains at all, gravity will cause a net shift downslope and seaward transport will result. (2) In the middle zone, beneath shoaling waves that are feeling the bottom, the decreasing depth causes a progressive change in both the shape and orbital motion of the waves. Here, for waves of relatively long wavelength, a short rapid onshore motion (beneath the steep wave crests) is followed by a longer and slower offshore return (beneath the broad flat troughs). Under these conditions, which are most pronounced close to shore and beneath low waves that are far apart, the onshore current can advance many grains that the offshore flow is too

Fig. 177. *Upper:* Silt clouds and rip currents along the Southern California coast. *Lower:* Suspended fine sediment stirred up by waves along the north shore of Lake Erie.

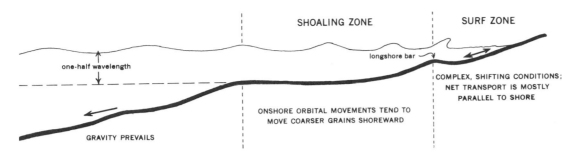

Fig. 178. Generalized nearshore profile indicating probable relations between waves and movement of sand (arrows). The widths and positions of the zones informally identified here, the number, shape, and positions of longshore bars, the depth of water, and accordingly the shape of the profile, are all continually changing on natural beaches.

weak to return, resulting in net shoreward transportation of grains that exceed some critical size. Thus sand may be moved toward the beach, or at least held on this slope, while finer material under two-way agitation may still move seaward because of gravity. (3) Where the waves break, these processes abruptly give way to the turbulence, swash, and backwash of the surf zone. Movements within this third zone are complex, but their net effect is the development of a slope or profile of equilibrium on which the shoreward transport by vigorous swash up the beach is balanced by seaward transport accompanying the backwash, which, although weakened through loss of water by percolation, is aided by gravity. A submerged longshore bar is commonly found between the second and third zones—under the outer edge of the surf zone.

A possible explanation for beach changes during local storms stems from the fact that storm waves are generally higher and closer together than swells of more distant origin. Under storm waves the on- and offshore orbital movements near the bottom are more nearly equal, there is less net sand migration toward the beach, and accordingly the profile assumes a gentler slope. Since the beach is the inner end of the profile, reducing the supply of sand and the slope will depress the beach. Some of the sand lost from the beach may be washed inland onto tidal flats, but most of it is generally spread offshore, whence it is returned to the beach when the waves are again low and far apart.

Oceanographers using diving equipment near La Jolla, California, have observed that at least some of the sand thus removed from the beach in winter is spread offshore to depths of at least 180 feet in the form of a very loosely packed blanket. In some places this sand seems to be creeping downslope and, occasionally, perhaps when triggered by a small earthquake or other disturbance, it apparently flows rapidly out to sea by way of small steep-sided gullies. Probably such sand will never return to the beach.

The overall similarities of beaches can be attributed to the fact that wherever sandy material is available to them, waves work it over according to the same rules, and in the same relatively narrow nearshore zone. Once sand escapes from this zone it is out of their reach. But within this zone the rules are complex because the processes are highly sensitive to many minor variables: the sand includes grains of a variety of densities and sizes, the waves vary in size, shape, and angle of attack, and the tidal cycle shifts all the wave processes back and forth over the profile. The inevitable result is continuous modification in endless pursuit of fleeting adjustment to conditions that are constantly changing within rather narrow limits.

Longshore Transportation of Sand

Pounding waves on a sandy shore probably keep more sand grains in suspension more of the time than most rivers do. The almost constant agitation in the surf zone keeps the sand mobile, allowing it to shift in sensitive response to many complex influences. In a few places measurements have been made from which we can estimate the magnitude of some of these changes. Along the outer shore of Cape Cod, for example, the exposed beach has been cut down as much as 9 feet during one tide and has lost over 500 cubic feet of sand per foot of shoreline during a single storm.

The problem of where this sand goes leads directly to consideration of sand movements parallel to the beach. Accurate measurements of such *longshore transport* are difficult to make, but good estimates have been obtained near newly constructed jetties which, by interrupting the flow, cause measurable accumulation on one side and depletion on the other. At Santa Barbara, California, for example, the average daily longshore transport of sand is about 800 cubic yards (equivalent to more than 1,000 tons, dry) and four times this figure during storms. Thirty five miles away, at Port Hueneme where the coast is less protected, the transport is about 3,000 cubic yards per day. Near Atlantic City, New Jersey, the daily longshore movement of sand is estimated to be about 1,100 cubic yards. Whether there is any movement at all past pocket beaches protected by rocky headlands is generally unknown.

One way to demonstrate longshore movements is to compare maps or photographs showing the same active beach at different times. The upper view here is northward along Nauset Beach, the point of sand that forms the "elbow" of Cape Cod, east of Chatham, Massachusetts. Two former positions of the tip of this large *sand spit*, taken from U.S. Coast and Geodetic Survey charts of earlier date, have been indicated on the photograph to show what has happened. The average rate of growth, based upon these positions, is about 265 feet a year for the period 1890–1958. More frequent observations would undoubtedly show that the changes are quite irregular—as might be inferred from the figures just cited on the effects of a single storm. The general explanation seems to be substantially as follows.

Fifteen to twenty miles north, near the center of the bow of the outer Cape, the waves are actively cutting back a 60 to 100 foot cliff in sandy glacial deposits; the general area is out of sight at the upper left in these scenes but Figure 214 affords a close view. Between 1953 and 1958 this cliff retreated at an average rate of about 1 foot a year; later a single storm (early April, 1958) cut it back 4 feet, and lower bluffs nearby lost as much as 50 feet. Large waves rapidly redistribute this sand. When their approach is oblique to the shore, those coming from the east and southeast shepherd the sand along the gentle curve to, and around, the north tip of the Cape. Waves approaching from the northeast, the source of the most severe storms, work the sand southward. Since the uprush of the oblique wave is slanted to north or south, the path of each grain as seen from above is a series of arcs, resulting in net migration north or south along the beach. In addition, even weak alongshore components of flow in the turbulent surf zone can account for very large quantities of longshore transport because so much sand is kept in suspension there; speeds as high as 50 feet per minute have been measured with marked sand on some California beaches.

Apparently southward longshore transport of sand tends to prolong the gentle outer curve of the Cape by building the narrow submerged spit over which waves are breaking (at the right in all these views). In the two lower views, which show conditions 14 and 30 months after the upper one, either the same spit or its successors can be seen emerging and advancing on Nauset Beach while the point of attachment moves south. During the same period the short ridge of sand which formed the near tip of Nauset Beach in 1958 (lower left in the upper view) has similarly moved onto the point, a change which probably represents the last stage in the history of an earlier submerged spit. Thus the addition of sand to Nauset Beach may be, in part, a two-stage process in which a submerged spit is formed by longshore transport and then driven onshore. Probably this sequence is in some way related either to seasonal changes or to occasional severe storms. Whatever the mechanism, it is estimated that a quarter of a million cubic yards of sand have been added to the south tip of Nauset Beach each year since 1940.

Fig. 179.
Upper: Looking north over the southern part of Nauset Beach; Chatham, Massachusetts, at the left. Low tide, June 29, 1958. *Lower left:* Same scene at high tide on August 23, 1959. Note oblique incidence of swells breaking over submerged spit at right.
Lower right: Same scene at low tide on December 9, 1960, from higher altitude. Note the many changes in the tip of Nauset Beach since 1958.

Shorelines Arising from Relative Submergence

A *shoreline* is the line along which land and water meet. In geology the term is also commonly used for any mark left on the land that records a former lake or ocean margin. Initially any shoreline must be level and, if marine, at sea level. Because of the general restlessness of the crust, however, shorelines of the past have been uplifted, depressed, and warped; where recognizable today these once level lines provide important records of deformation.

The shorelines surrounding the continents have unique importance in earth history and the study of geologic processes. They separate a well-known land realm covering less than three-tenths of the earth's surface and dominated by erosion from a little-known marine realm covering more than seven-tenths of the surface and dominated by sedimentation. At the shoreline there is a significant discontinuity in the surficial geological processes. Rock decomposition, downslope movements, erosion, and sedimentation are all active both on land and under the sea, but in rather different ways.

The position and pattern of the shoreline is, of course, very sensitive to any change in the vertical relations of land and sea. This is particularly true along lowland coasts; on the Gulf Coast of southeastern Texas, for example, a change of 50 feet in the relative levels of land and sea would shift the shoreline about 35 miles either way. If we arrived on the scene after such a change had taken place, would we be able to recognize what had happened?

In the upper view at the right the land is low and composed of resistant crystalline rocks. Seawater is backed up into the valleys and surrounds some hilltops. Along with the absence of recent shallow-water marine deposits on the land these facts point to relative submergence or drowning of the land. There are five kinds of changes that could have brought this about: (1) the land sank beneath a stable sea, (2) the sea rose and flooded a stable land, (3) the land subsided and the sea rose at the same time, (4) both sea and land went down, but the land was depressed more than the sea was, (5) both sea and land rose, but the sea rose more than the land did. The last is probably most applicable here.

In the lower scene the land slopes gently down to the left, and the surface of weak sediments is drained by a series of small parallel valleys whose lower ends are now flooded with seawater. The drowned valleys and fingers of land that separate them are truncated at the left by a *barrier beach* that sweeps away in the gentle curve characteristic of shorelines shaped by uninhibited waves and longshore currents (cf. Fig. 179).

The barrier beach converts what would otherwise be an extremely irregular shoreline into an almost straight one. The conversion takes place for several reasons. Refraction causes waves to converge on points of land that extend into the sea, and to spread out in bays (see Fig. 174). The headlands thus draw more than their share of the wave attack and unless they consist of rocks more resistant than other parts of the shore they will be eroded more rapidly than the land around the bays. At the same time either of two processes may contribute to the formation of barrier beaches. One is longshore transportation, which often builds spits downcurrent from the headlands, across the mouths of the bays. The other is the reaction of waves to a seaward slope of submerged land that is more gentle than their profile of equilibrium; in such circumstances they may build an offshore bar, whose outer slope is the profile of equilibrium for the prevailing waves and sediment, and then gradually drive this shoreward to become a barrier beach.

The lower scene demonstrates the normal tendency of waves to straighten an irregular shoreline, a process retarded by resistant rocks, shallow water and small waves in the upper view. Both views are evidence of a geologically recent change involving a rise in sea level relative to the land.

Fig. 180. Looking offshore east of Brunswick, Maine.

Fig. 181. Looking west along the south shore of Martha's Vineyard, Massachusetts.

Marine Terraces

Along many coasts it is possible to pick out features which suggest former shorelines that are now a few tens or hundreds of feet above sea level. The most obvious are benches or steps back of the shore. The view on this page is toward the north along a bench on the central California coast. The elevation of its inner edge near the highway at the right is about 190 feet, but it seems to descend slightly to the north and rise to the south. As may be seen at the top of the cliff in the foreground, parts of the bench are mantled with a blanket of weak deposits and everywhere its gentle seaward slope cuts across the structural patterns in the sea cliffs. The many coves and rocky points demonstrate considerable variation in the resistance of these rocks, yet the surface of the bench is smooth. It is thus a truly erosional surface whose flatness and inclination must have been established by the agent that cut it rather than by the structure and hardness of its rocks.

Such a displaced wave-cut bench, along with any beach deposits remaining on it, is a *marine terrace;* its inner margin marks an ancient shoreline.

What is the age of this marine terrace? How much of its emergence was caused by lowering of sea level and how much by uplift of the land? The steep cliff at the present shoreline proves active erosion by waves, and the many exposed rocks outside the breaker zone suggest what soundings prove: there is also a wide wave-cut bench underwater. The five-fathom line (30 feet of water) is 1,000 to 2,000 feet offshore, indicating that the average seaward slope of the submerged bench is even less than that of the exposed one. The underwater platform means, of course, that there has been a corresponding landward advance of the shoreline; indeed, the cliff is evidence that this encroachment is still active. The cutting of such a bench could take place during stable conditions or slow submergence, but would be highly unlikely during a period of emergence.

The exposed platform is obviously older than the offshore bench now eating into it. Also, because sea level was lowered at least 300 feet during the last Ice Age, it is almost certain that

Fig. 182. Looking north along the central California coast near Gorda, between Carmel and Morro Bay; a well-defined marine terrace. (*U.S. Geological Survey*, Cape San Martin Quadrangle; 1:24,000.)

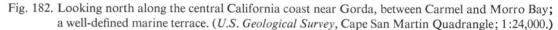

Fig. 183. Looking northwest along the southern California coast at Encinitas; a marine terrace striped with low ridges of sand that probably represent ancient beaches or bars. Note refraction of obliquely incident waves.

the offshore bench has been cut since sea level recovered and is therefore younger than the last large accumulation of ice on the continents. If the emerged terrace had been cut at the lower levels of this time the post-Ice Age rise of the land would have to have been more rapid and greater than the rise of sea level in order to put the old shoreline 190 feet above the sea today. Though not impossible, such a rate of crustal uplift (about 5 feet per century) is high. It seems likely, therefore, that the exposed terrace is older than the last major accumulation of ice on land. But does it represent a change in sea level, or local deformation of the crust? If it was produced by a higher stand of the sea before the last Ice Age there should be corresponding terraces along the stable parts of the coasts of all the continents. That such terraces have not been recognized favors a local origin. So does the possibility that this terrace has been warped, as suggested by the fact that it apparently stands at different heights along this coast.

The view above includes part of a terrace system that is 4 to 12 miles wide along the southernmost coast of California. Of special interest are the low ridges that trend approximately parallel to the shoreline. These rise from 20 to almost 100 feet above the terrace and are composed of well-sorted sand. Although these ridges are cut through by streams flowing to the sea, their segments are perfectly aligned, implying that they were once continuous and are probably older than the stream valleys. There can be little doubt that the terraces are of marine origin; the sand ridges probably originated either as beaches or as bars reworked by the waves of the receding sea. In the background is a long lagoon produced by the drowning of a major stream valley: there are others beyond this view. (See also Fig. 153.)

The very simplest history indicated by all these features is: (1) uplift of the land, which exposed these terraces and ridges; (2) Ice Age lowering of sea level, which brought about deepening of the valleys; (3) post-Ice Age recovery of sea level, which drowned the valleys and produced the lagoons. The cliffs again imply relative stability or continuing slow submergence as the present state.

Analysis of another part of this coast is undertaken on page 334.

17 WIND, DUST, AND SAND

Blowing Dust and Sand

Transportation and deposition of sediment by wind reaches important proportions only in relatively dry climates and along some sandy shores. But these very limitations add to the geological value of rocks composed of *eolian* (wind-deposited) material because of their implications regarding ancient environments. Let us turn, then, to the movement and accumulation of wind-borne sediment.

Hourly wind velocity averages ranged from 50 to 70 miles per hour at the airport (far right) during the afternoon the photo below was taken. *Dust* is being picked up from scattered fields and raised several thousand feet above the ground while it is carried toward the upper right. Note in the upper left how large quantities of dust are derived from some fields while others yield none—a graphic demonstration of the effectiveness of vegetation and dampness in inhibiting wind erosion.

In Figure 185, at the right, we are looking into a wind that is blowing chiefly *sand*, whose particles, being heavier than those of dust, stay within a few feet of the ground. (For a better look at the resulting crescentic *dunes*, see Figure 188.) Figure 186 is a closer view of a smaller amount of sand in motion. In the foreground, within an inch or two of the surface, a thin sheet of sand is moving to the right, blurring the image of the ripple marks. A little farther away a plume of sand is being blown over the crest of a small dune whose symmetrical shape implies frequent reversals of wind direction. In the background are some dunes partly anchored by bushes.

Fig. 184. Dust storm raised by a northeast wind over the vineyard district of southern California. Viewed from about 16,000 feet; average wind velocity at airport (far right) was 67 miles per hour when this photograph was taken.

Fig. 185. Sand storm raised by west wind blowing toward the Salton Sea. See also Figure 188.

These scenes demonstrate that the transportation and accumulation of dust is quite different from that of sand. Composed of particles small enough (under about 0.07 mm, or 3/1000 inch) and light enough to be held aloft by turbulent air, dust is carried farther and spread more widely than sand. Dust is also more generally available, notable sources being soil unprotected by vegetation, the fine loose alluvium of some floodplains and playas, and occasional explosive volcanic eruptions. Even though a wind may be able to drag sand along the surface at the same time that it carries dust aloft, it will drop the sand long before it drops the dust. The two are seldom mixed in eolian deposits.

Some of the world's most fertile soils are found downwind from large sources of dust. Many millenia of winds blowing from the Gobi Desert have enriched the soils of eastern China, and prevailing westerly winds blowing across the high plains and Mississippi Valley during the closing stages of the last Ice Age produced rich soils in parts of Iowa, Illinois, Missouri, and east of the lower Mississippi River. Such deposits, known as *loess* (German: *lösen*, to loosen, release), are characteristically unstratified and tend to cover the topography like a blanket instead of being confined to the lowlands as water-laid sediment would be.

Fig. 186. Light sand movement from south wind, Death Valley. Looking west.

Barkhan on the March

One of the simplest and most graceful forms in which windblown sand accumulates is the crescentic dune or *barkhan* (also spelled barkan, barchan). Examples are shown in the views on these two pages.

The horns of the barkhan always point downwind, in the direction of sand transport. The upwind surface is usually circular to parabolic in plan and, with a slope of 5° to 10°, presents a streamlined form to the wind. More than half of the downwind face of the dune, between the horns, stands at the steepest slope possible—the angle of repose for dry sand, ordinarily 31° to 34°.

17
WIND, DUST,
AND SAND

Barkhans occur almost exclusively on hard or gravel-covered flat surfaces downwind from a source of sand over which the wind blows with nearly constant direction. This is evident in the barkhan fields shown here, both of which are fed from the same source, which is about 10 miles from the one below and 15 from that on the facing page.

It is notable that barkhans are characteristically surrounded by barren areas devoid of sand. The explanation lies in the fundamental way sand is moved by wind. Movement begins with a few grains rolling along the surface. If the wind speed remains above a threshold value (usually around 10 miles per hour), most of the rolling grains either bounce into the air or throw up the stationary ones they collide with. During its short flight, usually no more than a few inches above the ground, the airborne grain acquires forward velocity from the wind and comes in for its next landing with high speed and at a low angle (10° to 16° by actual measurement in wind-tunnel experiments). The resulting more vigorous impact may result in another bounce, the ejection of another grain, or in a splash-out of several grains; the process of *saltation* (movement in a series of jumps) has now begun.

On a hard surface, or one paved with pebbles too large to be moved by the wind, the grains bounce high from the immobile surface, have long trajectories, and keep moving. But if the

Fig. 187. Nearly vertical view of barkhan by Tule Wash, west of Salton Sea, southern California, in 1958. Retouched to show positions on other photographs taken in 1932 and 1950. Dune is 33 feet high, 400 feet between tips of horns. *Inset:* Slipface of same dune, immediately following a period of advance. (Also shown in Fig. 158.)

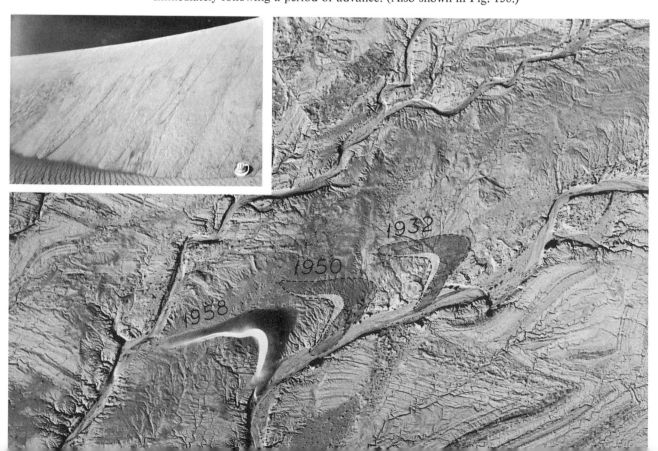

Fig. 188. Looking westward (upwind) over cluster of barkhans near west shore of Salton Sea.

same grains, driven by the same wind, encounter a patch of sand or a dune, many of them will bury themselves in its surface, dissipating energy in the splash. These become trapped and are added to the dune, and the splashed grains travel inches or less instead of feet. Thus stony areas tend to remain bare; sand patches and dunes tend to grow.

In time the grains added to a barkhan are bounced or splashed up the windward slope and over the crest, just beyond and below which they go over the brink into air so quiet that they simply rain down through it. The vast majority of them land high on this lee slope, because the average saltation jump is short. This tends to oversteepen the lee slope so that from time to time thin tongues of sand slide down it—in allusion to which the lee slope is also known as the *slipface* of the dune. The inset in the illustration at the left shows the slipface of a barkhan a few hours after the wind has stopped; note the many tongues of sliding sand.

The effectiveness of the slipface in trapping sand blown over the crest is clearly shown by the bare darker ground just downwind from the barkhans in the scene above. By the same token it is only from the tips of the horns that sand can leave a barkhan, as can be seen in the same view. Note how this influences the position of the next dune downwind, which uses the sand shed from the tips of those behind it.

The successive positions of the barkhan in the other view were selected from photographs which, taken together, show an average migration of about 50 feet a year between 1932 and 1962. In Peru, barkhans of many sizes have been studied as they were moved across the same plain by the same wind; the small ones almost invariably move faster than the large ones. Many factors influence the speed of migration, but height is usually dominant in any given environment because the higher the slipface the greater the volume of sand that must be added to build it forward any given amount. As a result small dunes sometimes overtake and merge with large ones. Large ones almost certainly can spawn small ones too; the process is probably taking place in the area shown on this page, and is implied by the fact that among the thousands in the Peruvian barkhan field, the dunes do not become progressively larger downwind.

Fig. 189. Coastal dunes south of Pismo Beach, California. Slipfaces indicate active migration inland. The form of the beach and the volume of sand it has contributed imply copious transport of sand by longshore currents.

The Accumulation of Wind-Borne Sand

Quartz, the most durable of the common minerals, is the chief constituent of most beach and desert sands and probably of most dunes. But there are notable exceptions. The Peruvian barkhans mentioned on the previous page are derived almost entirely from quartz-free volcanic ash, and the gray color of several dunes illustrated here results from abundant dark volcanic particles. In contrast to these, the dunes of White Sands National Monument in New Mexico are composed chiefly of the soft white mineral gypsum.

In most dune areas it is easier to find the source of the sand, and winds competent to transport it, than to determine why it accumulates where it does. Because the two principal sources of sand are beaches and deserts, most of the world's dunes are either *coastal dunes* or *desert dunes*. The scene above (Fig. 189) includes an example of the former. Prevailing winds off the ocean have swept beach sand inland, depositing it in two or three broad ridges, or waves, parallel to the shoreline. Dark vegetation occupies the low spaces between them. Such parallel ridges of sand, transverse to the prevailing wind, are found all over the world where the wind is constant in direction and there is an abundant supply of sand unencumbered by vegetation. Exactly why this should be so is not known. Note too that the individual dunes become larger inland; this may again be a result of smaller dunes overtaking and joining larger ones downwind.

The upper view at the right (Fig. 190) includes part of a 40-mile belt of dunes on the downwind side of an arid desert basin that for thousands of years has alternately been a dry playa and an undrained shallow lake. Although sand is generally available in the basin, at various times lake waves have concentrated it in beaches, and these have probably been the principal source of supply for the dunes. Indeed, this entire sand mass may have originated along a

Fig. 190. Part of the 40-mile long Algodones dune belt near the Mexican border in southeastern California. Looking north over the All American Canal; highest dunes rise 200 to 300 feet above the desert floor.

Fig. 191. Looking westward across dunes near the southern end of Death Valley. Highest dune is about 400 feet above its surroundings.

persistent shore and been migrating downwind since. Neither the large barren areas within the dune belt, nor its overall shape, are fully understood.

The scene in Figure 191 includes part of one of the patches of dunes in Death Valley, California, a long narrow trough that constrains most winds to blow along its axis, sometimes from the south (left) and sometimes from the north. Since the sand is alternately, perhaps seasonally, swept up from opposite directions, there is no net migration and the dunes grow much higher than isolated barkhans.

Like snow, sand forms drifts around obstacles that retard or disturb the transporting wind. If the obstacle is pervious, like the bushes in the foreground of Figure 191, a tail of sand forms downwind in the slackened current. If it is solid, like a boulder, and not streamlined, there is a little stagnant air in front which may cause deposition there as well. The larger such drifts become, the greater is their chance for survival as self-extending dunes; the critical stage in one-direction winds is the development of a slipface, whose existence is proof that some traveling sand grains are being captured.

The dunes in Figure 192 cover more than 25 square miles along the west base of the Sangre de Cristo Range near Alamosa, Colorado. The high proportion of volcanic particles in the sand helps confirm its derivation from the broad flat San Luis Valley to the west, an area largely floored with lava, ash, and alluvium derived from these and other rocks. The localization of these dunes seems to be related to the passes in the mountains beyond them, which, although serving to funnel the prevailing southwesterly winds out of the valley, are too high, steep, and forested to be crossed by many saltating sand grains. Occasional very strong northeast winds also help; the transverse crests of the dunes move at an average rate of less than 3 feet a week toward the mountains but occasionally travel more than 7 feet a day in the reverse direction when a northeast wind is blowing.

The small barkhans in the foreground show that the prevailing wind is toward the mountains. Under the present climate, valley-floor dampness and vegetation are trapping sand along their margins, which accounts for the low, irregular ridges trailing from their sides. In several places, especially at the right, pairs of these ridges converge and join in the downwind direction, indicating that a few dunes have been sapped to death by this process. No doubt the climate has been both drier and wetter at different times in the past, so the history of accumulation of the main sand mass may be long and complex.

The dune clusters in the lower scene (Fig. 193) are on the Snake River volcanic plain north of Idaho Falls. They are clearly moving to the left (northeastward) despite the handicap of clutching vegetation. Note how the bushes crowd in behind the dunes. The heavy losses from the horns are clearly shown by the long ridges that correspond to those in Figure 192; in addition (out of view at the left) there are many of these ridge tracks that no longer have dunes at their heads, apparently demonstrating the fate of others that marched by in advance of these. In the other direction, some of the tracks can be followed back for 50 miles toward a shallow intermittent lake which may have supplied much of the sand.

It is apparent from these samples that patches of dune sand occur under a wide variety of conditions. Among those pictured on these pages, those in the Salton Basin and Death Valley receive 3 inches or less of rain per year; these are in true desert environments. Around the Colorado and Idaho dunes the annual rainfall is 12 to 20 inches. About 45% of the Oregon shoreline is backed by stretches of dune sand, some of them reaching inland more than three miles, in a region where the average annual rainfall exceeds 60 inches. Similar conditions prevail on parts of Cape Cod, Massachusetts.

As we shall see on the next page, dune sand makes a very distinctive sedimentary rock, but in view of the distribution of these modern accumulations of wind-borne sand it is obviously not safe to assume that all such deposits reflect desert conditions in the past. We will face a problem of this sort in connection with the history of the Grand Canyon region, page 283.

Fig. 192. Looking northeastward at the dunes of Great Sand Dunes National Monument near
Alamosa, Colorado. Mosca Pass through the Sangre de Cristo Range in the background.

Fig. 193. Looking southeast over far-travelled dune clusters about 10 miles northwest of Ashton, Idaho.

The Anatomy of Sand Dunes and Recognition of Ancient Eolian Deposits

Inasmuch as sand may accumulate on the sea floor, under fresh water, and as dunes on land, there must also be sandstones that originated in each of these environments. In order to reconstruct the geologic histories of areas in which sandstones occur we must be able to distinguish between such varieties as eolian (wind deposited), fluvial (river deposited), and marine sandstones. With this objective in mind, let us look more closely at dune sands.

An ancient eolian sandstone and a marine sandstone probably deposited by a turbidity current were compared in Figure 45. These magnified views illustrated two valuable distinguishing clues in the textures of these rocks. First, because wind-transported sand grains move almost entirely by saltation (rather than in suspension) and collide in air more violently than is possible in water, they usually have smoother, more-rounded shapes. The occasional water-deposited sandstone whose grains are similarly well-rounded has probably been through several cycles of erosion and redeposition—or was derived from older eolian deposits to begin with. Second—and this is probably the more useful distinction—eolian sand grains display a very small range in size. Those larger than a few millimeters in diameter (about 1/10 inch) the wind seldom moves at all, and those smaller than about 0.07 millimeters (about 1/350 inch) are carried away as dust. This remarkable uniformity of grain size in dunes is easily demonstrated. In the Peruvian dunes, 98% of the sand grains (by weight) lie between 0.10 and 0.42 mm; in a sample from Libya 98% lie between 0.2 and 0.9 mm; more than 98% of the Oregon Coast dune sands fall between 0.12 and 0.50 mm. Furthermore, these figures apply to the whole dune, and thus to potential sedimentary rock at least tens of feet thick. By contrast, a lesser thickness of water-deposited sand is likely to include a variety of textures and even beds of conglomerate or shale (as in Figs. 42 and 73), neither of which is ever found within purely eolian deposits.

An even more conspicuous characteristic of dune deposits is their sweeping large-scale cross-bedding. Thin layers, differing very slightly in texture and composition, occur in groups, each of which usually truncates those on which it lies and is truncated by those that succeed it. (This relationship can often be used to determine the original top of the sequence if such a sandstone is later turned on edge.) Such cross-bedding is clearly shown in all the views at the right. In Figure 194 the two upper photographs show pits dug, when the sand was damp, into a steep (*left*) and a gentle (*right*) slope within the dune complex shown in Figure 192 on the preceding page. Note that in both the highest laminae conform, as they must, to the surface of the dune. Deeper down are groups of laminae with different orientation, denoting different, earlier surfaces. As the winds shift the dunes, or parts of them, sand is removed from one surface and added to another—whose dip may be anything from the 33° or so on the slipface to horizontal in the hollows between dunes. Thus laminae of one orientation are partly cut away and then covered with a new and different set.

On the lower left of Figure 194 is shown part of the windswept crest of a large dune in the same locality; the new slipface growing out to the left truncates the laminae on the windward slope at the right, from which, at the moment, sand is being removed by the wind. The junction of the two sets is a conspicuous surface of angular unconformity. If later, through another wind shift, new laminae should be deposited on this crest, they would truncate both the sets whose edges are now visible.

The two other views show natural outcrops about 10 feet high of two different ancient sandstones, respectively about 150 and 250 million years old. Each provides a large vertical exposure of the kind of internal structure revealed in the walls of the pits above. Because of the characteristic cross-bedding so beautifully displayed and the even, fine texture of the sand in the rock, there can be no doubt that these are ancient dune deposits.

Fig. 194. *Upper left:* Excavation near base of slope of large dune in Great Sand Dunes National Monument near Alamosa, Colorado. See also Figure 192. *Upper right:* Excavation in hollow between large dunes, same locality. *Lower left:* Structure along crest of active dune, same locality. *Lower right:* Exposure of Navajo sandstone near Kanab, Utah. (*Photo by Tad Nichols.*)

Fig. 195. Exposure of De Chelley sandstone, 20 miles north of Kayenta, Arizona. (*Photo by Tad Nichols.*)

18 GLACIATION

Introduction

The changes in the landscape which can be produced by large masses of moving ice hold a special fascination for geologists. Not only do they include highly distinctive shapes resulting from erosion and deposition, but the materials left by such ice are quite unlike those deposited in or by water. Furthermore, one of the greatest insights ever to enrich our understanding of recent geologic history emerged between 1820 and 1850 when a number of observers in Europe and Scandinavia realized that these tell-tale features exist far beyond the limits of present-day glaciers. Out of their findings grew the whole concept of former Ice Ages as times of greatly expanded mountain glaciers and continental ice sheets. With this concept came a host of implications regarding climate, changes in sea level and the circulation of sea water in the ocean basins, shifting loads on the crust, and the influence of these on drainage, land sculpture, and on both marine and terrestrial life. Glacial geology provides a superb example of the value of studying the present in order to learn the language in which the past is recorded.

The arrangement of the scenes on the pages that follow more or less parallels this historical development. The first concern the origin and behavior of modern glaciers; the others, using as clues the more enduring erosional and depositional features produced by ice, present some of the evidence for past Ice Ages.

Origin of Modern Valley Glaciers

The most widely known and thoroughly studied glaciers today are those occurring in mountains. In order to support glaciers, the mountains must be high enough to penetrate the zone of permanent snow. The *snowline*, or lower limit of perennial snow, is determined by the balance between accumulation and melting. In general the elevation of the snowline rises toward the equator and inland from windward coasts. Although highly variable in detail, it is around 20,000 feet in the Andes, 13,000 in the Sierra Nevada of California, 8,000 to 10,000 in the Cascades of Oregon and Washington, 3,000 on Mt. St. Elias in Alaska, and at sea level just inside the Arctic Circle. In the Olympic Mountains near the coast west of Seattle the snowline is around 5,000 feet; at Glacier National Park, less than 500 miles inland and a little farther north, it is around 10,000 feet because the total precipitation is less and the summers are warmer and drier.

Above the snowline, the accumulation of snow exceeds the loss through combined melting and evaporation and thus would pile higher and higher over the years if it did not ooze away as slow-moving streams of ice like those pictured here. Because these ice streams move down the major valleys, often joined along the way by smaller tongues coming down tributary valleys, they are known as *valley glaciers* (or mountain or alpine glaciers) to distinguish them from such relatively flat plateaus of ice as those that cap Greenland (background of lower scene at right) and Antarctica.

Although valley glaciers are nourished by the permanent snow fields of their source areas, as clearly shown in these views, there is little similarity between the porous, feathery texture of new-fallen snow and the solid ice in the heart of a moving glacier. However, cores taken from boreholes in glaciers show that there is a complete gradation between snow and ice, the change taking place gradually with time and increase in depth. Under the microscope this change can be seen to be one of progressive compaction and recrystallization. Within a few days the delicately branching snowflakes are transformed into rounded granules—like the corn snow of skiers—as a result of crushing, compaction, local melting, and recrystallization. As the packing of these grains improves they produce a permeable granular mass known as *firn* or *nevé*. Firn becomes increasingly solid as continued compaction expels air and as higher pressure causes individual grains to melt at points of contact and the resulting water freezes again in the spaces between them. In the transformation from fresh snow to glacier ice the porosity changes from more than 90% to less than 20%, the specific gravity from as low as 0.05 to between 0.85 and 0.91, and the texture from that of lightly tangled snowflakes to an impermeable mass of solidly intergrown crystalline ice grains that may be more than half an inch across. Further growth, and changes in grain shape and arrangement, are often associated with the deformation accompanying the flow of the ice.

The change from snow to firn to ice requires burial under 100 to 200 feet of accumulating snow and the passage of one or more seasons. The process is retarded by very low temperatures and slow accumulation. By analogy with other rocks, snow is an eolian sediment, firn is somewhat like a sedimentary rock, and glacier ice, having recrystallized under conditions of confining pressure and shearing stress, is a metamorphic rock.

Fig. 196. View westward over the source of the Aletsch glacier on the slopes of the Jungfrau, Monch, and Eiger in the Swiss Alps. (*Photo by Swissair, Zürich.*)

Fig. 197. Part of a perennially snow-covered plateau and two of the valley glaciers that drain it, as seen from the air; about 125 miles southwest of Scoresby Sound at 69°N latitude near the east coast of Greenland. (*Photo by Geodætisk Institut, København, Danmark, Eneret.*)

Fate of Modern Valley Glaciers

In the simplest possible terms, the fate of glaciers today is—melting. The ice will persist until it descends to climatic zones where it melts as fast as it is supplied from above, or it will push out into the sea until pieces break off and melt as they float away. Direct evaporation probably accounts for less than 5% of the wastage of most modern glaciers.

In principle this is analogous to the fate of deep-seated rocks. Just as it requires certain temperatures and pressures to produce glacier ice, and the resulting body is unstable and changes when moved into a realm of significantly different conditions, so it requires certain temperatures and pressures to produce a deep-seated rock such as schist or granite, and it too becomes unstable when moved into other environments. Both may be deformed and recrystallized in the solid state. Both may be melted. Exposed to the atmosphere, which is foreign to the environment in which either was born, the deep-seated rock weathers to sand and clay, and the glacier ice liquefies and evaporates.

In the upper view at the right the *terminus*, or *snout*, of a valley glacier is prominent in the foreground. Although many glaciers push out into the sea and there break up into small icebergs, the majority end as this one does—in or near the foothills of the mountains that cradle them. In Scandinavia and the Alps, where they often come down close to summer pasture lands, some glaciers have been observed for centuries. Not only do the positions and shapes of their termini change from year to year, but, like most of the world's glaciers, they have become thinner and shorter during the past century. Carefully prepared maps of Switzerland issued in 1877 and 1932 show that during this interval the area under ice was reduced 25%; because of thinning, the loss in volume of the glaciers has been even greater. Since annual measurements were begun in 1918, the terminus of Nisqually glacier on Mt. Rainier in Washington has melted back fairly steadily at an average rate of about 21 feet per year.

What determines the position of the lower end of a glacier? Basically it is the balance between accumulation above the snowline and *ablation* (= evaporation + melting) below it. In general, melting is accelerated at progressively lower altitudes until, at some point, ablation can just dispose of the volume of ice delivered by the moving glacier. Here one might think that the terminus would be stationary, even though the rest of the glacier was creeping forward, but seasonal and longer variations in both velocity of flow and rate of melting result in erratic fluctuations of the front. Further, there is mounting evidence that pronounced variations in accumulation near the head of a glacier do not simply ride down it at the rate of ice advance; heavy avalanching or excessive snowfall, for example, may produce a kinematic wave or bulge that travels down the glacier faster than the ice is acually moving. The arrival of such a *glacial surge* at the snout can cause anomalous fluctuation in the shape and position of the terminus.

In the upper view here note that both to right and left of the dark snout there are faint ridges, or embankments, running parallel to the valley floor and extending beyond the terminus off the bottom of the picture. These are composed of debris that collected along the sides of the ice and prove that this tongue of ice was once thicker and reached farther downvalley than it now does.

In the lower view the small glacier at the left has obviously shrunk within its valley, since the valley is smooth, broad, and straight for some distance below the terminus. The same reasoning and conclusion apply to the larger glacier, on the right, the only difference being that the lower part of its valley is blocked and flooded to form a meltwater lake. The source of the prominent medial moraine (page 212) is well shown in the background.

It will be noted in these, as well as in many other views, that the ice of glacial termini may be dark and almost indistinguishable from the surrounding bare ground. The explanation, of course, is that the rock debris frozen within and carried on top of the glacier becomes concentrated as melting reduces the volume of the ice but not of its load.

Fig. 198. Looking eastward up Yanert glacier, toward Mt. Deborah, in Alaska.
(*Photo by Bradford Washburn.*)

Fig. 199. The Finsteraarhorn, Upper and Lower Aar glaciers, and the Grimselsee in the Swiss Alps.
(*Photo by Swissair, Zürich.*)

Flow in Modern Valley Glaciers

There are many ways to prove that glacier ice moves. In the upper view here, for example, note how the rock debris that falls from the steep valley walls onto the edges of any glacier becomes a dark longitudinal stripe in the ice stream when two glaciers come together. This could not happen if the ice were not flowing downvalley. Each such median stripe is a *medial moraine*, produced by the union of two *lateral moraines* riding along the edges of confluent ice streams. The graceful curves of the medial moraines in the lower part of the Yanert glacier on the preceding page and the tightly folded moraines in the lower view at the right demonstrate the plastic nature of glacial movement.

It is a little more difficult to learn exactly how glacier ice moves. In the middle reaches of most glaciers stakes driven into the surface in straight rows from bank to bank lose their alignment in a few hours or days because the central stakes move downstream faster. As in rivers, the current is evidently swiftest near midstream—at least at the surface. Velocities range from almost imperceptible to more than 150 feet a day; a few inches to a few feet is most common. Friction retards the flow at the sides and probably along the bottom, but the internal velocities are not well known. Meager evidence indicates that slippage between the basal ice and the underlying rock probably accounts for less than half the observable movement of most valley glaciers. The details are complex because the ice responds both to gravity and to differences in pressure. Thus irregularities in thickness and in the shape of the rock floor can cause ice near the sole of the glacier either to flow uphill or to accelerate, as it slowly squirts over an obstacle or through a constriction. Around small obstructions on its bed, if the temperature is not too far below the melting point, the ice probably melts under the pressure on the upstream side and refreezes on the downstream side, where the pressure decreases. As noted above, now and then one or more glacial surges, somewhat like flood crests on a river, may move down a glacier with speeds several times that of the ice itself. Some surges seem to have been set in motion by earthquakes and the avalanching they cause, but the Muldrow Glacier in Alaska has surged without such provocation (e.g., 1956–1957), as though after many years the accumulation in its source area had reached some threshold amount that suddenly permitted the release of excess ice to begin.

The deep gaping cracks (*crevasses*) frequently found on the surfaces of glaciers demonstrate that this ice is brittle. But crevasses seldom extend more than about 100 feet below the surface, and the association of cracks and flow patterns unmistakably shows that crevasses are caused by adjustments in the brittle surface ice as it accommodates flow in the deeper part; the effect is particularly noticeable at any sharp bends and steps in the valley. The distinction between these zones is recognized by referring to the surface carapace, probably less than 200 feet thick, in which the ice behaves as a brittle solid, as the *zone of fracture*, and to the mobile zone beneath this, in which the ice behaves as a plastic mass, as the *zone of flow*. In the downstream parts of many of the glaciers illustrated on these pages ablation has removed so much ice from the upper surface that the zone of fracture now consists of ice whose contorted structures were produced when it lay deep enough to be in the zone of flow.

These zones illustrate again the fundamental fact, first discussed in connection with metamorphic rocks, that materials which are brittle in small masses when on the surface and subjected to rapid deformation, may be quite plastic and deform by flow when in large masses, in a deep environment, and subjected to slow distortion. In this sense glacier ice is a weak rock that deforms near the earth's surface in the course of a few years, in ways that gneisses do only at depths of several miles and over periods of millions of years. Empirically, ice stands at the weak end of what might be termed a rigidity series of the common rocks, followed by rock salt and gypsum, limestone, schist, and granite, each of which is more rigid than the last and would therefore require deeper conditions (higher temperature and confining pressure), and longer time before it would exhibit plastic deformation.

Fig. 200. Looking northeast up Barnard glacier, near the head of Chitina River, Alaska. (*Photo by Bradford Washburn.*)

Fig. 201. Contorted medial moraines between Steller and Bering glaciers, 40 miles northwest of Yakataga, Alaska. (*Photo by Bradford Washburn.*)

Transportation and Deposition
by Modern Valley Glaciers

Glaciers are among the most impressive and distinctive agents that sculpture the landscape. Individually they are more efficient than rivers; a given volume of water as glacier ice can transport a considerably larger load than the same volume of water as a flowing stream. In terms of world-wide effectiveness, however, this advantage is more than offset by the enormously greater discharges and lengths of most streams; in a given length of time, streams will move more rock debris farther.

As dramatically shown in the upper view at the right, glaciers carry part of their load on top of the stream of ice. As with streams, most of this material has been delivered to the margins of the glacier by frost wedging and downslope movements on the steep valley walls—mostly as talus and rockfalls. Generally there is not enough of this debris to account for the amount appearing at the snout; evidently a larger load is carried and dragged along at the base of the ice, somewhat like the bed load that is rolled along the bottom of a river. Occasionally some of this internal load can be seen in the broken snout of a glacier. From such observation it is inferred that most glacier ice is charged with rock debris wherever it is in contact with its containing valley, the lateral moraines being continuous, under the ice, with the bed load, or *ground moraine*. When a small glacier joins a larger one it sometimes rides on top of it, or becomes partly embedded in it so that its bed load travels at an intermediate level in the trunk stream of ice. By the time many tributaries have joined, as in Figure 200, the glacier is laced with curving sheets of rock debris that connect the medial moraines with one another or with the bed load of the main mass. These are an important source of the debris that becomes concentrated in termini like the one shown in Figure 198.

The form in which this load is deposited and the shape it assumes after the ice has disappeared depend upon the part of the glacier from which it was released and the extent to which it has been reworked by meltwater. The all-embracing term for rock debris of glacial origin, regardless of how, where, and in what shape it was deposited, is the noun *drift*. Generally the greatest concentration of material is left near the terminus of a glacier where, unless it is rapidly advancing or retreating, the annual contributions of many years may be heaped in approximately the same place. The result is to build up the characteristic arcuate *terminal moraine* that marks any position long occupied by the snout of the ice. Pauses and minor readvances during the general melting back of the ice front may produce a succession of irregularly spaced terminal moraines, sometimes referred to as recessional moraines. (See, for example, Fig. 208.) In the lower scene at the right the terminal moraine is a low, dark, arcuate ridge reaching to the left margin of the view, separated from the present terminus by a belt of irregular lakes. In three or four places meltwater streams have cut through the terminal moraine and, flowing toward us with markedly braided pattern, are spreading their load in an *outwash plain* resembling the coalescing alluvial fans of arid regions (cf. Fig. 148).

Terminal moraines are commonly more or less continuous with the lateral moraine embankments left along the valley sides. Medial moraines, however, are usually not so easily recognized after melting, because as the terminus melts back upstream they merge irregularly with the ground moraine, and their central position in the valley exposes them to maximum reworking by late-stage meltwater streams.

Fig. 202. Contorted moraines on lower Susitna glacier southwest of Mt. Hayes, Alaska.
Some of the larger loops may be related to surges in the advance of tributary ice streams.
(*Photo by Bradford Washburn.*)

Fig. 203. The snout of Allen glacier and its associated terminal moraine and outwash plain. Looking
down the Copper River, 40 miles northeast of Cordova, Alaska. (*Photo by Bradford Washburn.*)

Erosion by Valley Glaciers

Mountains must rise above the snowline if they are to have glaciers. At most latitudes they cannot reach such elevations without undergoing vigorous stream erosion, and glacial erosion is therefore generally superimposed on topography already sculptured by fluvial processes.

There are a number of characteristic modifications that mountain glaciers tend to impose on pre-existing stream-cut valleys. *Valley straightening* (Fig. 200), occurs because it is difficult for the highly viscous glacier ice to turn sharp corners. As part of this process the lower ends of ridges are trimmed off—the spurs become truncated and their steep blunt terminations contrast strongly with the tapering tips of ridges that normally project into stream-cut mountain canyons. Viewed along its axis a canyon modified by a mountain glacier acquires a characteristic flaring U-shaped cross-profile (Fig. 207) whose steep sides and relatively flat floor differ from the normal V-shaped cross-profile of stream-eroded mountain canyons. (Where such straight, steep-sided, U-shaped glacial valleys have been invaded by the sea they are called *fjords*.) The erosive action of a valley glacier deepens the valley and leaves a *scoured floor* from which all loose weathering products have been removed; the resulting fresh bedrock surface is usually mantled with a combination of moraine and outwash left by the melting ice. The deepening may so lower the valley floor below the mouths of its tributaries that they enter far up on the steep valley walls, a process that accounts for some of the world's most beautiful waterfalls; the valleys thus cut off are called *hanging valleys* (Fig. 364). Where the mantle of moraine and outwash are absent, as on mounds of resistant rock or in the upper ends of many glaciated valleys, one can usually find surfaces of *striated and polished rock* (Fig. 209) resulting from abrasion by the debris locked in the moving ice.

Erosion beneath moving ice takes place in two principal ways. One is this *abrasion*, the scraping effect of rocks of all sizes and shapes that are embedded in the ice and dragged along the floor and lower sides of the containing valley. This is sometimes referred to as the glacial rasp. The other is *plucking* (or quarrying), which consists of the removal of blocks of bedrock which the glacier ice freezes onto and then drags along as it moves down the valley. Despite low temperatures, some ice is undoubtedly melted by the pressure beneath most glaciers; upon refreezing, this water may pry rocks loose and help bind them to the moving ice.

Followed upstream, most glaciated valleys end abruptly in a semicircle of high cliffs, often with a small rock-bound lake nestling at its foot. This headwall is a *cirque*, one of the most easily recognized and distinctive products of glacial erosion. The steep walls of a cirque are craggy rather than polished and clearly more the product of quarrying than of rasping. Indeed, as shown in the upper view at the right, it is sometimes here that the glacier is born—as an independent product of local avalanches and drifts of snow and firn instead of outflow from an extensive snowfield. The firn and ice of the upper end of the glacier rest but lightly against the headwall. At times the ice pulls away, leaving a deep narrow gap between ice and headwall. Angular rock debris, loosened by the freezing of water in cracks, falls into this gap and immediately becomes part of the glacier's load. Thus an abundance of blocks with sharp edges unblunted by erosion is available at the very head of the valley for rasping the floor of the cirque. The result, minus ice, is visible in the lower view at the right, which includes more than a dozen cirques deeply cut into the flanks of a smooth mountain summit. This picture suggests another factor that might promote the formation of some cirques: It is probable that, because of winds, very little snow would accumulate on the summit surface of this mountain; initially it would form drifts in such protected places as valley heads. As the latter enlarged, they would trap nearly all of the snow that crossed the summit, thus creating conditions conducive to severe glacial erosion in the valley head while virtually none was taking place on the adjacent summit.

Fig. 204. Three small tributary glaciers heading in cirques. Looking northwest across Harvard glacier, about 80 miles east of Anchorage, Alaska. (*Photo by Bradford Washburn.*)

Fig. 205. Cirques cut into the summit of the Uinta Range during the last Ice Age. Looking west-southwest over the 12,000-foot crest, immediately west of Leidy Peak.

Evidence of An Ice Age I:
Moraines and the U-Profile Left by Valley Glaciers

Glaciers leave deposits of telltale shape. Fluvial erosion may alter this shape in a comparatively short time, but even after this has happened, polished and striated boulders in the deposits may provide a clue to glacial origin. Coupled with the erosional effects of glacier ice itself, such features make it possible to prove the former existence of glaciers in places where no such ice remains today.

In the Alaskan scene below the moraines typically associated with the lower end of a mountain glacier are clearly shown. Dark embankments of talus-like rubble along the sides form lateral moraines and loop around the end of the glacier as terminal moraines. The terminus seems to have melted back a short distance from its maximum advance, for a braided meltwater stream has already cut away parts of several loops.

In the upper view at the right we see a similar valley in California. Its open U-shaped cross-profile and gently curving plan contrast strongly with the smaller narrow and crooked stream-cut canyons at the right. This glacial shape continues beyond the mountains between well-developed embankments of lateral moraine (foreground). In the lower right part of the scene this broad trough, many times wider than the stream now flowing in it, becomes a narrow tree-lined gully with proportions befitting the stream. The point at which the change begins marks the limit reached by the glacial snout, the gully being the work only of meltwater and later runoff. If one imagines the ice tongue restored down to this point, the scene is geologically almost identical to the one on this page.

The lower view shows the mouth of another glaciated valley on the east slope of California's Sierra Nevada. More than a dozen nested loops of terminal moraine are discernible between the foreground and the morainal dam of the lake in the background. Some of the nearer ones correspond with the multiple lateral moraines at the far left. Such piles of unsorted rock debris, arranged along the sides and around the mouth of a U-shaped valley, can have been produced only by a glacier, the several crests marking successive halts in its shrinkage.

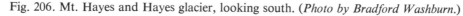

Fig. 206. Mt. Hayes and Hayes glacier, looking south. (*Photo by Bradford Washburn.*)

Fig. 207. Lateral moraines of Green Creek, 10 miles south of Bridgeport, California. View is westward, up east slope of Sierra Nevada.

Fig. 208. Terminal and lateral moraines below Convict Lake at the east base of the Sierra Nevada.

Evidence of an Ice Age II:
Glacial Polish and Striations

The rock fragments dragged along in the base of a moving glacier range from blocks many feet in diameter to pulverized rock flour. The result of abrasion with such particles is combined grooving and polishing. Three samples are shown here.

The view below is across the summit of the Devil's Post Pile, a small mass of columnar lava already seen from the side in Figure 27. Only the tops of the nearly vertical columns, scraped and ground off to form a smoothly rounded surface, are visible here. The overriding ice has left a characteristic combination of striations (grooves and scratches) and polish, produced respectively by blocks and rock flour held in the ice. Weathering has destroyed patches of the polish, but it is a slow process; probably at least 10,000 years have passed since the last glacier occupied this area.

In the upper view at the right, grooved and polished granitic rock protrudes from beneath a cover of glacial deposits. Note the irregular shrinkage cracks at the upper right, which imply a high proportion of clay, probably produced by weathering of rock flour. There are also blocks several inches across in the foreground. This combination would be almost impossible in a water-laid deposit, but is very characteristic of glacial *till*, the variety of drift that is deposited directly from glacier ice and not subsequently reworked or sorted by running water. (See also Fig. 223.) Moraines are composed of till.

The lower view at the right shows a till-like deposit (upper right) resting on a smooth, striated rock surface. Such a combination is interpreted by most geologists as evidence of glaciation, despite the fact that the overlying rock is very much older than the other examples on these pages. Tills deposited during the last Ice Age are generally loose, unconsolidated sediment, but this is a hard rock (tillite) probably at least 200 million years old. Such occurrences probably mean that the earth has witnessed more than one Ice Age.

Fig. 209. Glaciated top of the Devil's Post Pile in the eastern Sierra Nevada, California. (*Photo by Robert C. Frampton.*)

Fig. 210. Glaciated granitic rock beneath unconsolidated till near the border between Sweden and Norway.

Fig. 211. Exposure of the Dwyka tillite resting on glacially smoothed and striated igneous rock near Kimberley, South Africa. (*Photo courtesy of R. F. Flint.*)

Evidence of an Ice Age III:
Kettles in Outwash

Till, as we have just observed, is rock debris deposited directly by ice. It is a special variety of sediment, most readily recognized by the shapes of the deposits in which it originally accumulates and by its internal characteristics, particularly the association of rock flour and striated boulders. Because of its origin in contact with ice, till is of limited occurrence; its initial distribution can have been no wider than that of the ice itself. But much of the load of any glacier is ultimately picked up, transported, and finally redeposited by streams flowing on, within, beneath, and beyond the ice. Closely associated with till, therefore, is another variety of drift which, although derived from glaciers, was deposited by streams composed principally of meltwater. These *glacial outwash deposits* (for example, that forming the outwash plain in Fig. 203) are readily distinguished from till because they are stratified; typically, lenses of gravel alternate with cross-bedded sands as they might in the deposits of braided and occasionally torrential streams. (See also Figs. 224 and 227.)

Clearly, glacial outwash may be difficult to distinguish from ordinary stream deposits. The farther the sediment is carried from its glacial birthplace, the greater the likelihood that its striated stones will become rounded and all its ingredients weathered and worn to resemble ordinary sediment. But there are clues.

The stagnant end of a large glacier enters from the left in the upper scene at the right. Two meltwater streams, characteristically milky with rock flour, are flowing into the sea, where their load of fine sediment produces conspicuous silt clouds. (Note the effect of a longshore current.) Between the gray ice at the left and the beach at the right is a forested area dotted with irregular pits and depressions, some of which contain ponds. These have no outlets and are not connected by streams. Somewhat similar ponds occur near the margins of the bare ice. However, there are none on the central part of the glacier or beyond the ice at the upper left. We must conclude that they are related to the ice and have been produced during the melting of its stagnant margin.

Obviously the pits in the ice itself will disappear when the ice melts. But trees require soil, which means that the forested area, with its ponds, is underlain by till or outwash. Where these deposits have buried or surrounded an isolated block of ice, the later melting of the ice produces a depression in the sediments; this is the explanation of the ponds in the forested area. Such depressions are known as *glacial kettles*, and a generous sprinkling of kettles is good evidence that the deposit in which they occur is glacial outwash rather than some other kind of alluvium.

The lower scene is a comparable view of the outer shore of Cape Cod. Here again is an area of uncemented sands, clays, and pebble beds, topped by an uneven forested surface pitted with depressions, many of which contain ponds. These depressions are evidence of glacial outwash, but why are the ponds more nearly round than those in the upper scene? The almost circular one just right of center seems to be in the process of cutting itself off from its rambling bays, one of which on the near side is partly filled in. Evidently it once had a more irregular shape, more like those to the left of it. Recalling the tendency of wave action to smooth out shorelines (page 192) by attacking headlands and building spits and barrier beaches, we may be reasonably sure that there is a similar explanation for the well-rounded shapes of the larger ponds; in loose sandy deposits there would be little resistance to the effect produced by shifting winds driving waves against all sides of each depression. The larger and deeper the pond the larger the waves, which is probably a major reason for their more regular shapes.

The probability that this part of Cape Cod is glacial terrain is examined further on the next page.

Fig. 212. Looking southward over the terminus of Fairweather glacier, 80 miles southeast of
      Yakutat, Alaska. (*Photo by Bradford Washburn.*)

Fig. 213. Looking northeastward over the central part of the outer shore of Cape Cod, east of Wellfleet.

Evidence of an Ice Age IV:
Extensive Pitted Outwash Plains

The large pond in the upper scene here is the same one that was left of center in the last view (Fig. 213). Here, in the sea cliff, we can obtain some idea of the materials on which this uneven surface of hummocks and hollows has developed.

Note that even though there is much loose sandy debris along the cliff, none accumulates at its base—evidence that the cliff is being actively eroded by the waves. This is the source area for much of the sand that drifts southward to form the spits and extend the beaches shown in Figure 179. Yet despite active undermining at its foot, there are only a few places where even a suggestion of stratification can be seen in the cliff; it must, therefore, be composed chiefly of loose, uncemented sediment. It is particularly interesting that the meager indications of bedding tend to parallel the undulating surface at the top of the cliff. This is difficult to explain in any other way than by differential settling of the sands and gravels. The gradual melting of buried slabs and blocks of ice provides a likely mechanism.

Consider some alternative origins for this kind of hummocky surface on this kind of unconsolidated sediment: The combination cannot be eolian because the deposits contain lenses and beds of gravel and show water-current, not eolian, bedding. Well records show that the area is not underlain by limestone, salt, or gypsum, so this cannot be a blanket of sand accommodating itself to irregular dissolving of a deeper layer. (See *karst topography*, page 238.) The depressions cannot be craters from a meteorite shower because they show neither explosive disruption of the bedding nor rims of material thrown out of the holes.

Great numbers of depressions in a thin deposit of loose stream-laid sand and gravel are characteristic only of kettles in glacial outwash. Such a pock-marked blanket of meltwater deposits is often referred to as a *pitted outwash plain*. Geologically, it is an almost unique combination of features and very good evidence of former glaciation.

To conclude that this landscape is of glacial origin, even though there are no lofty mountains nearby and it is 150 miles to the nearest ridge even 5,000 feet high, is to pose another question: Where did the ice come from?

If we examine Cape Cod back to its connection with the mainland, we find that it consists throughout of pitted outwash plain modified only by some small and recent stream valleys, some low bouldery ridges that are undoubtedly terminal moraines, and some belts of recent sand dunes. Continuing the examination to the northwest, along the mainland coast of Massachusetts shown in the lower view at the right, it is clear that pitted outwash plain here covers hundreds of square miles. Indeed, it was just over 100 years ago that geologists began to realize that virtually all of the lowlands of New England are mantled with outwash and till, and that all the bedrock hills which protrude above this have been scoured or smoothed by overriding ice. Principal credit for this insight goes to Louis Agassiz, who arrived at Harvard University in 1846, fresh from studies of similar evidence in Europe, and who first marshalled the evidence for the former existence of a North American ice *sheet* that was not dependent upon nearby mountains.

Of course the logical direction in which to look for the source of such an ice sheet was northward, especially because the glacial deposits terminate a little farther south, between New York City and Philadelphia. On the next six pages we will examine more of the evidence for this *continental ice sheet*, as distinct from valley or alpine glaciers.

Fig. 214. Closer view of the cliff along the outer shore of Cape Cod shown in Figure 213; a pitted outwash plain.

Fig. 215. General view northward along the pitted outwash of the Massachusetts coast; Plymouth Bay in the background.

Evidence of an Ice Age V:
The Glacial Limit

As a test of the hypothesis that an ice sheet once covered extensive lowland areas in the northern states and adjacent Canada, let us examine some samples far west of Massachusetts.

The upper view here is toward the southeast on the Columbia Plateau of east central Washington. Two distinct kinds of topography are apparent. At the right and in the background the land surface is smooth and almost flat as far as the eye can see—a patchwork of rectangular cultivated fields. In contrast to this the surface at the left is rough—a jumble of hummocks and hollows with an average elevation 50 to 100 feet above the smooth farmlands. The rough ground is further distinguished by scattered boulders on its surface, many of which can be seen to equal the size of the house and barn in the central foreground. The local farmers refer to these huge blocks of dark basalt as "haystack rocks."

Geologically, the smooth land consists of rich soils developed on a discontinuous thin blanket of loess that overlies horizontal basalt flows. Embedded in the rough ground, in addition to the haystack blocks, are many smaller boulders of granite, gneiss and other rocks whose only possible source is beyond the edge of the underlying lavas, at least 10 miles away. All of these pieces of rock have been carried to their present positions on the lava plateau by some agent capable of transporting them for many miles over almost level ground. The large haystack boulders almost certainly came from the limiting cliffs at the edge of the plateau and must therefore have been lifted up onto it. Granitic and gneissic rocks are exposed only under the lavas and north of the plateau, 10 to 30 miles away, and thus boulders of these rocks must either have been lifted up onto the lavas or have been carried across the 1,000-foot deep canyon of the Columbia River. Only ice could have done any of this—and deposited such a load on top of *un*eroded lavas. (Such isolated blocks of displaced rock, transported into foreign surroundings from which they could not have been derived, are termed *erratics*. This noun is used chiefly for glacially transported blocks left on or near bedrock of distinctly different composition. (See also Figure 228.) And finally, as might be expected, there are places on the plateau, 35 miles to the northeast, where exposed basalt still retains a grooved and polished surface of the kind characteristically produced under moving ice. In short, the uneven ground was once under ice, the smooth ground was not, and the somewhat irregular boundary between them marks the *glacial limit* in this area—the outer edge of the terminal moraine left by an ice sheet.

The lower view is toward the west in the same area and shows in somewhat greater detail the contrast between morainal topography in the foreground and unglaciated plain beyond. The particular value of these two scenes is that each permits direct comparison, in one view, of glaciated and unglaciated landscapes in a region where there is no complication from other landforms.

It is worthy of note that there are almost no outwash deposits on the lava plain immediately beyond the moraine in these views. This probably means that there was not a great deal of meltwater coming from this ice front, and that most of what there was flowed more or less along the ice margin, cutting gullies that are still plainly visible. Some of these are still functional and some abandoned. The principal drainage line in these scenes follows the glacial limit fairly closely, and is the ice margin stream of Figure 160. All this implies that the ice sheet here was thin, and perhaps melted very slowly.

(This ice was part of the Okanogan Lobe shown in Figure 315 and discussed on pages 338–351.)

Fig. 216. Looking southeastward along the margin of the moraine left by the Okanogan Ice Lobe on the Waterville Plateau between Chelan and Grand Coulee, Washington. (Also shown in Fig. 160.)

Fig. 217. Closer view of another segment of the same glacial limit, looking west.

Evidence of an Ice Age VI:
Landscapes on Glacial Deposits

The most obvious clues to extensive continental glaciation are the peculiar details which an ice sheet imprints on the topography of its own deposits. Here are three examples.

The view below (Fig. 218) is a typical landscape on ground moraine. Notice the random pattern of small irregular hills and the many ponds in various stages of conversion to bogs. These and the absence of an organized stream system eliminate all possibility that this scene could be the result of normal processes of stream erosion. It is not even well drained, much less eroded by running water. Some of the depressions are kettles, but probably most of them are simply incidental to the uneven thickness and distribution of the till.

Less extensive, but often somewhat resembling the scene below, are the many low ridges of terminal moraine left by the ice sheets. The terminal moraines of ice sheets are much more difficult to see and photograph than are those of valley glaciers because they are usually several miles wide and less than 200 feet high. The small example at the upper right (Fig. 219) is readily identifiable because it runs into the sea, where its subtle form is largely responsible for a chain of low islands. In the foreground its rough and wooded surface at the right stands in clear contrast to the flat outwash plain at the left, which is dotted with houses and larger ponds. The ice came from the right.

The lower scenes (Fig. 220) show a special variety of glaciated landscape—elongate, parallel, smooth hills known as *drumlins*, which represent one of the many aligned patterns produced under moving ice. They are generally less than 100 feet high, up to half a mile long, and tend to be steep at one end and tapered at the other. In both these views, which are toward the north, it is clear that the steep end faced the oncoming ice, thus demonstrating both the fundamental principles of streamlining and, indirectly, the fluid behavior of the ice. Some drumlins are composed entirely of till, some also contain stratified sand, and a few seem to be shaped partly from knobs of bedrock. They usually occur in clusters or fields; the one centered east of Rochester, New York, of which these views are a sample, is estimated to contain about 10,000 drumlins in all.

Fig. 218. Sample of morainal topography near Roslyn in eastern South Dakota.

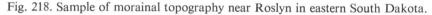

Fig. 219. View southwestward over the Buzzards Bay moraine from Falmouth
to the Elizabeth Islands, Massachusetts.

Fig. 220. General (*left*) and close (*right*) views of drumlins east of Rochester, New York.

Evidence of an Ice Age VII:
Drainage Changes along the Ice Front

When testing a hypothesis it is valuable occasionally to interrupt the systematic examination of evidence with a leap into creative speculation. Imagination plays a vital role in research, especially in suggesting things to look for. For example:

If an ice sheet once extended into the northern plains region, we may be quite sure it came from the north, both because that would be consistent with present-day climatic zones and because the signs of prehistoric ice disappear southward. But if ice sheets crept down from the north what might they do, for example, to the drainage system? And what changes would this make in the landscape? Probably the most marked changes would result from direct interference with streams and rivers. Those flowing northward would be blocked by the ice front. They might be immobilized by freezing part of the time, but during warming trends, when the ice sheet was melting, there would be times when they would be ponded against the ice front. The shallow lakes thus formed would receive large amounts of meltwater with its load of sand, silt, and clay. If a lake rose high enough it might overflow through a divide where streams had not flowed before; for a time the level of the lake would be controlled by the level of this spillway. Under favorable conditions we would expect the waves to engrave one or more shorelines, the highest corresponding to the level of the spillway. Remembering the spreading of mud on the floor of Lake Mead and the ability of waves to agitate shallow bottoms, we would expect the floor to consist of a nearly flat blanket of fine sediments with deltaic tongues of sand opposite the mouths of streams that flowed into the lake. Perhaps when the ice melted sufficiently to uncover the old northward channel the outlet would change, abandoning the temporary high-level spillway in favor of the old lower one. Sudden lowering of the lake level and trenching of the lake-floor deposits should follow.

Streams flowing southward, away from the ice, would be changed too. At the very least there would be times when they would be swollen by quantities of water and sediment greatly exceed-

Fig. 221. View southward over the flat floor of glacial Lake Dakota south of Aberdeen, South Dakota.

Fig. 222. View southward along the spillway of glacial Lake Agassiz at Browns Valley
on the South Dakota–Minnesota boundary. Big Stone Lake now occupies part of
the channel in the background.

ing their normal discharge and load. These conditions would modify the shapes of the valleys, probably by widening and the addition of extensive floodplain deposits; subsequent normal flow would trench these, leaving stream terraces in the glacial fill. Lakes and the changes that accompany them could develop if these valleys were obstructed downstream—as by terminal moraines.

That this is not all idle speculation is shown by abundant evidence from Montana to Indiana to Maine. The view at the left shows part of the extraordinarily flat plain in South Dakota across which the James River (in the foreground) meanders today. Over an area of more than 1,800 square miles this surface has a local relief (except among a few sand dunes and along the river) of less than 3 feet, and although it is 105 miles long, the south end is only 10 feet lower than the north. On the east and west it is bordered by gently rising land, and at the southeast this is subtly terraced by the remains of old shorelines. At the south it is closed off by a complex of terminal moraines through which the river has cut a long postglacial trench. Clearly, a lobe of the ice sheet pushed its way down the ancestral James River valley and left broad compound terminal moraines to choke its southward course in the southern part of the state. These served as a dam which retained for a time, as the ice sheet melted, shallow glacial *Lake Dakota*, whose old bottom deposits account for the flat surface shown here.

Northeast of glacial Lake Dakota was an arm of the much larger glacial *Lake Agassiz*, which was centered in Manitoba and had an area of over 200,000 square miles, more than twice that of all the present Great Lakes combined. This greatest of all known North American lakes was produced when the ice front blocked several rivers flowing northeastward toward Hudson Bay. The photograph on this page shows part of the giant spillway toward the south that for a time accommodated the outflow of Lake Agassiz—a trench 200 feet deep and 1 to 4 miles wide cut across flat ground which today can supply it with only a trickle of water, and on a gradient so gentle that long shallow lakes now occupy its low places. Dozens of shorelines and the largest truly flat area in North America lie north of it, attesting to the great ice-margin lake responsible for this abandoned river bed. (A tributary to this spillway is shown in Figure 154.)

231

Some Characteristic Features of Glacial Deposits

The six views on these two pages portray some of the distinguishing characteristics of glacial drift and also illustrate their use in reconstructing local conditions during the last Ice Age.

Fig. 223
Glacial Till.—At the left is a close-up of glacial till in Vermont. Note particularly the absence of size sorting or bedding and the combination of boulders, some of them angular, with a clay-rich matrix (revealed in the vertical scrape at the left of the photograph.)

Fig. 224
Glacial Outwash.—Shown here, for comparison with glacial till, is glacial outwash exposed in an excavation on Cape Cod. The cross-bedding and abrupt changes from sand to gravel imply torrential water currents, but there is no proof here that they were derived from melting ice. However, there is a good terminal moraine a few hundred feet away, and this deposit contains many kettle holes and smaller structures that are most reasonably ascribed to the melting of buried ice blocks.

Fig. 225
Lateral Moraine.—A cross-section through a narrow ridge more than 10 miles long that lies along one side of a valley at the north base of the Alps is shown in this figure. The shape of this ridge and its position near the mouth of a valley possessing all the classical features of mountain glaciation leave no doubt that it is a lateral moraine. Being deposited directly by the ice, it should be composed of till. A boulder from this moraine is shown in the next figure.

Fig. 226

Striated Boulder.—This 12-inch slab of rock was taken from the exposure in the preceding view. The slab's surface is grooved and scratched from being dragged along in different positions or inscribed by passing stones. Looking again at Figure 225 we can see that this slab was part of an unstratified deposit that includes blocks 10 feet or more across, randomly distributed in finer material. These facts and the abundant clay in the matrix leave no doubt that this is till.

Fig. 227

Esker.—A section through one of the many sinuous ridges on the lowlands of central and southern Sweden is shown for comparison with Figure 225. Some of these ridges are more than 100 miles long, and most trend perpendicular to the associated terminal moraines. Their constituent materials, crudely stratified sand and boulders, are outwash, not till. Known as eskers, the ridges were almost surely deposited by torrents of meltwater flowing in confined channels on, within, or beneath stagnant ice.

Fig. 228

Erratic.—Because it lacks the sensitive relation between velocity and transporting power typical of water and wind, only glacier ice can transport and deposit together sediment ranging from fine clay to huge boulders. Blocks like the one at the right, composed of rock types not found nearby, must have been delivered by moving ice. If their source is known, such erratics can be used to determine the direction of movement. Near Orebro, Sweden.

19 GROUNDWATER

Introduction

We have seen that in the long run only a small part (usually less than a third) of the average precipitation on the lands appears in streams and rivers as runoff; the rest is returned directly to the atmosphere through plants or by evaporation. It is also clear that some of even this small part is delayed in becoming runoff simply because it soaks into the ground, partly or wholly filling the accessible cracks and pore spaces between soil particles, mineral grains, and rock fragments, to become what is known as *groundwater*. In some places groundwater rejoins surface runoff in a few hours or days; in others it may remain underground for hundreds, or perhaps even thousands, of years.

It is groundwater that supplies wells, issues from springs, and keeps streams and rivers flowing through dry seasons. It is groundwater that spurts out in geysers and flows into mines. Water vapor far exceeds all other gases erupted by volcanoes, and most of this is derived from groundwater that has seeped into the volcanic conduits. Under some conditions groundwater dissolves large quantities of limestone to produce caverns like Mammoth Cave and Carlsbad Caverns; under others it may reverse the process, adorning caves with spires and columns of deposited *dripstone*. A number of rich mineral deposits, including much of the iron in the Lake Superior district, some copper in Arizona and Utah, and most aluminum ores, owe a large part of their value to the combined effects of weathering and groundwater, which have selectively removed unwanted minerals and redeposited ore in higher concentrations.

In temperate climates groundwater is replenished by downward percolation from rain and snow. Normally there is a lower zone in which all pore spaces are filled with water, the *zone of saturation*. Above this is one through which the water filters on its way down; here the rock particles are more or less moistened but the cracks and pores also contain air. This is the *zone of aeration*, in which weathering processes are highly active. The boundary between the two is the *water table*—the top of the zone of saturation. The water table is an uneven and sometimes interrupted surface whose shape and depth below the land surface depend both on the physical properties of the rocks and the fluctuating supply of water.

In the relatively dry climates found over much of the southwestern United States the water from light rains often evaporates before reaching the water table. In the mountains rainfall is relatively high and a large part of the groundwater in the lowlands is replenished by mountain streams that lose themselves completely by percolation in the alluvium of the valleys.

Springs may exist where the water table comes to the surface, and wells, of course, must penetrate the zone of saturation if they are to serve their purpose. Continuing yield in both cases is possible only because groundwater moves laterally, albeit slowly. Rates range from a few inches to a few miles per year and depend primarily upon the *permeability* of the rocks—the ease with which they transmit fluids. This in turn depends largely on the size and connectedness of the pore spaces and is surprisingly independent of the *porosity* or total amount of pore space in proportion to rock. A clay or shale with porosity of 30 to 40% may be far less permeable to water than a sandy gravel with porosity of 20 to 25%—because the interstices of the clay are so minute that friction and capillary forces inhibit the flow.

The direction of groundwater movement is generally down the slope of the water table, but because it is subject to many fluctuating influences, such as pressure gradients, the actual flow patterns may be quite complex.

The next nine pages are devoted to brief examination of a few of the more conspicuous geological changes (other than ordinary weathering) wrought by groundwater, and to a summary of man's use and abuse of this vital resource.

Effects of Solution

The most spectacular accomplishment of groundwater is the dissolving of certain kinds of rock. Naturally occurring rock salt, like ordinary table salt, is highly soluble in water; even when quite impure it is so susceptible to removal by rain and groundwater that it almost never forms outcrops at the land surface. Indeed, the rocks overlying buried salt deposits often display irregular structural depressions caused by collapse where underlying salt has been dissolved away. (An example is explored on page 372.) To a lesser degree the same is true of gypsum.

Less soluble than either of these, but far more widespread, is limestone. Rainwater is very slightly acid from the carbon dioxide it has picked up in falling through the atmosphere; its effectiveness in dissolving exposed surfaces of limestone is illustrated in the view below. This etched surface, with conspicuous rills running parallel to the paper clip, is solely the result of rain falling on impure limestone. Large quantities are not required—the fretted surface shown below developed on a boulder in the desert near Palm Springs, California, where the rainfall today is less than 6 inches per year. In humid climates decaying vegetation often makes the water in and beneath the soil even more acid. The result is that in almost any climate limestone shows evidence of slow removal in solution.

By the same token, water taken from wells, springs, and some surface streams in limestone country is rich in calcium compounds taken into solution from the rock. Such water does not cooperate well with soap; householders call it "hard." Removal of the calcium compounds is one of the main objectives of water "softening" processes.

The upper photograph at the right was taken at one of the excavations for dams in the Tennessee Valley. In order to place the dam on a solid footing it was necessary to remove the

Fig. 229. Detail of solution effects of rainwater falling on limestone in the desert of southern California. (*Specimen collected by A. O. Woodford.*)

Fig. 230.
General view of joints enlarged by groundwater solution. Limestone uncovered during excavation for dam, Tennessee Valley. (*Photo courtesy of Berlen C. Moneymaker.*)

muds and gravels of the river bed at the site. This view shows the underlying horizontally bedded limestones as they were revealed in the process—traversed by sets of straight trenches representing solution along joints by circulating groundwater. In some places these openings were found to extend several hundred feet below the water table.

Some of the details of such a trench are shown in the lower view. Beneath the dried mud and rounded pebbles of the river bed are thin horizontal layers of nearly white limestone. Note the delicate sharp points with which these project into the trench—a characteristic product of solution and one which helps to distinguish it from normal surface erosion. Study of hundreds of openings like these show that the pattern and shapes of the cavities are controlled by cracks, which allow circulation, and by variations in the solubility of different layers in the limestone. These examples, of course, have been cleaned out. Most such solution cavities are filled with clay and silt—the insoluble residues from the limestone—plus material that has fallen into them from overlying strata.

Fig. 231. Detail of a similar solution cavity in limestone beneath river gravels.
(*Photo courtesy of Berlen C. Moneymaker.*)

Karst Topography

Limestone caves are numerous in many parts of the world. Some, like Mammoth Cave in Kentucky and Carlsbad Caverns in New Mexico, are visited by more than half a million people each year. But from the geological point of view these are only local curiosities, compared to the thousands of square miles of landscape that show unmistakable evidence of solution below the surface.

The upper scene at the right is eight miles south of the entrance to Mammoth Cave. The characteristic feature of this lowland underlain by limestone is the profusion of small depressions, or *sinks*, that dimple its surface. Some contain ponds, some bogs, and some are merely shallow, cultivated swales. A normal drainage system is conspicuously lacking; there is not a single creek for tens of miles along this belt of dimpled ground. Yet the annual rainfall is around 50 inches and several small streams flow into the lowland from higher areas to the right and left. Where does the water go?

Of course, part of this moisture is returned to the atmosphere directly by evaporation and part by transpiration through plants. But when, as in this scene, hollows without surface outlets do not hold water, while neighboring ponds prove there is enough precipitation to fill them, it is obvious that many of the sinks leak. This is more dramatically shown at the occasional point where a stream flows into a hollow and disappears; such sinks are appropriately referred to as "swallow holes." The water that enters them emerges again in springs at lower elevations.

Sinks, and drainage that goes underground, are elements of *karst topography;* the name is used for any landscape conspicuously influenced by subsurface solution and comes from such a district in Yugoslavia. The development of karst is favored by soluble rocks (usually limestone) cut by systems of joints that provide initial and localized water passages, and plenty of water to circulate through them. In the early stages the ground sags into shallow sinks over points where rock has been dissolved, but as solution progresses more and more water finds its way into the growing subterranean passages. Then the entire flow of a stream may go underground for a short distance, reappearing a mile or so away, creating the anomaly of a streamless segment of a valley in a humid climate. In time the sinks enlarge, sometimes by collapse into subterranean caverns, and a residual blanket of the less soluble ingredients of the rock accumulates at the surface. Most soils on such material are relatively rich in clay and many are reddish in color.

The lower view shows a group of sinks along the east side of the valley of the Pecos River near Roswell in semi-arid New Mexico. The visible rocks in the walls of the sinks are chiefly gypsum and rock salt with interbedded layers of red silt and sand. This sequence dips very gently to the right (east) and is covered, at the left, by alluvium of the Pecos valley. Underlying the gypsum, salt, and red beds is a limestone so riddled with small solution openings that it is highly permeable. This limestone comes to the surface along the base of the Sacramento Mountains west of the valley and there absorbs some of the greater rainfall on this higher ground. The water percolates down the dip of this *aquifer* (= water-conducting rock) to the area under the valley, where other less permeable strata of silt and clay (possibly including the lower part of the blanketing alluvium) inhibit further movement. The water thus confined is under hydrostatic pressure; when wells reach it, at depths of a few hundred to more than 1,000 feet, it rises almost to the surface and in some cases overflows.

The sinks are clearly the result of groundwater solution of the gypsum and salt and the accompanying collapse of the surface, but why they are localized along the edge of the valley is not known. Perhaps they are sites of old springs through which, when the rainfall was a little heavier (as late in the Ice Age), water under hydrostatic pressure escaped upward, either along joints or along the edge of the less permeable blanket of alluvium.

Fig. 232. Karst topography on lowland underlain by limestone, just west of Park City, Kentucky.

Fig. 233. "Bottomless Lakes," a group of sinkholes less than 100 feet deep near Roswell, New Mexico.

Fig. 234. General view of the caliche deposit capping Mormon Mesa west of the
Virgin River in southeastern Nevada. For a wider view, see Figure 152.

Precipitation of Minerals from Groundwater

What becomes of the material taken into solution by groundwater? This includes not only the limestone and other soluble rocks removed to make caverns and karst topography but also enormous amounts of dissolved mineral salts derived from ordinary weathering of common rocks. These salts are of complex composition because, over long periods of time, no mineral can wholly withstand the attack of percolating natural waters as they seep downward through the zone of aeration to the water table and then slowly move toward some spring, well, stream, or shore.

This great chemical migration can be partly traced by observing differences in the natural *salinity* (concentration of naturally dissolved salts of all kinds) of surface waters. The salinity of rainwater is very low, usually less than 10 parts per million (ppm). The natural salinity of the Great Lakes is about 100 ppm, increasing slightly downstream from Lake Superior to the St. Lawrence River. The average for the Mississippi River is about 200 ppm; that for the Colorado is about 600 ppm, and for the Gila 1,000 ppm, reflecting the greater aridity of their basins. (Drinking water should not exceed 500 ppm, comparable to about a quarter teaspoonful of salt per gallon.) The average for sea water is about 35,000 ppm or 3.5%, and water in small enclosed basins in dry climates, like Great Salt Lake and the Dead Sea, goes above 200,000, or 20%. Clearly, terrestrial waters pick up soluble salts on their way to wherever they are going and leave most of this dissolved load behind when they return to the atmosphere by evaporation: this is one reason the oceans are salty. By the same mechanism seawater is nearly saturated with calcium carbonate—which makes it easy for clams to have shells.

Although most of the dissolved material in groundwater is carried into streams and ultimately reaches the sea, a small but geologically important part is deposited along the way. Somewhat as evaporating "hard" water will leave a crusty scale of calcium carbonate inside a teakettle or in the spout of a slowly dripping faucet, so evaporating groundwater deposits its dissolved load near the surface in dry climates. The result is a calcareous crust or "hardpan" known geologically as *caliche* (Spanish, from the Latin word for *lime*).

The platform above (Fig. 234) is capped with a resistant layer of caliche formed long ago when its surface was approximately the level of the valley floor (see page 158 for a general discussion of this history.) In this arid climate most of the downward-moving water evaporates within the zone of aeration, and during long dry spells deeper moisture is drawn upward by blotter-like capillary action but evaporates before reaching the surface. In either case caliche is deposited, its depth and thickness depending on the degree of aridity, permeability, and time. The result in this example is a zone in which the otherwise unconsolidated sediment is firmly bound with a characteristically spongy cement of calcium carbonate derived principally from the abundant limestone fragments occurring throughout the deposit. Although from a distance it looks like a different and harder layer deposited in sequence with weaker ones beneath it, it is apparent from its composition and the closer view in Figure 235 (*left*) that it is not a stratum and is distinguished only by its somewhat irregular cementation.

The limestone pebble conglomerate in Figure 235 (*right*) demonstrates both solution and deposition. Most of the spaces between pebbles are filled with precipitated calcium carbonate, but many of the pebbles are conspicuously dented. These dents have sharp rims and are perfectly fashioned to accommodate neighboring pebbles; they must have formed *after* the sediment arrived at this place. In all likelihood solution by groundwater was most active at (or even limited to) the points of contact between pebbles. This is plausible, for the solubility of limestone, like that of most substances, is increased by pressure: groundwater could be depositing carbonate in voids at the same time it was dissolving it from pressure points. With such a delicate balance slight differences in the texture or composition of adjacent pebbles might determine which member of any pair would penetrate the other.

Groundwater helps produce and transport the iron oxide that forms rusty nodules and hard layers in and below some soils. It also contributes the calcium carbonate which, added to minute fragments of shell or limestone that act as nuclei, causes them to grow into the lumpy *concretions* found in some sandstones. Silica in groundwater is precipitated in many forms, including opal, chalcedony, and agate, and, by replacing the carbon in buried logs, produces petrified wood.

Most important of all, the traces of water that are associated with deeply buried detrital sediments play a vital role in converting them into sedimentary rock. The Coconino sandstone (page 280) was once loose dune sand yet today it stands as a vertical cliff near the top of the walls of the Grand Canyon. Under the microscope it is quite clear that this is possible because the sand grains have been cemented together by the deposition of calcium carbonate (and a little silica) in the pore spaces. The Coconino sandstone is overlain by marine limestones: both these and the seas in which they formed could have contributed calcium carbonate to the groundwater from which the sand later derived its cement.

Fig. 235. *Left:* Close view of edge of Mormon Mesa caliche cap. *Right:* Detail of limestone-pebble conglomerate showing effects of groundwater solution and redeposition of calcium carbonate. Width, 12 inches. Nagelfluh, near Zürich, Switzerland.

Fig. 236. Looking southeastward along the trace of a branch of the San Andreas fault in the desert northeast of Palm Springs, California. Note effect on groundwater in middle distance.

Man's Use of Water

The average rainfall on the 48 contiguous United States is about 30 inches per year. Most of this, equivalent to about 21 inches, is returned directly to the atmosphere through the transpiration of natural vegetation and by evaporation, leaving about 9 inches available for our use in the form of runoff and groundwater replenishment. At the present time, approximately 6 inches of this flows unused to the oceans. Of the 3 inches withdrawn by man, about half is put to nonconsumptive use (washing, cooling, and other domestic and industrial uses which ultimately return the water to sewage and streams), and half is consumed, chiefly by irrigation.

In broad terms, irrigation and industry are the two giant users of water, each accounting for more than 46% of the total. The remaining 6 to 8% includes all water supplied to cities, households, and to farms for purposes other than irrigation—an amount equal, on the average, to about 150 gallons per day per person.

The basic problems connected with man's use of water arise from sharp upward trends in the needed quantity and in demands for water of greater purity. Of course growth in population will necessitate ever larger municipal and rural-domestic supplies, but these are only a small fraction of the total. It is the expected trebling of industrial use and possible doubling of irrigation use that will require huge volumes of water not now being withdrawn from natural sources. Both industry and agriculture already have problems.

Industry depends mostly on surface waters; such manufacturing cities as Minneapolis, Chicago, Detroit, St. Louis, Cincinnati, Buffalo, and Pittsburgh are situated on rivers or lakes, or both. Even when these industrial centers supplement their supply with groundwater, the waste

242

Fig. 237. View northwestward along the San Andreas fault between Valyermo and Palmdale, California. Clump of trees marks place where fault crosses dry bed of Little Rock Creek, causing groundwater to rise.

is released to the streams. The problem of polluted surface waters in such areas is becoming acute at an accelerating rate, for while the volume of industrial discharge increases, thereby lowering the quality of the water in the streams (and compounding the difficulties of the next user downstream), industry is looking for ever-larger amounts of *better* water. The processing and reuse of water is an absolute necessity along these rivers, but it probably cannot fully meet the demands of such burgeoning fields as processed foods, beverages, and synthetics, all of which require water of high quality. In the petroleum-refining industry about 20 gallons of water are required for each gallon of product; about 65,000 gallons of water are required for the production of one ton of steel. In each case, over 90% of these requirements can be met with used water, but such recirculation of processed water is usually not practiced until economic considerations demand it. Where the quantity and quality of surface waters have reached their usable limits, the next step is to turn to groundwater.

Groundwater now supplies about 20% of the nation's withdrawals of fresh water. It is likely that this proportion is increasing; but even if it were not, the situation would be alarming because withdrawals already exceed replenishment in many areas. The greatest problems here involve expanding irrigation in areas dependent upon groundwater.

In the seventeen contiguous western states, which contain 90% of our irrigated lands, most of the groundwater is held in the pore spaces of alluvial deposits filling the valleys. Because most of the bedrock floor beneath this fill is uneven, the thickness of the water-bearing sediments is irregular. In some places movement on steep faults has crushed and sheared these sediments, reducing the permeability along nearly vertical planes until groundwater movement is inhibited. One effect of this can be seen in Figures 236 and 237. In the first we are looking along the trace of a fault cutting through alluvial deposits that slope gently to the right. Mountains out of sight

243

to the left are the source of the alluvium, of the intermittent runoff that transported it, and of the steadier seepage of groundwater beneath its surface. The fault acts as a barrier, impeding the flow enough to cause the water table to rise until it can be reached by the roots of natural vegetation. The result is the dark areas along and on the upslope side of the curving fault trace. The view in Figure 237 shows a part of the San Andreas fault zone that separates the mountains on the left from a desert plain on the right. The dry and almost barren stream bed that crosses the scene from left to right supports a conspicuous clump of trees at the point where it is crossed by one of the faults in the zone. A parallel fault crosses the same stream bed (foreground, far right) with no apparent effect, probably because it has been inactive since the stream course was established.

Such faults as these serve as boundaries between *groundwater basins*, each of which has its own water level, and in each of which the water table rises and falls with recognizable individuality in response to replenishment and pumping. Groundwater basins may also be set apart by differences in the shapes of their floors, original variations in the thickness or porosity of the water-bearing sediments, or buried faults without surface expression. Although many basins are vaguely defined, and most are leaky, the water level in wells sometimes differs sharply in adjoining basins and the pattern of seasonal fluctuations tends to be characteristic in each.

The water levels in most southwestern groundwater basins have been declining in recent decades. Since about 1890 the fall in southern California has averaged 3 to 8 feet a year during dry periods—which have ranged from 11 to 17 years long—interrupted by insignificant recovery during the shorter wet periods. The cumulative result in many wells has been a lowering of the water level by hundreds of feet. Current withdrawals are so high that no ordinary period of wet years could restore the original water levels in these wells.

An important distinction should be made here. From the geological point of view groundwater is mobile and temporary; very little of it remains in the relatively shallow basins of western America for more than a few hundred years. It has been estimated that the entire supply of groundwater in the United States to a depth of half a mile is equivalent in volume to the total replenishment during the past 160 years. But from the point of view of most living persons this is a resource ready and waiting to be exploited. Every landowner feels he has a right to extract all the water he wants to from beneath his property, often forgetting that it is the only significant reserve we have, the only large quantity of fresh water that is "in the bank," so to speak. Wherever groundwater withdrawals exceed replenishment we are mining an accumulation inherited from the past.

Large withdrawals of groundwater may also create more immediate problems. In some areas the drying and shrinking of clays in the alluvium that has resulted from lowering the water table by overdraft pumping has caused the land surface to subside by as much as nine feet. On Long Island and along parts of coastal California pumping of wells has lowered the water table below sea level, allowing seawater to seep inland and render many near-shore wells useless through saline contamination.

Most problems of water use can be solved. The water now withdrawn can be used more efficiently, the distribution can be improved, and seawater conversion may become economically feasible. The greatest gains can come from more and better processing and reuse by industry, and, especially in the realm of groundwater, more careful and effective use in irrigation. About one-fourth of the water destined for irrigation is now lost by evaporation and seepage, and in many areas the water applied to the ground is three or four times that required by the crop. The problem of distribution can be partly solved by expanding agriculture in areas where water is abundant—for example, the Snake River plain—and by transporting water from regions of excess to those of shortage. Seawater conversion can probably assist only a few coastal cities at present, but this development is in its infancy.

IV   Time

How geologic events can be arranged

in chronological order, and some

insight into the magnitude of geologic time.

20 SEQUENCE AND DURATION

Up to this point we have primarily concerned ourselves with the physical processes at work on and within the crust. In so doing we have encountered many different kinds of evidence of deforming movements in the crust, ranging from such obvious things as displaced shorelines, deformed sedimentary layers, and volcanism, to the more subtle but often more impressive implications of unconformities, metasediments, exposed plutonic rocks, tectonic landscapes, and the very existence of mountains. Each of these, in its own way, embodies evidence of a restless earth. But only a few contain even a hint of the *time* involved in their creation.

Geologists make two rather different uses of time. The first is to establish the order, or *sequence*, of events by determining their relative positions in time. For example, we know that a dike or volcanic neck is younger than all the rocks intruded by it. Similarly, metamorphic rocks are older than any sediments in depositional contact with them, faults and folds are younger than the rocks they deform, and the erosion represented by a surface of unconformity must have taken place after the deposition of all the rocks truncated by it but before the deposition of any of the strata resting upon it. By means of such relations we can work out the sequence of geological events recorded in a region.

In some cases we can also establish approximate contemporaneity between events, as when it can be shown that erosion in one area was contributing directly to sedimentation in an adjacent one. When it is established that two separated geological events took place at the same time, or that two separated rocks—possibly even on different continents—are of the same age, they are said to be *correlated*. The process of establishing such contemporaneity in geology is called correlation. Most geological correlation is accomplished with the aid of fossils, according to principles which will be discussed later in Part IV.

Before the discovery of spontaneous disintegration in some chemical elements (radioactivity), contemporaneity and sequence were the only relationships in time that could be firmly established in geology. In the geological study of an area, large or small, they continue to be absolutely fundamental to the untangling of the incomprehensibly long history of the earth's crust.

The second use of time is as a measure of *duration*—the length of time represented by a given geological event. Until recent decades this was possible only in very special cases, notably the deposition of certain thinly laminated fine-grained sediments (varved clays) whose rhythmic bedding is controlled by seasonal changes and may therefore be used to count years in the same way that annual growth rings in trees are used. (Varves are illustrated and discussed further on page 304.) With the discovery of radioactivity just after the turn of the century, and especially with the rapid development of sensitive instruments in the Atomic Age, it has become possible to use the natural radioactive disintegration of certain elements in making age determinations. We are now able to determine the approximate age in years of a growing number of mineral occurrences in the range from a few hundred thousand to more than three billion years. Carbon-bearing remains of plants and animals that ceased living within the past 50,000 years can also be "dated," utilizing the decay of the radioactive isotope Carbon-14, which seems to be present as a fixed fraction of all the carbon in living organic tissue but decreases at a definite rate by spontaneous disintegration after the plant or animal dies. The present rate of refinement of these dating methods arouses the hope that some day it may be possible to determine the age of almost any rock; certainly if such a day should come geology will have completed one of the most significant forward steps in its entire history.

A man's life, and indeed all of recorded history, is so short, compared to the age of the earth, that almost every geological event or story we study is *pre*historic by a wide margin. The earth has possessed a crust about 4,000 times longer than man has lived on it. Yet we have had considerable success in deciphering these ancient and obscure events by first studying the present. As already pointed out, examination of modern glaciers and their effects on the landscape led to recognition of prehistoric Ice Ages, and study of the details of modern sediments has made it possible to reconstruct the environments in which very ancient sedimentary rocks accumulated. The geological maxim, "the present is the key to the past," has had many fruitful applications, as will be further demonstrated on the next forty-two pages.

It is equally true that the past is the key to the present, for a sense of history is part of the fabric of geology. It is the succession of geological events over past millions of years that has determined nearly every detail of the position and shape of our lands and exactly where and in what state we find our mineral deposits. The industrial revolution of the past two centuries would not have been the same if blind searching for iron ores and accidental development of coal deposits had not been replaced by some understanding of the geological events that have produced these resources. It is first a question of where to look (there is no use looking for oil in exposed plutonic and metamorphic rocks, for example); then of exactly where to dig or drill, and finally of what to do as production declines or where the ore is cut off by a fault. Intelligent answers to such problems involve an understanding of what localized the deposit in the first place and what has happened to it since—which, in turn, requires the ordering of these events in time.

Let us begin by examining the sequence of geological events in a specific area. Using the region of the Grand Canyon of the Colorado River, we will combine the evidence from rocks and from fossils to reconstruct as much as possible of the story they record—the *geological history*. Following this we will investigate the more general challenge of relating such histories in different parts of the world, and finally review the methods of actually measuring geologic time.

21 EVENTS IN THE VICINITY OF THE GRAND CANYON OF THE COLORADO RIVER

Introduction to the Structural History

The fame of the Grand Canyon of the Colorado River is amply justified by its enormous proportions and colorful scenery alone. Many visitors see only these aspects and go home feeling fully rewarded for their journey to northern Arizona.

But there is a grandeur in the Grand Canyon scene which is less obvious to the casual observer, one that can be appreciated only by those possessing a little geological insight, and which, added to the scenic beauty, makes the Canyon an even more soul-stirring place. This more subtle feature is the long and remarkably eventful history recorded in the rocks so magnificently exposed there.

To the geologist, whose constant challenge is to decipher and understand what has happened beneath the surface, the earth is all too smooth. In a subject inherently three-dimensional he is presented with 50 million square miles of horizontal information (not counting the ocean floors) —and a vertical view limited to a few thousand feet. Anything that reveals rocks in the vertical dimension has special value: which accounts for the geologist's keen interest in mountains, mines, wells, and boreholes. But mountains are usually geologically complex, reflecting the deformation that produced them, and mines, wells, and boreholes provide extremely small samples. The extraordinary value of the Grand Canyon is that it affords such an extensive look, to such a depth, in a relatively uncomplicated part of the crust.

There are three sources of information on the geological history of the Grand Canyon region. The first, the structural relationships of the major rock bodies to each other, is presented in bold terms. Viewing this grand succession of events—each of which deserves pages of elucidation in its own right—is somewhat like reading an outline of a book. The second source is

Fig. 238. View of the Grand Canyon, looking downstream (northwest) near the mouth of Shinumo Creek. The small numbers in squares on the drawing (right) correspond to those used on Figure 258, where the rocks are more fully identified and described.

the nature of the rocks themselves, their compositions and textures and all that these tell about where they came from, how they arrived, and what has happened to them since. The third source is the fossils contained in the rocks, the record of some of the life that witnessed and participated in some of these events. Together, the last two supply the details of the story; whatever is not recorded in them is, as we have seen (page 72), largely lost, so that to obtain a full account of all the time involved is generally impossible.

The scene below, 20 miles downstream from Grand Canyon village, provides a key to the three principal and easily distinguished groups of rocks. These are (1) the horizontal layers, which for reasons that will be explained later (page 300) we will call the *Paleozoic strata*, (2) the tilted beds, which are known as the *Grand Canyon* series, and (3) the steeply foliated dark *Vishnu schist* along the walls of the inner gorge, close to the river. A few faults of small displacement may be seen in the last two groups of rocks but the first is remarkably undisturbed; these conditions prevail throughout the canyon.

The drawings that follow are based on this scene and represent the most obvious of the major geological events recorded and implied in these rocks. They are so arranged that we begin with the familiar scene and reason our way back through time, step-by-step, employing fundamental concepts that have been developed on preceding pages. In this way we will arrive at each stage aware of the logic by which it is reached and the justification for postulating it. The geologist must always start with what he sees and reconstruct the past from this as a basis, even though he may later tell the story in the conventional manner, beginning with the earliest event.

On the next ten pages, the text and associated drawings, which deal with this unraveling of the story, should be followed *without regard to the legends above the pictures*. After Step Ten has been reached, the procedure should be reversed and *only the legends should be read*, thus recounting the story from the earliest time to the present by examining the scenes in order from Ten to One: this will provide a summary, with emphasis on the earth movements involved, of the steps that led to the present scene.

Remember: To avoid confusion, *ignore the legends* when first reading the text; they make sense only when read in sequence from Ten to One.

Fig. 239. Step One: After the Colorado River has established itself in a winding course across what is to become the western Colorado Plateaus, further *uplift*, ultimately totalling more than 8,000 feet, raises all these strata well above sea level. The resulting erosion removes nearly all of the weak upper strata (above 13) from the vicinity of the Grand Canyon and the entrenched river cuts the Canyon in the more-resistant lower layers.

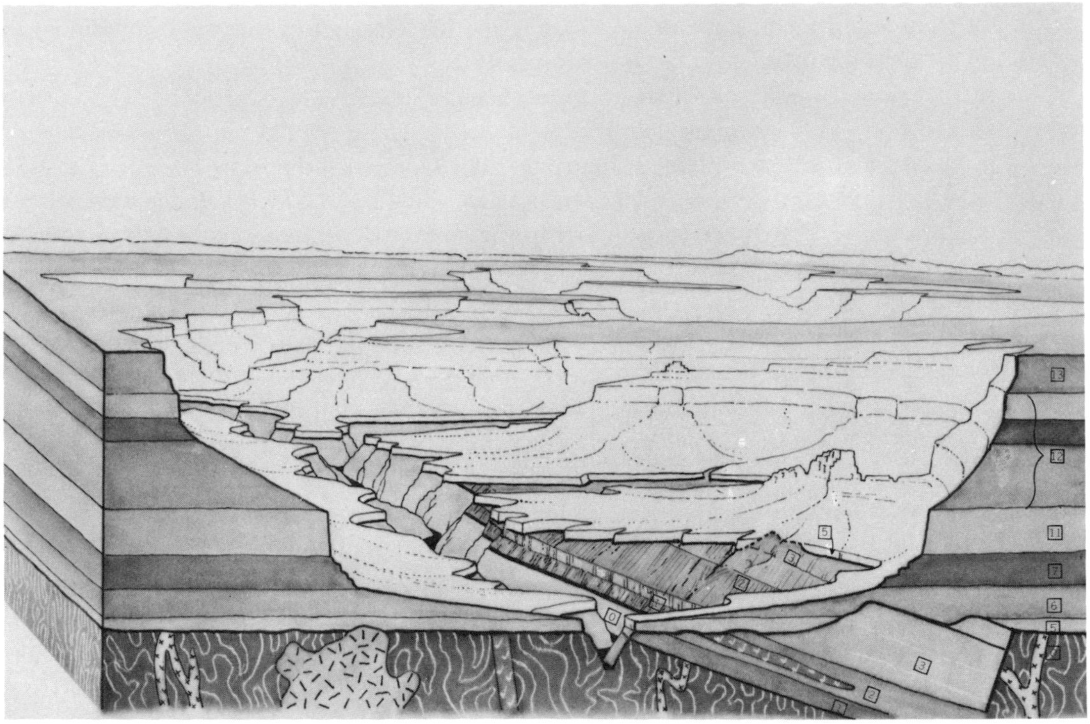

Step One: Analysis of the Present Scene

First, let us back away from the view on the last page and present the same scene as part of a block diagram that includes a larger area, seen from a higher altitude. The same tilted Grand Canyon series can be recognized to the right of the river, with the hill (marked by a dashed line) produced by resistant beds (in unit 3) protruding into the overlying horizontal Paleozoic strata.

On the vertical front of the block the essential relations between the different rocks are shown somewhat diagrammatically. Within the Vishnu schist, represented by a pattern of contorted lines, are a few thin pegmatite dikes marked with x's, a dike of basaltic rock marked with v's, and an irregular body of granite marked by scattered short straight lines. To the right of the river is a tilted wedge of Grand Canyon series sedimentary rocks whose lower boundary is the nonconformity visible on the preceding page (Fig. 238) and featured in Figure 113. The strata in this series are cut off at the right by a steep fault along which the relative movement was up on the right and down on the left. The two faults within the Grand Canyon series, visible in Figure 238, are omitted here because they are not essential to the story.

The small numbers within squares consistently identify the same rocks on any diagram where they are used. Thus here, as in Figure 238, the Paleozoic strata carry numbers from 13 at the top to 5 at the bottom, beneath which the Grand Canyon series strata are numbered from 3 to 1 in descending order and the Vishnu schist is numbered 0. Where numbers are missing in any drawing (here 4, 8, 9, and 10) it means that strata belonging in their positions, and present nearby, are missing from the section illustrated.

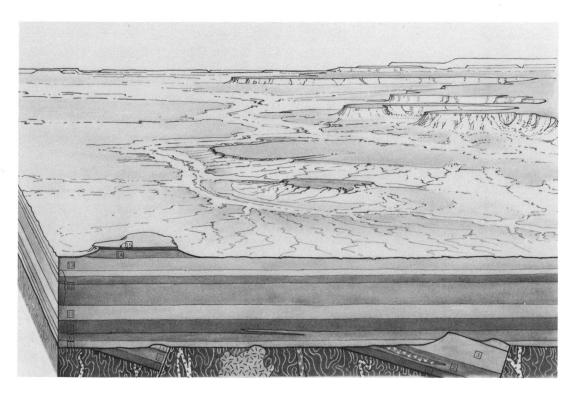

Fig. 240. Step Two: *Submergence* of this second peneplain induces deposition of the Paleozoic strata, whose base marks the second unconformity (Figs. 238, 252, and 257). Some of these layers accumulated on land, which requires the fluctuating conditions discussed on pages 266–286 but not shown here. The total thickness of the Paleozoic strata today is a little over 4,000 feet, they were deposited in the period from 600 to 200 million years ago. Later, *uplift* initiates erosion which is here beginning to strip off some of the higher strata.

Derivation of Step Two

The Grand Canyon provides a spectacular example of the meaning of Steno's third observation: sedimentary layers do not accumulate with their edges showing. The perfect match of the strata in the north and south walls of the Canyon throughout its 200-mile length, even to details measured in inches, renders it virtually impossible that they could have accumulated separately. It is safe to conclude that these strata were once continuous across the chasm, and that the canyon must have been *cut* into these very widespread and remarkably uniform layers.

Some might wish to argue that the river must have had some assistance—perhaps a system of faulting by which a section of the crust was dropped a few thousand feet, after which running water so sculptured the shape as to obscure or remove the evidence. But faulting dislocates rocks, and no amount of erosion can make them match again. Actually, faults in the Paleozoic strata along the Canyon walls are rare, strike across rather than along the Canyon, and at most have displacements of only a few hundred feet.

So, to take the first step back in time, we undo the most recent event by restoring these rocks to their condition before the Canyon was cut. A few layers (14 and 15) are added above those of the present rim (13) because scattered remnants of these higher strata can be found in several places near the Canyon today and their general nature is well known from extensive exposures on plateaus in the vicinity of Zion and Bryce canyons to the north. These higher plateaus are shown in the background at the right, extending farther south than they do today because, during the cutting of the Canyon, downslope movements and erosion undoubtedly caused them to retreat northward. We also have backed away again, so that our view of the front of the block can include a second faulted wedge of Grand Canyon series strata (at the left). Its position is hypothetical, but its existence is not; many such are known from different parts of the Canyon.

The fronts of the blocks are strictly comparable in this and the next eight drawings.

Fig. 241. Step Three: A long period of erosion follows, accompanied by enough *uplift* to bring the base of the Grand Canyon series above sea level in some places. Toward the end of this period the levels of land and sea are stabilized and a second peneplain is developed in which the only remains of the Grand Canyon series are small down-faulted wedges, all the rest having been removed by erosion.

Derivation of Step Three

In the pre-canyon stage just reached, one step back in time, we recognize that the highest layer is the youngest and that its accumulation represents the most recent event recorded in that scene. To take the next step back, therefore, this stratum must be removed, or "undeposited," whereupon the same argument applies to the next lower layer. (This is an application of Steno's first principle, the law of superposition.) Continuing this process, we remove all the horizontal Paleozoic strata, assuming for the moment that their accumulation was essentially uninterrupted.

Their removal exposes an old erosion surface with very low relief which truncates both Vishnu schist and the upturned edges of the tilted blocks of Grand Canyon series. Only the edge of this erosion surface can be seen today, but at least 95% of the more than 200 miles of such exposures in the Canyon has a relief of less than 150 feet and only a few hills of resistant sandstone rise as much as 600 to 800 feet above their surroundings.

The edge of this surface is well shown at the base of the Paleozoic strata in Figure 238; note the hill several hundred feet high just right of center (dashed line in the drawing) produced by beds of resistant sandstone ([3]) in the tilted Grand Canyon series. The lowest Paleozoic beds lap out against this hill, which probably, therefore, was an island during their accumulation.

From this evidence we postulate, as shown in Figure 241 above, a peneplain with a few low strike ridges of resistant strata and small hills rising as ancient monadnocks above its otherwise featureless and soil-covered surface.

Fig. 242. Step Four: The Grand Canyon series is *uplifted* and, probably at the same time, *broken* along faults between which the blocks are *tilted* northeastward. The fault-block mountains thus produced are, of course, immediately attacked by erosion.

Derivation of Step Four

Contemplating the pre-Paleozoic peneplain of the preceding scene, we realize that the latest event recorded there is the erosion that produced the peneplain. To take the next step back, therefore, it is necessary to undo this erosion by replacing the eroded rocks. This includes restoring the eroded extensions of the tilted Grand Canyon series strata (Steno's law of concealed stratification). Farther east in the Canyon some of the larger faulted wedges of these strata include many layers above those visible in Figure 238. Using this information, it is possible to reconstruct these blocks to stratigraphic thicknesses of at least 12,000 feet (by adding the strata marked 4).

When this is done the result is a set of tilted blocks, separated by faults, like leaning books on a shelf, the strata being represented by the printed titles on their backs.

Fig. 243. Analogy for tilted fault-block mountains.

There are many regions west of the Rocky Mountains today with this kind of structure; the resulting mountains are sometimes referred to as *fault block* ranges. Accordingly, we postulate a landscape similar to those of these western areas. The front face of the drawing shows the relation of the underground structure to the topography.

253

Fig. 244. Step Five: *Submergence* of this peneplain induces deposition of the Grand Canyon series sediments. As these are more than 12,000 feet thick the total subsidence must be at least this amount—three times the thickness of the horizontal Paleozoic strata exposed in the Canyon walls today. The base of the Grand Canyon series marks the lowest and oldest recognizable unconformity in the Canyon walls; segments of it are shown in Figures 113 and 238.

Derivation of Step Five

As originally deposited, the strata of the Grand Canyon series were neither tilted nor cut by faults. Applying Steno's second and third laws (original horizontality and continuity of strata), we see that the most recent event in the previous scene is the tilting and rupturing of these strata. The next step back, then, is to restore them to horizontality while undoing the faulting—to straighten up the books. This leaves us with the Grand Canyon series strata in the orientation in which they were deposited, but probably above the level at which they accumulated.

The rocks of this series are chiefly shales and fine sandstones showing ripple marks (Fig. 253, *right*) and mud cracks, plus a few beds of limestone and a sill intruded between the layers near the bottom. Near the middle of the full sequence, too high to show in Figure 238, are lava flows about 800 feet thick. The mud cracks prove intermittent drying out of the surface, so the water must have been shallow, and the limestones almost certainly mean that some of it was marine. There is no evidence of vegetation.

Thus we postulate a scene (Fig. 244, above) resembling the present head of the Gulf of California, in which arid lands shed sediment into a shallow and fluctuating sea and occasional volcanic eruptions occur near the shore. Under these conditions some beds in the sequence may be marine and others fluvial, as the shoreline shifts with changing levels of land and sea. Possibly the sill near the bottom was injected as part of the same igneous activity that produced the lavas near the middle of the sequence; they are of similar composition.

Fig. 245. Step Six: *Uplift*, probably amounting to about 10 miles, maintains the mountains for a time. Ultimately erosion wins out and, as the crust becomes more stable, a near-sea-level peneplain is produced on rocks that were once in the roots of the mountains. A small piece of this peneplain, seen on edge, is visible as the surface of nonconformity in Figure 113.

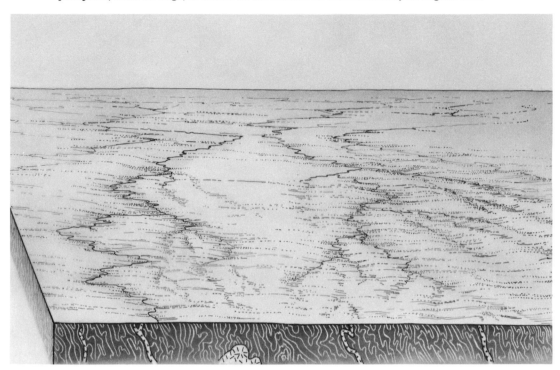

Derivation of Step Six

We have now reached a stage that superficially resembles that of Step Two, having reasoned our way back to another, but earlier, time when there was a thick sequence of undisturbed strata. In the last scene (Step Five) we recognize, as before, that the most recent event was the accumulation of the highest stratum. Applying the law of superposition to successively lower layers in this sequence, and again assuming that there are no important events hidden within or between layers, we work back through time by removing all of the Grand Canyon series. This exposes a remarkably flat surface cut across the Vishnu schist, a surface that truncates the foliation, the pegmatites and granites—all the structures and variations within it.

Judging by the fragments of this surface that are exposed to view in the present Grand Canyon (for example, Figs. 113 and 238), this was an almost perfect peneplain. It can be traced, in remnants, for many miles and is exposed in areas tens of miles apart, yet its relief is everywhere less than 50 feet. Both this and the peneplain of Step Three still retain patches of deeply weathered rock on their surfaces.

No scene quite like this peneplain is known today; the closest in this book might be Figure 146 with the monadnock removed. To explain the deep chemical weathering we guess that the climate was fairly humid. The drawing above incorporates these considerations but the details of such features as the stream pattern are, of course, conjectural.

Fig. 246. Step Seven: *Deformation*, arising from movements in the crust, crumples the Vishnu sediments and lavas. Under the pressures and temperatures existing at a depth of around 10 miles this results in metamorphism that converts them to schist. Bodies of granite and pegmatite form in the schist, either as injections from below or as a result of partial melting and recrystallization of the sediments themselves. The deformation almost certainly results in mountains at the surface. These events probably occurred about 1,700 million years ago.

Derivation of Step Seven

Schist, being a rock produced deep in the crust, is visible at the surface only in areas from which vertical miles of rock have been eroded away. In the stage we have now reached (Step Six), the Vishnu schist is not only visible at the surface; it has been leveled to a peneplain. This imples two earlier stages: first, uplift of something on the order of 10 miles, which was required to raise the schist from the zone where it was produced to one where it could be exposed; and second, a long period of stability during which the peneplain reached near perfection. Erosion, of course, was active throughout both stages.

To reverse these episodes and take the next step back in the story, we must undo this erosion by restoring the vertical miles of rock under which the schist formed. It is highly probable that the 10 miles or so of uplift necessary to bring about this amount of erosion would have produced mountains—perhaps several generations of them. And it is almost certainly in the deeper levels of this mountain-making deformation that the Vishnu schist was produced and that the pegmatites and granites were formed or intruded.

So we postulate a mountainous landscape on folded and faulted sedimentary rocks, which grade downward, through tens of thousands of feet, into the metamorphic Vishnu schist.

Fig. 247. Step Eight: Further *subsidence* results in burial of the Vishnu sediments, shown at the lower right, under several miles of superincumbent strata. The earlier Vishnu deposits are now more than 5 miles below sea level.

Derivation of Step Eight

Clues to the nature of the rocks from which the Vishnu schist was derived can be found in its composition and textures. Some layers are clearly derived from impure sandstones; the sand grains can still be distinguished with a microscope. In others, calcareous lumps suggest former concretions. Still others have chemical compositions that resemble those of certain sediments and volcanic rocks more than those of any other plausible ancestors. Some of the banding in the schist resembles sedimentary stratification, even to such details as relict cross-bedding (Fig. 251).

From such evidence we conclude that the Vishnu schist (excluding the associated pegmatites and granite) consists largely, if not entirely, of metasediments and minor metavolcanics. Before these materials can come up to view as schist they must go down in the crust as sediments and volcanics and there undergo metamorphism.

The next step back, then, is to undo this metamorphism, which takes us back to the stage when the ancestral Vishnu sediments were on their way down to the zone of metamorphism, probably in response to downwarping of the crust. Since subsidence on such a scale cannot take place without inviting sedimentation, we reconstruct a stage in which the Vishnu sediments are subsiding (at the right) and being buried by thousands of feet of overlying material. The character of these overlying rocks is unknown, but their existence is required as a prelude to the metamorphism.

Fig. 248. Step Nine: *Subsidence* of the crust submerges a large area of this floor and induces deposition of sediments which, with some associated volcanic rocks, are to become the Vishnu schist. Since some of the sediments are types found in shallow or moderately deep water, it is likely that sedimentation more or less kept pace with the subsidence. Some of the volcanic rocks may have originated as submarine eruptions or as intruded sills. The total thickness probably exceeds 5 miles.

Derivation of Step Nine

Before the Vishnu sediments could be buried, they had to be deposited. So the next reconstruction takes us back to the time of accumulation of the sediments that are to become the Vishnu schist.

It is fortunate, for our purpose, that in some of the layers of the schist the recrystallization and contortions have not wholly destroyed the original textures of the deposits. As has already been pointed out, chemical composition and textures show that many layers were once fine sandstones or shales.

Fine sandy and shaly sediments are most abundant today in shallow to moderately deep water, relatively near land—along the Gulf Coast of North America, for example. Accordingly, at this stage our reconstruction somewhat resembles the Gulf of Mexico, to which rivers like the Mississippi and Rio Grande bring abundant supplies of mud and sand to be spread by waves and offshore currents. We add a little volcanic activity to the scene to account for the meta-volcanics, not knowing whether the original rocks were erupted on land or on the sea floor, or whether some of them were intruded as sills. All we know is that they were of basaltic composition.

It has been estimated that there are more than 25,000 feet of sediments and more than 4,000 feet of volcanics represented in the Vischnu schist; this is more than seven times the thickness of the Paleozoic strata now exposed in the walls of the Canyon.

Fig. 249. Step Ten: The oldest rocks exposed in the Grand Canyon are the Vishnu schists. As they were largely derived from sediments, the earliest event we can infer is the accumulation of these deposits. But what were they deposited on? There must have been a land surface which was near sea level and which, to receive the first layer of the Vishnu sediments, was submerged by an advancing sea. This land surface, the floor on which the Vishnu sediments began to accumulate, is the earliest scene we can postulate, although we do not know what it actually looked like.

Derivation of Step Ten

Because in their present metamorphosed condition the original Vishnu strata are wrinkled and folded and standing on end, and because the walls of the inner gorge expose only a small fraction of the total volume of this rock, we know nothing about the character of the top and bottom of the original sedimentary sequence. But of one thing we can be sure: there was a first layer, and this layer was deposited on something.

Consequently, the very earliest event to which the evidence from the Grand Canyon points is the existence of a floor under the original pile of Vishnu sediments. We know absolutely nothing about its character, but as it must have been there a place is provided for the scene without prejudicing the imagination of any who would like to reconstruct it.

Now, knowing why it begins where it does, you should review the history of the Grand Canyon region in the conventional order. Read the legend at the top of this page, then that for Step Nine and the others in sequence back to Step One, studying each accompanying drawing as you do. In so doing you will move forward through time and the order of the events will be the real one. Perhaps the most impressive thing is the great amount of crustal movement that is involved, even without considering changes that may be recorded within, as distinct from between, the major rock units.

The times given are estimates based on radioactivity, which is explained on pages 305–310.

Fig. 250. General view of the Vishnu schist west of Kaibab bridge in the Grand Canyon. Note steep
foliation. Most of the light streaks (e.g., left foreground) represent sills and
dikes of pegmatitic granite.

Information from Lithology and Fossils I:
Vishnu Schist and Grand Canyon Series

The geologic history of the Grand Canyon region outlined on the preceding pages concerns only
the major physical events, the sequence of episodes involving uplift and subsidence, erosion and
deposition, crustal deformation and crustal stability, that produced the three major rock units
and the Canyon. The evidence for these episodes is inherent in the very existence of a schist
nonconformably overlain by a group of sedimentary layers, both of which are unconformably
overlain by a second group of strata. Let us now see what additional information can be obtained from the character of the rocks themselves: i.e., their *lithology*.

The steeply foliated *Vishnu schist* (Fig. 250, above) is the most challenging. Metamorphism,
by modifying the texture and mineral composition of the rocks, has blurred much of the record
we would like to have of these earliest deposits. It is uncommon but significant details, like the
cross-bedding shown here (Fig. 251), that provide reasons for postulating a sedimentary origin.

Fig. 251.
Relicts of sedimentary structure in the Vishnu schist.
Scale is 6 inches long. Lone Tree Canyon, about
3.5 miles upstream from Kaibab bridge. (*Photo
courtesy of John H. Maxson.*)

Fig. 252. The lower part of the Grand Canyon series as exposed at the mouth of Shinumo Creek in the Grand Canyon (cf. Fig. 238). Key to numbers is in Figure 258.

We reason that if fairly definite metasediments are found here and there throughout an estimated 25,000 feet of schists whose overall character is quite uniform, it is probable that the entire mass was derived from sediments. If so, the much less abundant dark schists of appropriate composition, probably represent lavas, tuffs, or sills.

There is no direct evidence in the Vishnu schist to indicate that its ancestral sediments accumulated in the sea; in fact, we should not overlook the possibility that oceans as we know them did not exist in that early time. About the best we can do is to observe that the fine-grained texture and rather uniform character of the rocks would be most likely to result from sedimentation in a large body of water, either the sea or a large lake.

Fossils would be helpful in this connection, but they rarely survive metamorphism and none have been found in the Vishnu schist. Furthermore, fossils from this very primitive era, if found, might be so unlike later forms that it would be impossible to determine whether they were marine or nonmarine.

It is important to realize that the nonconformity between the Vishnu schist and the Grand Canyon series strata may represent quite a long gap in the record at the Canyon. As already pointed out, during this time rocks formed deep in the roots of mountains were lifted above sea level, a process which would probably take at least millions of years, during and following which the mountains were eroded down to an extraordinarily flat peneplain, a process probably requiring tens of millions of years. Of course the sediment produced by the weathering and erosion of these mountains was deposited somewhere; if the place could be found, and the deposits are still there, we might learn more about this interval.

The *Grand Canyon* strata that rest on the post-Vishnu peneplain are normal sedimentary rocks whose details can profitably be compared with modern types. The remnant shown here (Fig. 252) is divisible into three easily distinguished parts. The lowest forms a series of ledges and cliffs near the river (marked ⬛1) and rests directly on the peneplained Vishnu schist (⬛0⬛). It consists of resistant beds of silty dolomitic limestone alternating with weak beds of shale

261

Fig. 253. *Left:* Typical exposure of impure sandy limestones and interbedded shales of the Bass limestone. Kaibab trail, Grand Canyon. *Right:* Ripple-marks preserved on slabs of muddy Bass limestone. Near Kaibab trail, Grand Canyon.

(Fig. 253, *left*). In some places there is a thin conglomerate at the base. A few mud cracks and ripple marks (Fig. 253, *right*) in the shale plus the muddy character of these strata point to accumulation in very shallow water, which occasionally, as on a tidal flat, receded or evaporated so that the surface of the deposit dried out from exposure to sun and air.

Some of the limestones contain faint casts or impressions of symmetrical objects, some that resemble primitive sea-pens (related to corals) and others that were possibly algae. The latter occur in large numbers, some singly and some in groups of two and three as shown here (Fig. 254). The close similarity among all of them, and the fact that the individual forms in a cluster seem to have existed simultaneously rather than one after another, probably rule out the possibility that they were made by rain drops or bubbles. If these are indeed of organic origin, as seems likely, they are among the oldest known direct indications of life on earth.

This 250-foot sequence of limestone ledges and associated shales in the lower part of the Grand Canyon series is known as the *Bass limestone*, the name having been chosen because a typical cross-section of the strata is exposed in Bass Canyon, a side canyon less than a mile from the scene in Figures 238 and 252, and because limestone is the most characteristic rock type in the sequence.

Fig. 254. A pair of algae-like fossils(?) from the Bass limestone in the Grand Canyon. (*Specimen in the Raymond M. Alf Museum, Webb School of California.*)

The only place where extensive bedded limestones are forming today is on the sea floor. All ancient limestone bodies of any size seem also to be marine; most of them contain marine fossils and nearly all are associated with other rocks, such as shales, that are typically marine. Extensive dolomite strata are always marine. Thus, the uniform solid beds of the Bass dolomitic lime-stone, interbedded with shale and devoid of sandstone, are much more like known marine than known nonmarine deposits. The Bass limestone is much stronger evidence for the existence of an ocean than was the Vishnu schist.

Overlying the Bass limestone is 600 to 800 feet of brown to red sandy shale (2). The upper part is bright vermilion, an unusual color for any kind of rock and easily the most conspicuous color in the Canyon, although it can be seen in only a few places. In Figure 252 the intrusion of a dark sill of basaltic rock within the shale adds several hundred feet to the thickness. The sill is about as weak as the shale and therefore does not stand out in this view, although its dark debris masks the distinctive red color. Many bedding planes in the red shale exhibit ripple marks and mud cracks, as well as cross-bedding (Fig. 255) and raindrop prints, from which we

Fig. 255. Typical exposure of the Hakatai shale and one of its few included sandstone lenses; note cross-bedding in the latter. Kaibab trail, Grand Canyon.

infer that these sediments also accumulated under very shallow waters with frequent emergence to permit surface drying. The red color is caused by iron oxide and is essentially a form of rust coating the sedimentary grains. Since iron rusts more readily if it is occasionally exposed to air than if it is continuously submerged, the color also favors the idea of deposition in shallow water—unless it was produced long after the sediments accumulated. There are no fossils and no strong reasons for concluding that the original muds accumulated in the sea rather than in fresh water.

These red beds have been named the *Hakatai shale* because they are well exposed in Hakatai Canyon, the tributary in the middle background of Figure 238.

It seems probable that the Hakatai shale records a further shallowing, or filling up, of the sea in which the Bass limestone accumulated. Perhaps similar muds were being deposited throughout the time represented by both rocks, the change from impure limestone and shale to red sandy shale recording a change from marine conditions to those of an estuary, lagoon, delta or floodplain, which halted the production of limestone and increased the exposure to air.

The red Hakatai shale is overlain by a group of resistant layers that produce ledges and small cliffs, marked ③ on Figure 252. These, however, are composed entirely of sandstone, some of it very firmly cemented with silica. The rock is fine-grained, the larger grains are smooth and round, and there is very little clay or silt; most of the layers show cross-bedding (Fig. 256) and many are ripple-marked. It is clearly water-laid, yet rather clean and uniformly fine-grained for a deposit with such an origin. This could indicate that the sand was originally accumulated by the wind, after which the dunes were reworked into this deposit by the waves and currents of an advancing sea.

Fig. 256. Close view of very hard layer in the Shinumo sandstone, showing probable cross-bedding. Kaibab trail, Grand Canyon.

Because it is well displayed in Shinumo Canyon (right foreground of Fig. 238) this sequence of relatively resistant beds is known as the *Shinumo sandstone*. The most prominent layers are highly resistant to weathering and erosion because their grains are cemented with silica instead of the more common and more soluble calcium carbonate or iron oxide. The Shinumo sandstone is more than 1,100 feet thick and overlain (out of sight to the right) by more than 2,000 feet of shaly sandstones that are distinctly less uniform, less clean, less resistant, and are grouped under another name.

Let us digress at this point to consider the significance of such terms as Bass limestone, Hakatai shale, and Shinumo sandstone in relation to the fundamental methods of field geology. To repeat a basic truth: Almost everything in geology depends on our ability to tell one rock from another. Folds are delineated by tracing a stratum, or group of strata, through its contortions without losing track of it. Faults are located and evaluated by comparing rocks across them and recognizing displaced parts of the same rock on either side of the break. Deciphering the history of the Grand Canyon region depends on being able to tell schist from shale and lava from limestone and knowing the origin of each.

The basic tool for studying the relations among the different rocks of an area is a *geologic map*—one showing, usually by distinctive colors or patterns, the distribution of each kind. (Examples are show in Figs. 102, 289, 374.) In mapping it is generally impractical to plot the area of every different rock separately; many are too thin or too small to show individually on maps prepared at scales that are otherwise useful. Consequently, the field geologist subdivides the totality of rocks he is mapping into groups, each of which is large enough to show on the map and distinctive enough to allow easy tracing of its bounding contacts on the ground. Each of these mappable subdivisions of the rocks in a given region is a *formation*. The Bass limestone, Hakatai shale, and Shinumo sandstone are formations; they are three of the recognized subdivisions of the Grand Canyon strata. The Vishnu schist is also a formation, and if we wanted to delineate an individual granite mass in the Vishnu, we could make it a formation too.

The name of a formation becomes formalized when a full description, giving details of its character, thickness, location, and the origin of its name, is published in a professional journal. Most formations are named for a physiographic feature, at or near which they are typically exposed, followed by such a word as "limestone," "shale," "conglomerate," or "basalt" if such a rock is prominent in the unit, or simply by "formation" if it contains a variety of rock types. A *type section* or *type locality* is designated so that other geologists can examine the original as an aid in making comparisons. (Similar procedures, involving type specimens—which are made available in museums or other collections—are followed in naming plants, animals, and fossils.)

Most formations, like the three mentioned here, are fairly homogeneous and are so defined that the contacts between them are obvious. The Bass limestone forms cliffs and ledges and its contact with the overlying Hakatai shale is drawn at the top of the highest limestone bed, separating dark resistant ledges below from weak slopes of sandy red shale above. (The whole Bass formation is visible as the dark horizontal layers in Figure 113, resting on Vishnu schist and overlain by weaker, lighter-colored rocks.) Similarly, the contact between the Hakatai and Shinumo formations is placed at the top of the weak, vermilion, sandy shales, a horizon that is also the base of the first resistant ledge of the overlying Shinumo sandstone.

Because of the way they are chosen, each formation tends to represent fairly uniform conditions of origin that differ from those of its neighbors. Thus the fluctuating shallow muddy sea of Bass time was replaced in Hakatai time by conditions unfavorable to the accumulation of limestone; either the sea withdrew or the quantity of mud increased. Both imply relative uplift of the land nearby and rejuvenation of streams. In a similar way, conclusions can be drawn about the environments in which the overlying formations were produced.

Before leaving the Grand Canyon series, it should be pointed out that Figure 238 includes two good examples of relatively datable faults, one of which is also visible in Figure 252. These nearly vertical faults displace the Grand Canyon formations (and, of necessity, the underlying Vishnu schist as well), but have no effect on the overlying Paleozoic strata. Clearly these displacements took place after the Grand Canyon series was deposited but before the development of the peneplain on which the younger strata accumulated.

Fig. 257. The "great unconformity" as exposed on the west side of Garden Creek, Grand Canyon. Vishnu schist nonconformably overlain by lowest layers of Paleozoic strata. Height of cliff shown, 40 feet.

Information from Lithology and Fossils II:
Tapeats, Bright Angel, and Muav Formations

One of the longest chapters in the history of the Grand Canyon region was written after the time of accumulation of the Shinumo formation and before the first Paleozoic stratum was deposited. At the very least this included the accumulation of over 10,000 additional feet of Grand Canyon series sediments, the breaking of these strata into tilted fault blocks, uplift of at least 12,000 feet to bring the lowest layers above sea level, and the removal by erosion of all but the lower corners of some of the tilted blocks, parts of which protruded as monadnocks on the resulting peneplain. It is not possible, at present, to say exactly how much time is represented by these events, but a reasonable guess, based in part on radiometric ages, would be several hundred million years. These events are included in Steps Five, Four, and Three, illustrated on pages 252–254.

This entire chapter is now represented by the younger surface of nonconformity, a short section of whose hundreds of miles of exposure in the Canyon is shown in the view above (Fig. 257). Here the Paleozoic sediments lie directly on the Vishnu schist, which is deep reddish brown owing to weathering that took place when it was a land surface, before the overlying sediments were deposited. Several of the earliest of the overlying strata have a similar dark rusty color, evidently caused by incorporation of sediment derived from this old weathered zone on the schist.

Note too that here the lowest beds lap out (toward the right) against a gentle slope—which was probably the flank of a low hill or monadnock on the Vishnu peneplain. Most such hills on the Vishnu surface are less than 50 feet high; the exceptions are composed of especially resistant rocks, such as hard Shinumo sandstone. As already noted, a few of the latter rise 600 to 800 feet above their surroundings—high enough to have remained as islands during the early stages of inundation and to be surrounded but not buried by the earliest layers of Paleozoic strata (Figs. 238 and 241).

The sedimentary record begins again with the Paleozoic strata, which began to accumulate

266

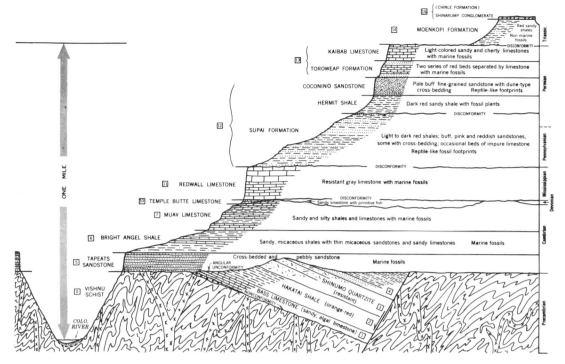

Fig. 258. Diagrammatic representation of one wall of the Grand Canyon summarizing the character and relationships of the different rocks. This drawing provides the key to the small boxed numbers used on several other illustrations in this book. Names at far right relate these rocks to the Standard Column and Geologic Time Scale discussed on pages 294–300.

when the second, or post-Grand-Canyon-series, peneplain was submerged. Most of the information we need to reconstruct an account of the time recorded in these rocks is summarized in the diagram above (Fig. 258), which is a more detailed representation of the stratigraphic section to the right of the river on the front of the block diagram in Figure 239.

The *Tapeats sandstone* (⑤), which makes the lowest horizontally layered cliff in the scene below (Fig. 259), consists of 200 to 300 feet of medium to very coarse brown sandstone in well-defined horizontal beds. The main part is firm enough to produce a prominent cliff, but the upper 20 to 100 feet is composed of finer-grained, less-resistant layers, which form a transition zone between the Tapeats and the overlying shales. These relations are clear in the closer view

Fig. 259. General view of the Paleozoic strata exposed east of Bright Angel Creek on the north side of the Grand Canyon.

Fig. 260.
Closer view of Tapeats sandstone resting
nonconformably on Vishnu schist.
Near Trinity Creek, Grand Canyon.

of the formation in Figure 260 in which the vertical schistosity and pegmatites of the Vishnu contrast conspicuously with the horizontal beds of the Tapeats. In a small tributary canyon (Fig. 261) we can see that the individual beds are only a few feet thick. A closer look at one of these (Fig. 262) shows the foreset type of cross-bedding that is characteristic of all but the lowest layers. (A close-up of the texture and cross-bedding was also shown in Figure 75, which accompanies the discussion of cross-bedding on page 74.)

Measurements made at more than five hundred outcrops of this sandstone throughout the Canyon show that the cross-bedding consistently dips toward the west and southwest, indicating that this was the direction of movement of the transporting currents and implying that the source of the sand was east of the Canyon. This interpretation is borne out by the fact that eastward the Tapeats sandstone disappears entirely where it abuts against hills of older crystalline rocks, and that westward (into Nevada) it thickens and earlier strata appear in sequence beneath it. The Tapeats also contains numerous worm trails and the tracks of bottom-dwelling marine animals (trilobites) whose fossilized remains occur sparingly in the transition zone and are fairly abundant in the overlying shales. From these observations, and considering its uniformity over thousands of square miles, we conclude that the deposit is marine and that the sea

Fig. 261.
Tapeats sandstone as exposed along lower
Garden Creek, Grand Canyon.

Fig. 262.
Cross-bedding in Tapeats sandstone,
Garden Creek, Grand Canyon.
(See also Fig. 75.)

in which it accumulated spread slowly eastward from southern Nevada to a shoreline in north-eastern Arizona. Under these conditions the earliest Tapeats sands in the west should be a little older than those in the east, a fact supported, as we shall see, by the positions of two distinctive fossil horizons. The difference in age represents the time required for the shoreline to advance about 120 miles; this was probably only a few million years at the most, long by human stand-ards but rather short as part of a series of events that began between 500 and 600 million years ago.

The Tapeats sandstone is overlain by weak fine-grained strata whose susceptibility to erosion has led to the development of gentle slopes and the widest bench to be found at any level in the Canyon. These strata, the *Bright Angel shale*, are 350 to 400 feet thick in most places. Although composed chiefly of fine-grained silty beds with layers less than an inch thick, there are occa-sional sandstone ledges and, especially in the upper half, prominent beds of sandy dolomite and silty limestone that are similar to the dominant rock in the overlying formation.

The lithology most characteristic of the Bright Angel formation is shown in Figure 263, a greenish-buff silty to sandy and often micaceous shale. (See also Fig. 46.)

Three general types of fossils occur in the Bright Angel shale. The most obviously marine

Fig. 263.
Typical exposure of Bright Angel shale,
east of Kaibab trail, Grand Canyon.

Fig. 264.
Brachiopods of genus *Lingulella* and *Paterina* and other fossil fragments from the Bright Angel shale. (*Specimens at Museum of Northern Arizona, Flagstaff.*)

Fig. 265.
A nearly complete trilobite, and the tail of another, probably belonging to the genus *Glossopleura*. From the Bright Angel shale, Grand Canyon. (*Specimens at Museum of Northern Arizona, Flagstaff.*)

are small thin shells of such extinct brachiopods as *Lingulella* and *Paterina* (Fig. 264). Most of these are fingernail-size or smaller, and most are broken, but they are quite abundant in some places.

Less abundant but equally distinctive are trilobites, extinct creatures with horny external skeletons, whose remains are numerous in early Paleozoic sediments in all parts of the world. Figure 265 includes an almost whole individual (*Glossopleura*) at the left and the tail plate of another at the right. The lithology of the strata in which they occur proves that many of them lived on muddy to sandy bottoms in shallow water. The ribbed midsection, which resembles the thorax of many modern insects, bore crustacean-like appendages with which the animal could walk and probably swim.

The third type of fossil comprises the tracks and burrows of trilobites and of what were probably worms. The track left by a trilobite crawling over the bottom consists of two parallel rows of closely spaced and rather faint indentations, somewhat like those left by a large beetle. Trilobite burrows, on the other hand, produce a real disturbance of the sediment. In Figure 266 we are looking at the underside of a fine sandy layer. The double rows of small ridges were formed by successive pushes of the appendages as the animal worked its way through the soft mud. It looks as though one individual entered from the right and after a few changes in depth climbed back up toward the surface (behind this layer) at almost the spot, by the paper clip, where another chose to burrow down to this depth and work his way toward the upper left corner. The modern horseshoe crab has similar habits on tidal flats today. In the wet sediment the burrow closes in immediately behind the animal.

There are many varieties of fossil trilobites in the Bright Angel shale. Some of them occur only in thin zones, a few feet thick, although they are widely distributed horizontally, extending for many tens of miles up and down the Canyon. If we make the reasonable assumption that

Fig. 266.
Collapsed burrow marks made by trilobites in a sandy layer of the Bright Angel shale. (*Specimen at Museum of Northern Arizona, Flagstaff.*)

these particular species came and went as coherent groups, each occupying the sea floor for a relatively short part of the period represented by the whole formation, then each thin fossil zone is essentially contemporaneous throughout its extent and constitutes a rough time plane within the Bright Angel shale. In this light it is significant that these fossil zones are not quite parallel to the top of the Tapeats sandstone: they are farthest above it near the western end of the Grand Canyon and the separation diminishes eastward. One, containing the trilobite *Olenellus* and marked xxx on Figure 269, is 13 feet above the base of the Bright Angel at the west end of the Canyon. Traced eastward it gradually descends into the Tapeats transition zone and ends, 30 miles east, at the top of the main mass of the Tapeats. Another, containing *Glossopleura* and other trilobites, is not quite so sharply defined but can be recognized for a greater distance (· · · on Fig. 269). It is about 420 feet above the base of the Bright Angel at the west and descends to about 120 feet from the base at the east end of the Canyon. In other words—accepting the fossil zones as time planes—by the time this assemblage of organisms arrived on the scene, the thickness of shale that had accumulated on top of the Tapeats was about 300 feet greater at the west end of the Canyon than at the east end.

These relations could be explained either by more rapid accumulation of the sediments in the west, or by a gradual decrease in the age of the Tapeats and Bright Angel formations as they are traced eastward. In view of the physical evidence for eastward transgression of the sea already mentioned and the fact that neither formation thickens westward within the Canyon (Fig. 269), an eastward change in age is believed to be the explanation.

As may be seen in Figure 258, and just above the center of Figure 259, the Bright Angel shale is overlain by an approximately equal thickness of beds that are only slightly more resistant. This formation, the *Muav limestone* (⁷), begins with a low cliff (Fig. 259) above which is a ledgy slope of lighter color than the Bright Angel (yellowish rather than greenish) and is topped by a second cliff about twice the height of the basal one.

These cliffs and ledges are formed by limestones (dolomite near the top) that is given a mottled appearance by the irregular lumpy inclusions of mud that constitute about 25% of the rock. The ledgy zone in the middle is made up of thin beds—alternately limestone, calcareous siltstone, and sandstone—that resemble some of the underlying Bright Angel strata. Most of the rock is intermediate between limestone and fine sandy siltstone. As may be seen in outcrop (Fig. 267, *left*) some of the ledges are actually groups of thin beds; the strata in these are a little more resistant than those in other parts of the formation because they contain a higher proportion of calcareous cement. This is also demonstrated in the view along the base of the upper cliff (Fig. 267, *right*).

Fig. 267. *Left:* Typical outcrop of sandy limestone forming less resistant parts of Muav formation. Kaibab trail, Grand Canyon. *Right:* Base of upper cliff in the Muav limestone. Kaibab trail, Grand Canyon.

Fig. 268.
Ripple marks on the surface of a bed in the upper part of the Muav formation. Kaibab trail, Grand Canyon.

The Muav beds contain fossil trilobites, and ripple marks occur at some horizons (Fig. 268). Calcareous muds of this sort usually accumulate in fairly shallow water (up to hundreds rather than thousands of feet deep).

When the full length of the Canyon is considered, the Muav limestone and Bright Angel shale interfinger with each other, as indicated in Figure 269. The upper part of the Bright Angel shale disappears westward as fingers or tongues that project into the limestone, and the lower part of the Muav limestone disappears eastward as tongues that pinch out in the shale. A similar but less noticeable relationship exists between the lower Bright Angel and upper Tapeats strata.

The relationships among these first three Paleozoic formations are interpreted to mean that at any given time as the Paleozoic sea moved eastward, Tapeats sands were accumulating near the shore, Bright Angel muds were being deposited farther seaward, and Muav limestone was accumulating still farther west. On the sea floor the lateral changes between these varieties of contemporaneous sediment will naturally be much less abrupt than the vertical changes between layers deposited at widely separated times. We reason that the Tapeats is a fairly clean sandstone because the currents were strong enough to keep mud (the principal ingredient of the Bright Angel) in suspension and carry it farther offshore, many miles beyond the limit to which they could spread the coarse sand. By the same reasoning the Muav limestone formed still farther from shore after enough mud had settled out to leave the ever-present calcium carbonate as the dominant ingredient. (It will be recalled that the Bright Angel shale is calcareous and the Muav limestone high in silt and clay; the bulk compositions of the two formations are not so different as the words "shale" and "limestone" imply.)

As the shoreline migrated eastward so did the three belts of dominantly sandstone, shale, and limestone deposition. Thus any given part of the peneplain was first dry land, then a shoreline attacked by waves, then a beach mantled with shifting sands, and then an ocean floor; as the shoreline moved on, the floor received a blanket of sand (Tapeats) which in the deepening water finally came to rest because it was out of reach of wave-generated currents competent to move it. With further eastward migration of the shore this same spot received mostly muds for a time and then, as the quantity of these diminished, muddy limestone. The last, being derived principally from compounds in the water itself, is relatively independent of detritus introduced from land.

The intertonguing indicates that the conditions of sedimentation, and therefore probably the marine transgression, were not steady. The westward-pointing tongues of shale represent either times when mud was carried farther out to sea as a result of invigorated currents or increased supply from land, or times when the shoreline temporarily halted or reversed its eastward migration. Likewise, the eastward pointing tongues of limestone mark times when the water was clearer and less turbulent nearer shore, either because of a reduction in the supply of mud or because of more rapid eastward advance of the shoreline. What is known today about crustal instability and sea level makes it seem probable that this sea spread intermittently, rather than at a uniform rate. No doubt the shifting balance between subsidence and sedimentation, with their opposite effects on depth of water and position of the shoreline, was largely responsible for the intertonguing relations.

Note that each formation, as a mappable unit, consists predominantly of one kind of rock, even though it changes a little in age from place to place. The lower part of the Bright Angel shale in the west was probably being deposited before the upper part of the underlying Tapeats sandstone in the east had accumulated. Certainly the lower half of the Muav limestone in the west accumulated at the same time as the upper half of the Bright Angel shale in the east. These relations give strong support to the interpretation of the fossil zones already given.

When strata are as well displayed and studied as these, a diagram like Figure 269 saves a lot of words. Variations in thickness, lithology, and age can be seen at a glance. It becomes obvious that the mutual relations of formations change when large distances are involved; we can see, for example, that the Muav will almost surely pinch out entirely farther east between the thickening sandstones that appear above and below it at the east end of the Canyon.

When, as in this case, it can be proved that different varieties of sediment were deposited simultaneously in different parts of the same basin of accumulation, then each variety is a *sedimentary facies* (Latin: *facies*, face or aspect; compare "facet"). The sandstone facies here accumulated nearest to shore, the shale facies and limestone facies successively farther seaward. At any one time all three were being added to the sea floor, and each facies represents a lateral variant of these contemporaneous deposits. Obviously, if the peneplain had not been so flat and nearly level and the shoreline had not migrated so persistently eastward, the three facies would not have been laid one on top of the other so evenly. The relationships between different facies, both horizontal and vertical, are sensitive to a host of factors, including the shape of the sea floor, rate of submergence, steepness of offshore profile, the kind and amount and rate of delivery of sediment and the changing complex of currents available to distribute it. The truly remarkable thing is that so many individual strata in the Grand Canyon can be traced as far as they can.

Fig. 269. Relations of the lowermost horizontal strata throughout the full length of the Grand Canyon, compiled from measured sections made at each of the positions marked by vertical lines. *L* and *P* denote occurrences of *Lingulella* and *Paterina*, respectively. (*Data from Edwin D. McKee.*)

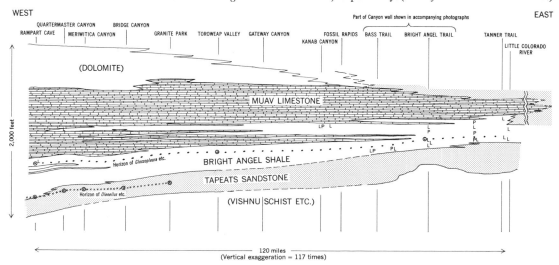

Fig. 270. Temple Butte limestone exposed in east wall of Pipe Creek, west of Kaibab trail, Grand Canyon.

Information from Lithology and Fossils III:
Temple Butte and Redwall Limestones

In the scene above (Fig. 270) we are looking at about 300 feet of the vertical south wall of the Grand Canyon. The well-bedded cliff in the lower half of the view is composed of the sandy dolomitic limestones that constitute the upper part of the Muav formation. The faintly banded rocks forming the massive cliff in the upper fourth of the scene belong to the Redwall limestone. Between them is the most inconspicuous formation in the Canyon, the *Temple Butte limestone.* In this view its upper contact is sharply defined and slightly wavy. Its lower contact is partly concealed by talus but is distinctive for its sag just below the center of the scene. No strata are bent, no deformation is involved: in this place the uneven, gnarly beds of the lower Temple Butte fill a channel about 30 feet deep cut into the top of the Muav limestone. A dozen or more such filled hollows can be found in the Canyon walls, most of them in relatively inaccessible places and partly obscured by loose debris. Between some of the channel fillings, and throughout many miles of the Canyon, the Temple Butte limestone is missing altogether and the Redwall limestone rests directly on the Muav formation (see Fig. 258).

The Temple Butte limestone contains few fossils. Poorly preserved corals, brachipods, and gastropods have been found, but most distinctive are occasional bony plates from a primitive armored fish. Figure 271 includes the inside surface of such a plate found in corresponding beds about 70 miles south of the Canyon.

Although the fossils and lithology leave little doubt that the Temple Butte is a marine deposit, the channels cut into the Muav prove that after Muav time and before Temple Butte time

274

Fig. 271.
Fossil bony plate from head of primitive fish (*Coccosteus arizonensis*) found near Flagstaff in beds approximately equivalent to the Temple Butte limestone. (*Specimen at Museum of Northern Arizona, Flagstaff.*)

there was erosion, which could have taken place only if the sea had withdrawn. The fact that Temple Butte beds exist only here and there proves that after Temple Butte time and before the overlying Redwall limestone was deposited there was another period of erosion, in which all but scattered thin patches of the Temple Butte formation were removed. Most of those that remain, as would be expected, fill depressions in the older rocks. Erosion surfaces like these, which separate parallel sets of strata, are called surfaces of *disconformity* (page 110); they record gaps in the sedimentary record. Those just described indicate that the accumulation of the Temple Butte was *not* the next event following Muav sedimentation; neither was it the event immediately preceding deposition of the overlying Redwall limestone. The Temple Butte may thus be a great deal younger than the Muav, or a great deal older than the Redwall, or both.

We have two clues that help in evaluating these possibilities. Since the sea in which the older strata accumulated encroached from the west it almost surely withdrew in that direction during the two periods of erosion. If, working from this hypothesis, we trace the Temple Butte westward into Nevada, we find that the erosion surfaces virtually disappear. The lower one is replaced by thousands of feet of sediment, the upper by a few hundred. In other words, in Nevada, where sedimentation was less interrupted, about ten times as much sediment was deposited in the interval between Muav time and Temple Butte time as in that between the latter and Redwall time. This implies that the lower disconformity represents the longer time and that the Temple Butte formation is closer to the overlying Redwall in age.

The second clue comes from the fossils in the Muav, Temple Butte, and Redwall formations. When these are compared with the total known record of life on earth and its evolutionary trends, it is clear that there is more of that record missing between the Muav and Temple Butte than between the Temple Butte and the Redwall formations. This also suggests that the lower disconformity represents the greater time interval. The fish plates in the Temple Butte are especially significant, for the earliest known fishes of any kind, anywhere, occur in rocks with ages much closer to the Redwall than to the Muav—and *Coccosteus* was not one of the first (see Fig. 292).

It is in recognition of these relationships that we have assigned the Temple Butte the number ⑩ on Figure 258, reserving ⑧ and ⑨ for the intervening rocks present to the west.

This reasoning still leaves us rather ignorant about what actually happened during these two periods of erosion in the Grand Canyon region—especially the long earlier one. There might have been a lowland there during the whole time, from which erosion removed only a few hundred feet of higher Muav rocks. Or there might have been post-Muav sedimentation thousands of feet thick, which, plus a little Muav, was removed before the Temple Butte submergence. Perhaps, someday, careful study of the corresponding strata to the west will provide some hints of how much deposition there was, how much erosion, and at what time or times either took place. (Note that these events are not apparent in the gross structural relations visible in the Canyon, which is the reason for their omission from the history on pages 248–259.)

Fig. 272. The south wall of the Grand Canyon, showing the complete Paleozoic section from Tapeats
sandstone (left foreground) through the Redwall limestone cliff in the middle to the
highest strata at the rim. Total thickness, approximately 4,000 feet.

Throughout the most-visited parts of the Grand Canyon the *Redwall limestone* forms a nearly unbroken 500-foot cliff almost exactly midway between the river and the canyon rim. It was prominent in Figure 259 and forms the lower of the two main cliffs seen in Figure 272.

The rock is so uniformly resistant that in most places weathering has failed to etch into prominence the subtle distinctions between individual beds, and the result is the most continuous massive precipice in the entire Canyon: this is clear in the close view of Figure 273. In strong contrast to the Muav, the Redwall is an almost pure limestone (dolomitic in the lower part) containing less than 1% sand and mud. Thin discontinuous layers of chert are common, especially in the lower part (Fig. 274). The fresh rock is mostly light gray, although weathered surfaces may be variously discolored; in many places a reddish stain has been imparted by seepage from the overlying deep-red shales—which accounts for the formation name.

Fossils representing at least twelve biological classes occur at a number of horizons throughout the Redwall. At the left in Figure 275 is a ribbed brachiopod and at the center and right two different coiled shells of extinct snail-like animals, all resting on a block of limestone from which protrude a dozen or more dark corals. All are marine and all occur in sequences of strata in many parts of the world, but only at levels well above any of the fossil forms found in the Muav limestone.

The extraordinary flatness of the disconformable contact at the base of the Redwall limestone, which generally lacks even the minor channelling seen beneath the Temple Butte, implies that the landscape that developed in the interval between the accumulation of the two formations possessed remarkably little relief. Some of the smoothing may have been accomplished by wave action as the sea advanced,

Fig. 273.
Typical Redwall limestone, near the
Kaibab trail. For a more general view in
this same area see Figure 39.

Fig. 274.
Lenses of chert (e.g., under head of hammer)
in Redwall limestone. Kaibab trail,
Grand Canyon.

for the basal Redwall beds contain recognizable detrital ingredients derived from the Temple Butte Limestone.

Limestones, because of their very low solubility in seawater, can be broken up by waves and distributed as clastic sediment by marine currents. In this way carbonate rock debris derived from older limestones exposed at a beach may be added to a new deposit forming offshore. The basal beds of the Redwall prove that this happens. But limestone is so soluble in most fresh water that, on land, as we have seen (page 236), it is often removed in solution rather than as clastic sediment, sometimes with the formation of karst topography. Since the known rocks immediately beneath the Redwall in the Grand Canyon region are predominantly limestone and dolomite, it is likely that a large part of the landscape surrounding the spreading Redwall sea was composed of these rocks. Such a landscape would probably contribute little more than water rich in dissolved calcium carbonate to the Redwall sea—a postulate supported by the purity of the Redwall limestone. If any parts of the lands contributing to this deposit were not dominantly limestone they must have been so low and flat that their streams carried no clastic sediment most of the time.

Perhaps it is significant that no karst topography has been found beneath the disconformity at the base of the Redwall. Does this mean that it existed and was erased by wave action, or that it never developed because of aridity or because the groundwater was already saturated through solution of more distant carbonate rocks? The answer may involve a combination of such factors. (There is karst, however, on top of the Redwall; see page 280.)

Today the Redwall limestone becomes uniformly thicker toward the northwest and thins in the opposite direction, but this was not necessarily true of the original deposit; some or all of the variation may be the result of post-Redwall erosion.

Fig. 275.
Fossil brachiopod, gastropods, and corals
from the Redwall limestone. (*Collected by
E. D. McKee and R. C. Gutschick*.)

Information from Lithology and Fossils IV:
Supai, Hermit and Coconino Formations

In the central part of Figure 276 is shown a steep ledgy slope that contrasts sharply with the high Redwall cliff beneath it and the vertical Coconino cliff above it. The rocks of this slope are the main source of reddish color in the Canyon and constitute the only thick and extensive sequence of red beds in the Paleozoic strata. The total sequence, about 1,200 feet thick, has been divided into two formations. The first 900 feet of alternating slopes and ledges above the Redwall limestone is the *Supai formation* and the uppermost 300 feet (immediately below the Coconino cliff), of slightly darker color and lacking resistant layers, is the *Hermit shale*. For our purposes these two red bed formations are enough alike to be discussed together.

The general lithologic character of these strata is clear in Figure 277. At the top and bottom are weak zones of deep-brick-red sandy shales that produce smooth slopes, and in the middle alternating weak shaly zones and small cliffs of resistant sandstone, the highest of which is the most prominent. The red color is most pronounced in the shales; from them it spreads, through weathering, to stain the exposed sandstones, which are usually cream colored where freshly broken. Most of the sandstones are fine-grained, distinctly cross-bedded (as shown in Fig. 74) and held together with calcium carbonate cement. A few beds of impure gray limestone occur in the lower part; these gain prominence westward until in southern Nevada the lower Supai is predominantly gray fossiliferous marine limestone with a few reddish sandstone layers and is known by another name (an example of facies change). The upper Supai and the Hermit beds thicken westward but retain their cross-bedded reddish sandstone and shaly siltstone lithology.

No marine fossils have been found in these red reds in the Canyon proper, and none would be expected, for the color of the beds implies a degree and kind of oxidation that almost certainly required at least intermittent exposure to air. Such exposure is proved by tracks of small animals, probably either amphibians or reptiles, on some layers in the upper Supai and the Hermit, and by fragments like that in Figure 278, which shows casts of mud cracks and raindrop prints as seen on the under side of a bed deposited on a dried surface. The Hermit shale also contains some fairly well-preserved plant fossils, including impressions of leaves of a primitive

Fig. 276. General view of south wall of the Grand Canyon emphasizing the ledgy slope of the Supai and Hermit formations between the Redwall cliff below and Coconino cliff above. The Muav formation shows below the Redwall on three spurs at the lower right.

Fig. 277. Closer view of Supai and Hermit formations resting on Redwall limestone in the north wall of the Grand Canyon. White cliff at upper left is the lower part of the Coconino sandstone.

seedfern (Fig. 279, *left*) and impressions of twigs that probably came from some cone-bearing tree (Fig. 279, *right*). The known fossil flora from the Hermit includes more than thirty kinds of land plants. The plants, the footprints, and the occasional mud cracks and raindrop prints clearly show that these strata accumulated on land. The lower Supai, which is definitely marine farther west, probably accumulated on the upper part of a delta or the lower part of a flood-plain. At least three surfaces of erosion have been recognized within these beds. The discon-formities, and the alternation of red beds and limestones in the lower Supai, suggest a low, near-shore environment in which small changes in level of either sea or land could cause wide

Fig. 278
Casts of mud cracks and raindrop prints, as seen on the underside of a slab of Hermit shale 7 inches long.

Fig. 279. *Left:* Leaf-impressions of the extinct fern-like plant *Callipteris conferta* from the Hermit shale. *Right:* Fossil twigs of probable cone-bearing tree as preserved on a slab of Hermit shale. (*Specimens at Museum of Northern Arizona, Flagstaff.*)

displacement of the shore line, thus occasionally interrupting the accumulation of nonmarine mud and sand with episodes of erosion or of marine deposition.

In some places the upper surface of the Redwall limestone projects 25 feet or more into the overlying Supai red beds; in others the limestone is cavernous to depths of several hundred feet. At many places the lowest Supai strata include pebbles derived from the underlying limestone. These evidences of karst topography and local erosion prove that this contact too is a disconformity—another interruption of the sedimentary record.

Full interpretation of the Supai-Hermit sequence is thwarted by some puzzles. For example, throughout the Grand Canyon region the typically aqueous cross-bedding in the Supai consistently dips toward the south and southeast, indicating that this was the general direction of the depositing currents. Yet it is difficult to find an adequate northern or northwestern source for such quantities of sand and mud; those that lie in the right direction seem either to have been under water or composed predominantly of limestone at the appropriate time. Perhaps the direction of the depositing currents was somewhat different from that of the transporting currents. Transportation of sediments from the northeast quadrant would agree better with the change to marine facies in the lower Supai, but this possibility is difficult to evaluate because in this direction most of the critical rocks (of Supai and older age) are now obscured by younger deposits. The problem may resolve itself when more is known about the subsurface geology of Utah.

The uppermost prominent cliff in familiar views of the Grand Canyon is formed by the *Coconino sandstone*, the most homogeneous and distinctive formation in the Canyon. It is about 300 feet thick at Grand Canyon village, which is near the center of its present area of 32,000 square miles; 100 miles south it is 1,000 feet thick; along the Utah border to the north and near the mouth of the Canyon to the west it thins to the vanishing point. Viewed from a distance, as across the Canyon, it is white or cream colored (for example, in the upper part of Fig. 277).

The Coconino is a fine-grained and almost pure quartz sandstone whose rather well-rounded and well-sorted grains are held together mostly by calcareous cement. (Its characteristic texture is shown in Figures 44 and 45, pages 40 and 41.) Most conspicuous of all is the bedding. Except for a few feet of horizontal stratification near the base in some places, it is, as shown here (Fig. 280), cross-bedded throughout with the great sweeping laminae characteristic of dunes. Many of these are 30 to 40 feet and some are 70 feet long. In some places they are ripple marked and

Fig. 280. Eolian cross-bedding in the Coconino sandstone as seen near the top of the Bright Angel trail, Grand Canyon. Tree trunk at lower left gives some idea of scale.

a number of the laminae bear fossil footprints such as those in Figure 281. Some types of footprints are so abundant, well preserved, and distinctive that their makers have been given generic and specific names even though they are known to us only by their tracks. At least 22 varieties have been recognized, all probably made by reptiles (or possibly amphibians), and all different from the tracks in the underlying Supai and Hermit formations.

Among the hundreds of examples of such tracks that have been observed all but perhaps four or five were made by animals ascending steep slopes (25° or more) in the dunes. Experiments with small lizards and chuckwallas on modern dunes have shown that the best tracks are made in dry sand and that the steeper the slope the greater the likelihood that only ascending tracks will be well defined—because the animals slide so much going downhill. Heavier animals make tracks in wet sand, and in any case the sand must be dampened while the tracks are fresh if good impressions are to survive until they are buried. Burial, in turn, is most likely to take place on slopes that face downwind, and these are generally steep (page 199); this may explain the one-way tracks in the Coconino. No bony remains of these animals have been found in this sandstone.

The preservation of raindrop prints on some laminae and many details of the ripple marks and slumping leave no doubt that the Coconino represents a great sheet of eolian sand. Its volume today is approximately 3,000 cubic miles, even after the loss of the unknown amount

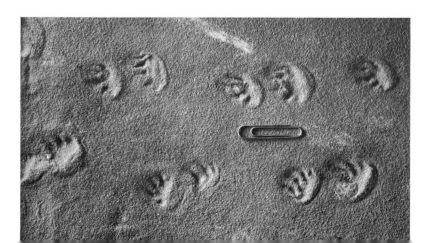

Fig. 281
Footprints of *Laoporus noblei*, an extinct four-footed animal, on a slab of Coconino sandstone. (*Specimen at Museum of Northern Arizona, Flagstaff.*)

Fig. 282. *Upper:* Contact between dark Hermit shale (below) and light Coconino sandstone (above) west of Bright Angel trail, Grand Canyon. *Lower:* Contact between Coconino sandstone (note inclined cross-bedding) and overlying Toroweap formation as exposed along Kaibab trail, Grand Canyon.

that has been eroded away along the thick southwest border. There is no trace of vegetation in the Coconino, although its animals must have had access to some. This evidence of completely barren shifting sands over such a large area leads many geologists to postulate an arid region analogous to the Saharan and Arabian deserts of today, but it must also be admitted that pure quartz sand, even though occasionally wetted, is not very hospitable even to today's diversified plants and may have been an impossible site, regardless of climate, for the more primitive plants of 250 million years ago.

Considering the definite and relatively simple requirements that have to be met to produce a sheet of wind blown sand, we know embarrassingly little about the origin of the Coconino. Both upper and lower contacts are in most places quite sharp (Fig. 282) and there is no evidence of prolonged weathering or extensive erosion at either surface. (Note, for example, the deep vertical crack in the Hermit, filled with Coconino sand, near the center of Figure 282, *upper:* such a sharply defined break is far more likely in fresh than in weathered rock.) Many hundreds of measurements of the cross-bedding, at widely scattered localities, show rather consistent dips to the south and southeast. Interpreted as slip faces these would indicate north and northwesterly winds and source areas. The well-rounded grains probably mean that the Coconino sands were derived from other, older sandstones that were exposed to weathering somewhere to the north. Perhaps the Coconino sand was partly derived from coastal dunes which traveled far inland. The thinning on the north and west is also a problem; it may be largely the result of redistribution of the sand by waves of the succeeding Toroweap sea, which probably spread over the dune area from the northwest while the sand was still fairly loose.

The three successive formations—Supai, Hermit, and Coconino—constitute the only large accumulation of definitely proved nonmarine sediments exposed in the Canyon walls. This is one reason they are lumped together as [12] on Figures 238 and 258. Note that the evidence proves erosion was dominant on the several earlier occasions when the Grand Canyon region was out of water; during the Supai-Coconino interval, however, there was sedimentation, and the strata provide information on the ancient environments that would otherwise be missing.

Finally, we may reason that the ultimate source of these clastic deposits must have been outside of the Plateaus region because (1) the horizontal sheet of Redwall limestone beneath them blanketed any source of sand within the area it covered, and (2) the absence of angular unconformities within the strata of the Plateaus proves that no prominent sources of clastic sediment were produced by local uplift there during the time represented by the Paleozoic strata. (The few monoclines and faults in the Canyon, many of whose slopes are now shedding clastic debris, involve the two formations overlying the Coconino and were therefore produced *after* Coconino time.) It is quite possible, however, that the gentle and broad warping suggested by the disconformities was sufficient, in the north, to induce reworking of nearly contemporaneous sediments which, in turn, may have been the immediate source of this sequence of nonmarine beds.

Information from Lithology and Fossils V:
Toroweap and Kaibab Formations

The top 500 to 600 feet of strata in the Canyon walls, from the top of the Coconino sandstone to the rim, form cliffs and steep slopes, as shown in the upper half of Figure 283. These strata are divided into two formations of approximately equal thickness (Fig. 258, 13). The lower one consists of two reddish sandy zones separated by about 75 feet of limestone and is known as the *Toroweap formation*. Being less massive and resistant than the Coconino it accounts for the steep slope and band of clinging vegetation immediately above the sheer Coconino cliff. Above this the more resistant *Kaibab formation*, consisting chiefly of sandy and cherty limestone, forms the topmost cliffs of the canyon and the surface of the plateau. Its creamy color is evident in boulders along the roadside, at the various scenic view stops, and in blocks used to construct many walls at Grand Canyon village. Many parts of the upper Kaibab are made reddish by thin, rusty, sandy siltstones that are interbedded with the limestone.

The limestones in these formations contain many fossils. Brachiopods, such as *Derbya*, *Meekella*, and productids (two specimens of each are shown in Figure 284) are typical. To judge by their occurrence, they were bottom dwellers. All three became extinct shortly after Kaibab time, at least 200 million years ago. The cephalopod, *Domatoceras* (Fig. 285), like its distant relative the modern coiled nautilus, was probably a free swimmer. It is not found anywhere in the world in rocks younger than the Kaibab. More than 80 genera of invertebrates are represented by the known fossils in these rocks. There are also a few fish teeth, including one from an extinct shark.

Eastward from the Grand Canyon both formations become increasingly sandy and the Toroweap disappears in less than 50 miles. Westward, the combined thickness increases slightly to 800 to 1,000 feet and the proportion and purity of the limestones rise steadily, even though the three red zones persist. Near and beyond the western end of the Canyon the red zones at the tops of the two formations contain gypsum; some of the deposits are large enough to be of commercial value.

The very sharp contact between Coconino and Toroweap (Fig. 282, *lower*) is so remarkably flat, wherever it can be seen, that any lengthy period of weathering and erosion between the

Fig. 283. The uppermost part of the south rim at the head of Pipe Creek, as seen from the Kaibab trail, Grand Canyon. Principal cliff is Coconino sandstone; below, slope on Hermit shale; above, horizontal beds and slope of Toroweap formation, overlain by layered cliff of Kaibab limestone.

284

Fig. 284
Fossil brachiopods from the Kaibab lime-
stone. Two views each of *Derbya* (left),
Meekella (center) and a productid (right).
(*Specimens at Museum of Northern
Arizona, Flagstaff.*)

deposition of the two formations seems extremely unlikely. As already hinted, this contact
probably means that the Toroweap sea spread across the Coconino dunes while they were still
loose uncemented sand, perhaps even still actively accumulating, and that wave action produced
the flat surface. Its preservation was insured because it remained under water and was soon
blanketed with Toroweap sediments. The slope of the sea floor must have been very gentle and
the wave action mild, because on open coasts today the offshore currents associated with vigorous waves are capable of cutting channels in near-shore sand deposits. If this interpretation
is correct the basal beds of the Toroweap should contain some reworked Coconino sand, which
is difficult to check because there are not many ways to distinguish one grain of quartz sand
from another. However, many of the basal Toroweap grains do have the size and shape of Coconino grains.

The facies changes in these two formations, as they are traced along the Canyon wall, show
that the Toroweap sea spread eastward, probably not far beyond the east end of the Canyon,
and then retreated to the west, leaving exposed reddish muds. On this surface the evaporation
of intermittent shallow lagoons or playa lakes precipitated gypsum; if more soluble salts were
also deposited they have since been dissolved away by percolating waters. The Kaibab sea
repeated this sequence of events, reaching farther east (where the Kaibab limestone now rests
directly on Coconino sandstone) and in its retreat likewise left, especially in the west, a series
of shallow basins in which evaporites were precipated. These two cycles of eastward advance
and westward retreat of shallow seas are the final events recorded in the Paleozoic strata of
this area.

Fig. 285
A fossil cephalopod, *Domatoceras*, from the
Kaibab limestone. (*Specimen at Museum
of Northern Arizona, Flagstaff.*)

Fig. 286. Cedar Mountain, a low flat-topped remnant of post-Kaibab red beds (shown as ⑭ and basal ⑮ in Figure 258). This view is southeastward from the upper end of the Grand Canyon, in whose wall (foreground) can be recognized the dark ledgy Supai formation and all overlying formations.

Of the strata laid on top of the Kaibab only scattered remnants of the first few hundred feet are left today in the vicinity of the Canyon. One of the most conspicuous is Cedar Mountain (Fig. 286). But to the north (Zion and Bryce canyons) and east (Painted Desert and Black Mesa) post-Kaibab beds form the basal part of a sequence more than 9,000 feet thick and composed entirely of sandstones, shales, and conglomerates. Red coloration, beds of low-grade coal, and vertebrate fossils indicate that most of these strata were deposited on land and therefore that after Kaibab time the Grand Canyon region was probably never again completely submerged.

Ever since the accumulation of these nonmarine strata ended, the dominating process in the Grand Canyon region has been erosion. First, beginning less than 100 million years ago, the 9,000+ feet of weak rocks (more than twice the thickness of the horizontal beds in the Canyon today) were stripped off the resistant Kaibab limestone throughout an area that by now exceeds 10,000 square miles. This erosion, often referred to as "the Great Denudation," must have been induced by widespread uplift of the Plateaus region. Then, beginning at an unknown time, but probably less than 20 million years ago, further uplift caused the Colorado River, which may by then have already been flowing in a westerly direction across the region, to become deeply entrenched. The Grand Canyon is the result of this last, relatively insignificant event.

Note that the Kaibab limestone, formed a little below sea level, now stands as much as 8,000 feet above it. This is a rough measure of this last and smallest uplift of the Plateaus (not counting whatever happened between Muav and Temple Butte time). It is a source of wonder that uplift of this magnitude over such an enormous area (considerably more than 100,000 square miles) was accomplished so uniformly that the result is a high tableland crossed by only a few faults and folds, all of them small in proportion to the total area involved. As already pointed out, the dominant feature of the Colorado Plateaus is the preponderance of essentially horizontal strata—Gilbert's "area of comparative calm." Theories to explain crustal deformation must be able to account for such gentle regional uplifts as well as for more localized mountain making—a problem we will touch upon in Part VI of this book.

Correlation to the West

The conditions and events deduced on the preceding pages greatly enrich the story of the accumulation of the Paleozoic strata. Close examination of the rocks and their fossils has made it possible to reconstruct a changing scene in which the sea came and went at least five times, interspersed with periods of erosion and of sedimentation on land. All of these events are included in Step Two (Fig. 240) and show how much was passed over when, in dealing with the Paleozoic strata (page 252), we injected the qualifying phrase, "assuming for the moment that their accumulation was essentially uninterrupted."

If, during the accumulation of the Paleozoic strata in the Grand Canyon region, shallow seas repeatedly advanced from the west and retreated to the west, there must have been a more persistent sea in that direction. The facies changes suggest the same thing: the Tapeats, Bright Angel, and Muav strata are increasingly marine westward, the Supai and Toroweap become almost completely marine in that direction, and the wholly terrestrial Coconino sandstone disappears near the western end of the Canyon. Furthermore, many hundreds of feet of marine strata were deposited in southern Nevada during intervals represented by erosional disconformities in the Canyon region; more than 20,000 feet of sediment was laid down there during the time that 4,000 feet accumulated in the Grand Canyon area.

We know that the deposits in these two regions represent approximately the same period of time largely because the top and bottom layers, the Kaibab limestone and Tapeats sandstone, can be traced from one area to the other. But there are some interesting developments when this is done. For example, consider the scene in Figure 287, near the Arizona-Nevada boundary, in which we are looking eastward toward the Colorado Plateaus, whose horizontal layers are visible in the background. In the foreground is a ridge made up chiefly of inclined strata. The

Fig. 287. Looking eastward across Whitney Ridge in the Virgin Mountains to the western edge of the Colorado Plateaus 15 miles beyond, in the northwest corner of Arizona. The tilted strata in the foreground correspond to the lower half of the Paleozoic strata in the Grand Canyon.

left end of this ridge is composed of rocks ([0]) that correspond closely to the Vishnu schist, and are overlain toward the right by recognizable Tapeats sandstone ([5]), Bright Angel shale ([6]), Muav limestone ([7]) and Redwall limestone ([11]). Between the last two are some strata that do not exist in the Grand Canyon, and directly above the Redwall are reddish sandy limestones— at the place in the sequence where, in the Canyon, we would find the Supai red shales and sandstones. Even though this sequence is thicker than its counterpart in the Canyon, and is separated from the Plateaus by an alluviated valley 15 miles wide, we can confidently correlate the two sequences because four of the formations are virtually identical in lithology and contain the same kinds of fossils. The only significant differences are the additional strata between the Muav and the Redwall (corresponding to the Temple Butte and its associated disconformities in the Canyon), and the limestone that here replaces the lower Supai sandstones above the Redwall in the Canyon. The thickness of the sequence from the base of the Tapeats through the Redwall as exposed here is 4,370 feet; in the Canyon it is 1,830 feet.

If we move still farther west, we may again see this part of the Paleozoic sequence exposed in the tilted block shown in Figure 288, which is 60 miles west of the edge of the Plateaus. Schist and pegmatite, strongly resembling the Vishnu of the Canyon, are exposed in a rough belt ([0]) at the base of the slope (far right), succeeded again by the first three Paleozoic formations; note the weak belt produced by the Bright Angel shale ([6]). Successively younger strata

Fig. 288. The west face of Frenchman Mountain 10 miles east of Las Vegas, Nevada.

occur toward the left in this concordant sequence; the crest of the main ridge is composed of massive Redwall limestone (⒒) and, to the left of the zig-zag road to the summit, bedded limestones occupy the position of the lower Supai red sandstones in the Canyon. Next comes a broad strike valley with some white (gypsum-bearing) layers along its axis; this is formed on red beds corresponding to the upper Supai formation and Hermit shale of the Canyon. To the left of these is a dark ridge composed of Toroweap and Kaibab limestones (⒀). Note that the Coconino sandstone is missing (cf. page 280).

In short, here is the whole Grand Canyon Paleozoic section, tilted about 55° to the left (down to the east). The top and bottom formations and the Redwall limestone near the middle are almost identical in lithology and fossil content with their counterparts in the Canyon walls, but the total thickness here is a little over 8,000 feet—twice that of the Paleozoic strata in the Canyon. Two changes are noteworthy. Between the Muav and Redwall, where in the Canyon we found less than 100 feet of Temple Butte limestone, there are here over 2,000 feet of limestone and dolomite, much of it undoubtedly deposited during the time represented by the two pre-Redwall disconformities in the Canyon. And above the Redwall, where in the Canyon we found the lower part of the Supai red beds, the presence of highly fossiliferous marine limestones here demonstrates again the westward change from nonmarine to marine facies.

In the next range of mountains to the west the thickness of the Paleozoic strata doubles again, and there is a new development. The same trilobites (*Olenellus* and *Glossopleura* and their associates) that in the Canyon are less than 600 feet above the Vishnu schist (Fig. 269), are found here as much as 10,000 feet above the base of the sedimentary sequence. Above them occur the same marine fossil assemblages that were above them in the Canyon, and although these are interspersed with many others missing from the Canyon because of the disconformities and nonmarine deposits there, their order is unchanged: for at least 200 miles these different fossil assemblages have retained their positions relative to each other. This is a compelling reason for believing that each fossil assemblage represents a definite time interval, probably only a small part of the duration of the sedimentation that produced the enclosing formation, and that each fossil zone may therefore be used to establish approximate contemporaneity throughout the region in which it occurs. Evidently each fossil assemblage is a sample of the life that flourished under a particular set of conditions in the sea at a given time. On succeeding pages we will further test this idea and discuss some of its limitations as well as implications.

If we conclude that these trilobites seen in the Canyon and in southwestern Nevada lived at approximately the same time, we must also conclude that sedimentation in southwestern Nevada began well before it did in the Grand Canyon region, because in some places almost 10,000 feet of sandstone had accumulated there before deposition began in the Canyon area. In other words, when *Olenellus* and associates appeared on the scene the sedimentary pile was already about two miles thick in southern Nevada but was just beginning to accumulate in the Grand Canyon region.

Examination of sedimentary sequences and facies and their associated fossils in various parts of the world has led to the discovery of many ancient and persistent marine sedimentary basins like this one. The sedimentary strata in them are much thicker and the geologic record is more nearly complete than in surrounding areas. Characteristically, they do not contain evidence of very deep water. Instead, they must have been slowly subsiding troughs that collected sediment from one or more of their sides and were kept shallow by these additions to their floors. The typical result is 20,000 to 50,000+ feet of strata, most of which seem to have accumulated in water considerably less than a thousand feet deep. And, as we have seen, although the axis of such a seaway might remain relatively fixed, its margins frequently shifted. The sea we have examined flooded its eastern borderland and then withdrew many times during the accumulation of the Paleozoic strata.

Such a persistent sedimentary trough, usually many hundreds of miles long, is known as a *geosyncline.* (The prefix "geo-" suggests its large size, and "syncline," as used here, refers to subsidence rather than downward folding.) One of the great generalizations of geology is that the belts of thick strata that accumulated in ancient geosynclines have nearly always been more severely deformed by later crustal movements than the thinner accumulations in adjacent areas. Certainly this has been true here, where the thick sediments are now the tilted and crumpled blocks of Nevada and the thinner sequence lies almost undisturbed on the Plateaus. A similar relationship on the eastern side of the continent is explored on pages 399–407.

pages 399–407

We have now found a way to relate the geologic history recorded in Grand Canyon to the geologic history of a nearby area to the west, even though the succession of events in the two places was not quite the same. Westward to the mouth of the Canyon this was possible because we could actually follow the strata almost continuously in the Canyon walls. Beyond that it was necessary to jump from one mountain ridge to another across valleys many miles wide in which only alluvium is exposed. Yet despite some marked changes we could correlate across these gaps because (1) some of the formations remained essentially unchanged and (2) each fossil assemblage seemed to be contemporaneous throughout its occurrence and therefore to mark a kind of time plane within the strata; this helped us to know whether certain events in southern Nevada took place before, during, or after known events in the Plateaus.

Many questions involving extensions of this process come to mind: Where were the other shores defining the shape of the land across which the Tapeats sea spread eastward? What was happening in the Rocky Mountains during the time these events took place in Arizona and Nevada? Many thousands of feet of sedimentary rock are exposed in the Appalachian Mountains; were they deposited before, during, or after the strata in the Grand Canyon? And what about similar sedimentary sequences in Europe?

To answer these questions we need to be able to establish contemporaneity over great distances, even across oceans. The example we have studied demonstrates that the lithology and thickness of some strata can change so markedly in less than 200 miles that, even when they are unusually well exposed, they are unreliable for establishing contemporaneity. A sandstone in one place is of the same age as a limestone in another; an erosion surface in one place corresponds to marine deposits in another. Although we obtained clues to these relationships from the persistence of such formations as the Kaibab limestone, neither it nor any other formation extends over a continent—much less across an ocean. How about fossils then? Can they be used for correlation over such great distances? The possibility is explored on the following pages.

22 GEOLOGIC TIME

The Importance of Fossils in Geology

Fossils were long ago recognized by many persons, including Herodotus and Leonardo da Vinci, as the remains and impressions of once living organisms, buried and preserved in sedimentary rock by natural processes. Even so, it was not until the eighteenth century, when open-minded observation began to replace dogma in the sciences, that this interpretation won general acceptance. A little later, in the early nineteenth century, fossils played a critical role in two developments of profound importance to natural science.

The first began with the work of one man, an English mine surveyor and canal builder named William Smith. His duties required that he travel throughout England and Wales, which, then as now, were more noted for their forests, fields, and gardens than for outcrops of bare rock. But since his work entailed examination of stream banks, sea cliffs, quarries, and excavations for canals, he had more than the usual opportunity to study the bedrock beneath the soil. He must have been a keen-eyed and diligent observer, for in the 1790's he recognized that different strata contained distinct groups of fossils and that with the help of these he could identify strata in isolated spots where only a small sample was exposed. Systematically he kept notes of his observations and recorded them on maps. After 24 years of such work he had compiled a map of England and Wales (published in 1815), a part of which is shown in Figure 289. Each group of exposed strata was represented by a distinctive color, deepest near the base, thus clearly marking the succession from oldest to youngest. He adopted this mode of presentation as a means of "generalizing the information" embodied in thousands of plotted observations so that it could be "seen and understood at a distance from the map, without distressing the eye to search for small characters."[1] In introducing this map Smith wrote:

> And I presume to think, that the accurate surveys and examinations of the strata . . . to which I have devoted the whole period of my life . . . have enabled me to prove that there is a great degree of regularity in the position and thickness of all these strata . . . and that each stratum is also possessed of properties peculiar to itself, has the same exterior characters and chemical qualities, and the same extraneous or organized fossils throughout its course. I have . . . collected specimens of each stratum, and of the peculiar extraneous fossils, organic remains, and vegetable impressions, and compared them with others from very distant parts of the island . . . and have arranged them in the same order as they lay in the earth; which arrangement must readily convince every scientific or discerning person, that the earth is formed . . . according to regular and immutable laws, which are discoverable by human industry and observation, and which form a legitimate and most important object of science.[2]

Thus did Smith, with great labor, accomplish the correlation of strata throughout soil-covered England. By means of it he was able to construct the first useful geologic map of a large area, a map which compares suprisingly well with modern ones. When we realize that this map was the work of one man who travelled on foot and horseback, without benefit of aerial photographs or other modern aids, it is a truly remarkable achievement. It is also eloquent testimony to the usefulness of his discovery that fossils can be used in the recognition and tracing of strata.

Events conspired to give prompt and practical importance to Smith's technique. James Watt's steam engine was hardly 50 years old and the demands of the burgeoning machine age were greatly stimulating interest in deposits of coal and iron. Geologic mapping with the aid of fossils was soon extended to sequences of strata in France, Germany, Italy, and Russia. Several

[1] William Smith, *A Memoir to the Map and Delineation of the Strata of England and Wales, with a Part of Scotland*, London, 1815, p. 11.

[2] *Ibid.*, p. 2.

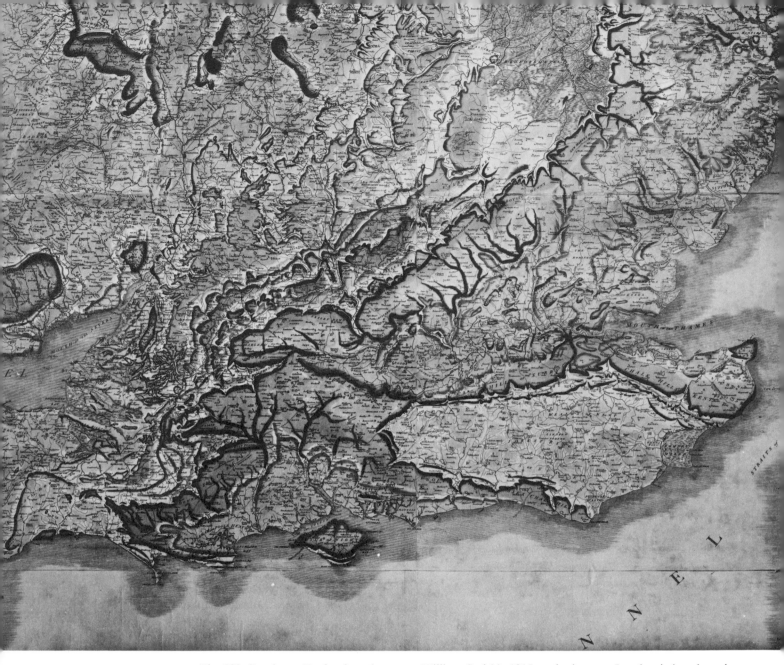

Fig. 289. Southeast England as shown on William Smith's 1815 geologic map. London is in a broad
syncline plunging gently eastward; south of this is the gentle arch or dome of the
Wealden district.

eighteenth-century pioneers in the study of the earth had already proposed standard sequences
in which they grouped all rocks according to presumed age; with this background Smith's
followers were quite prepared to find order among the strata. Perhaps, too, it was natural to
assume that the coal near Manchester was of the same age as that in Alsace, in the Ruhr, and
near Moscow.

Such studies made it possible by mid-century to compare, for the first time, detailed sedimentary sequences from widely separated localities, because in each place the order of the strata
and their enclosed fossils had been worked out, recorded, and reported. The comparison showed
that in each area studied there were some strata that contained essentially the same groups of
fossils. The succession of fossil assemblages both above and below the coal was in the same
order, even in areas 1,500 miles apart. This greatly strengthened the presumption that known
European coal deposits all formed at the same time.

It was a sound knowledge of the sequence of strata that enabled Godwin-Austen to predict in 1856 that coal would be found beneath the lowlands of the Weald, south of London, even though this area was more than 100 miles from the nearest natural outcrop of coal. To do this he had to know exactly which layers were at the surface in the Weald district, and he had to know their structure—essentials that could come only from a good geologic map.

The French naturalists Georges Cuvier and Alexandre Brongniart were studying the fossils and rocks around Paris at about the time Smith was mapping the geology of England and Wales. They produced a map that was inferior to Smith's, but because Cuvier, trained in anatomy, viewed his fossils as a biologist, he discovered an important fact that increased their usefulness: Not only do the assemblages differ from bed to bed, but the changes are systematic; those from the oldest strata are least like animals now living, and successively younger fossils increasingly resemble modern forms. This was clear evidence that some older species had become extinct and that new ones had appeared, a view generally rejected before Cuvier's time.

These discoveries helped set the stage for the concept of organic evolution. Darwin's *Origin of Species* was published in 1859, less than 50 years after the work of Smith, Cuvier, and Brongniart. From that time forward fossils have provided a major part of the evidence documenting the fact of evolution, although they help but little in explaining how and why it takes place.

The work of tens of thousands of geologists, probing rocks on every continent during the more than 150 years that have elapsed since the time of Smith and Cuvier, has demonstrated the worldwide validity of the principles they discovered. Pieced together these findings support a second profoundly important fact drawn from the fossil record: The kinds of organic remains at successive levels in the rocks are so markedly and systematically different that a fair sample from any particular level can be put in its proper place in the succession, even if the level from which it came is not known. This grand generalization rests squarely on the Law of Superposition, of course, for it is only by superposition that we can be absolutely sure of the relative ages of fossils in the first place. The generalization finds its most frequent application in the geological rule that *like assemblages of fossils indicate like geologic ages for the rocks in which they occur.*

The Standard Geologic Column

The use of fossils to identify strata of the same age in widely separated exposures unlocked the door to the greatest single advance in the history of geology. It made possible generalized geologic mapping of areas in which the soil limits observation of bedrock to scattered outcrops, and this, in turn, led to new discoveries. In preparing the maps it was necessary to group strata into formations, and the patterns these formed on the maps revealed the existence of such geologic structures as the basin around Paris and the elongate dome south of London (Fig. 289). The discovery of such structures was of prime importance, both to the fuller utilization of mineral resources and to the understanding of regional geologic history. The attention thus directed to sedimentary rocks gave birth to the branch of geology known as *stratigraphy*—study of the character, relations, succession, variations, and fossil faunas of strata, both locally and over large areas.

Long before Smith's map was published, there had been several proposals for grouping rocks (and mountains) into major categories. Notable among these was one advanced in 1759 by Giovanni Arduino, an Italian, who recognized such "divisions of the earth's crust" as:

1. *Primary or Primitive Mountains* composed of rocks in which ore deposits frequently occur (i.e., metamorphic and igneous rocks).

2. *Secondary Mountains* built up of limestone in which ore deposits are rare but marine fossils are abundant.

3. Hills composed of gravel, sand, clay, etc.; often highly fossiliferous. Volcanic rocks.

4. Alluvial materials washed down from the mountains by streams.

It is clear from his terminology that Arduino recognized the normally decreasing relative ages of (*a*) deep-seated crystalline rocks, (*b*) hard and resistant marine sedimentary rocks, (*c*) poorly cemented clastic sediments, and (*d*) unconsolidated alluvium.

Smith and those who succeeded him vastly improved upon Arduino's subdivision of the sedimentary rocks. They found that many of the strata of England and Europe fell naturally into larger groups, called *systems*, which included several formations and were, in many places bounded by unconformites. The first of these to gain general acceptance was the *Carboniferous System*, which embraced the well-known Coal Measures and the marine Derbyshire Limestone (= Mountain Limestone) underlying them. (Smith's names can be seen in Figure 290, a copy of the table of formations he published with his 1815 map; the major units in use today are shown in Figure 291, frequent reference to which will be helpful in the discussion to follow.)

Beneath the Derbyshire Limestone, Smith mapped a group of reddish deposits as Red and Dunstone which, to distinguish them from red beds occurring above the coal, have ever since been referred to as the Old Red Sandstone. These are largely nonmarine and barren of fossils, although primitive armored fish are locally abundant in interbedded shales, especially in Scotland. Southward, in Devonshire, the Old Red Sandstone interfingers with marine strata whose fossils make possible the correlation of both rocks with contemporaneous strata around the world. Since about 1840 these strata beneath the Carboniferous have been known as the *Devonian System*.

By the late 1800's three more systems had been established below the Devonian: in descending order, *Silurian*, *Ordovician*, and *Cambrian*. All three were defined in and near Wales and draw their names from the history of that region (see Fig. 291). The folding and faulting of these rocks is fairly complex in this area, they contain many local unconformities, and in some places they are mildly metamorphosed into such rocks as slate. The structure and stratigraphic succession were worked out rather slowly. For several decades only two systems—Cambrian and Silurian—were recognized, and the boundary between the two was in dispute. Finally, upon the recognition that there were really three distinct suites of fossils involved, the name Ordovician was proposed in 1879. Subsequent work in all parts of the world has shown that no strata beneath the Cambrian contain actual fossil remains of animals with hard parts, although some

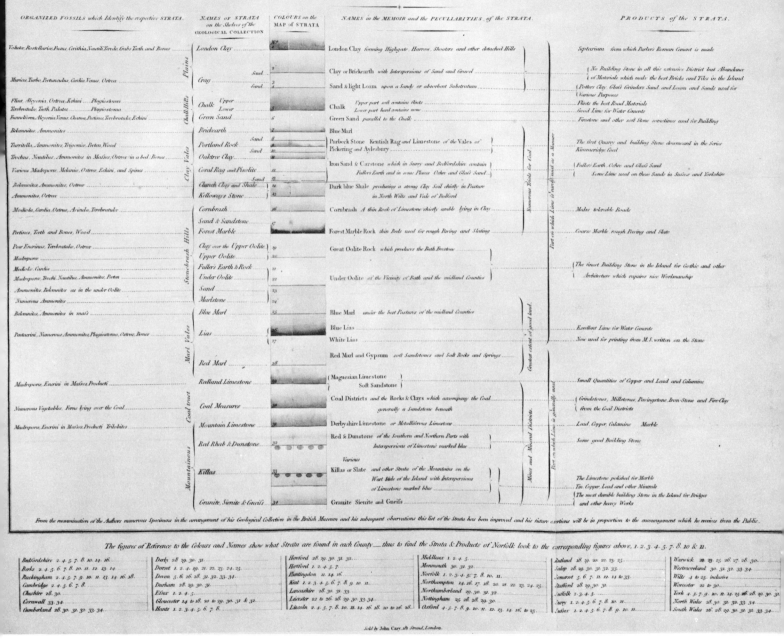

Fig. 290. Table of mapped formations with their associated fossils, characteristics, and uses, from the *Memoir* prepared by William Smith to accompany his geologic map published in 1815.

THE STANDARD GEOLOGIC COLUMN AND TIME SCALE

(As recognized by the United States Geological Survey)

Eras	*Systems (of Rocks)* *Periods (of Time)*	*Series (of Rocks)* *Epochs (of Time)*
CENOZOIC *(Recent life)*	**QUATERNARY** (An addition to the old tripartite 18th-century classification)	**RECENT** **PLEISTOCENE** *(Most recent)*
	TERTIARY (Third, from the 18th-century classification)	**PLIOCENE** *(Very recent)* **MIOCENE** *(Moderately recent)* **OLIGOCENE** *(Slightly recent)* **EOCENE** *(Dawn of the recent)* **PALEOCENE** *(Early dawn of the recent)*
MESOZOIC *(Middle life)*	**CRETACEOUS** *(Chalk)* **JURASSIC** (Jura Mts., Europe) **TRIASSIC** (From tripartite division in Germany)	

PALEOZOIC
(Ancient life)
	PERMIAN
	(Perm, a province in Russia)
	CARBONIFEROUS SYSTEMS
	(From abundance of coal in these rocks)
	    PENNSYLVANIAN
	    MISSISSIPPIAN
	DEVONIAN
	(Devonshire, England)
	SILURIAN
	(An ancient British tribe, the Silures)
	ORDOVICIAN
	(An ancient British tribe, the Ordovices)
	CAMBRIAN
	(Roman name for Wales)
		Note: Although local names are frequently used for series and epochs within Mesozoic and older strata, the generally recognized subdivisions are simply *Lower, Middle,* and *Upper* series *(Early, Middle,* and *Late* epochs) of the respective systems. Thus, where it is convenient to talk of smaller units, we may speak of the Upper Cretaceous series (of strata) which accumulated during the Late Cretaceous epoch (of time).

PRECAMBRIAN	Many local systems and series are recognized but no well-established worldwide classification has yet been attained.	

Fig. 291. The Standard Geologic Column. As explained on page 300, the same terms are used, as indicated, both for the rock sequences and for the time intervals during which these strata accumulated. Definitions in italic are from the Greek. (*Modified from Gilluly, Waters, and Woodford,* Principles of Geology, *W. H. Freeman and Company, 1959.*)

Precambrian cherts and limestones contain plant structures, and what may be the tracks, burrows, and impressions of primitive organisms are known from Precambrian sedimentary rocks in several parts of the world. The base of the Cambrian is thus everywhere drawn below the lowest actual remains of animals, among which such trilobites as *Olenellus* are characteristic.

Smith's Magnesian (or Redland) Limestone, lying next above the Carboniferous, is part of the *Permian System*, named in 1841 for the province of Perm along the west base of the Urals, where a more nearly complete sequence of fossiliferous marine strata occupies a similar stratigraphic position.

The next three systems above the Permian are all derived from studies of western European rocks. The *Triassic*, which overlies the Permian and includes the upper or New Red Sandstone of England, takes its name from the fact that in Germany it consists of two series of red beds separated by a dolomitic limestone. Although this threefold division is only local and more useful sequences of these rocks have been found elsewhere, the name persists on the very good grounds of priority and universal acceptance. Overlying this is the *Jurassic System*, named for the Jura Mountains along the French-Swiss border, in which these post-Triassic strata are richly fossiliferous limestones and shales. Jurassic rocks are widespread over southwestern Europe; in England they include the strata in which Smith first discovered the association of certain fossils with specific beds. In his "Geological Table," 13 of his mappable units (White Lias through the upper Blue Marl) now belong to the Jurassic System, whereas only two are Carboniferous and one Devonian. This many formations could be distinguished because the Jurassic of England consists of many relatively thin units, nearly all of them abundantly fossiliferous. Prominent in the strata overlying these is The Chalk, a massive deposit that is exposed in the white cliffs of the Dover Strait and may be seen in many other places in southeastern England. The upper part of The Chalk contains abundant flint nodules, which were first mined in the Stone Age. The chalk itself is a porous white deposit of almost pure calcium carbonate, much of it derived from the minute shells of microorganisms. From this distinctive variety of limestone comes the name of the system that includes it, the *Cretaceous* (Latin: *creta*, chalk).

As with human history, the record improves as we approach the present. The strata on top of the Cretaceous, up to and including modern alluvium, are divided into seven groups, even though the total time they represent is probably less than that of the Cretaceous. The first five of these (Fig. 291) are commonly referred to as the *Tertiary System* in the United States, the term having also been used by Arduino, and by extension of his categories the last two are considered the *Quaternary*.

Most of these seven youngest groups of strata were originally defined in terms of the progressive increases in the proportion of their fossils which are indistinguishable from living species. Strata were Eocene, for example, if 1 to 5% of their molluscan fossils consisted of species living today, or Pleistocene if over 90% were modern forms. Although useful in a general way, such definitions were not precise enough to permit determination of actual contacts in many areas, and international effort is now being directed toward finding the sequences of strata which, taken together, embody the best record of post-Cretaceous time.

The *Standard Geologic Column* (Fig. 291) is this master sequence of systems that makes possible worldwide comparisons of strata. Note that:

1. It represents actual rocks, real strata that can be examined in their type areas, such as Devonshire or Perm.

2. Since there is no known place on earth where sedimentation has been continuous throughout geologic time (if there were we could see only the top layer, and would probably have to go under water to do so), the Standard Column has been pieced together from sequences in different places.

3. The basic question of how well these pieces fit together has been under study ever since they were originally defined. In general, since in many of the original localities the systems were separated by unconformities, subsequent work has tended to fill the gaps between systems through the discovery of more nearly complete sequences as well as more richly fossiliferous strata.

4. Such refinement of the Standard Column raises the question: How precise can correlation by fossils become? In the present "state of the art" the answer is that correlation by comparison of fossil faunas is usually reliable for groups of strata as small as one-third to one-twelfth of a system (varying from system to system); rocks representing such large sections of the Standard Column (and periods of geologic time) as these can be placed in their proper order and correctly related to other such sequences with a high degree of consistency. Very small subdivisions of a stratigraphic sequence, such as single beds or tongues of sedimentary rock, cannot usually be correlated by this means. A major reason for this limitation, arising out of shifting environments, is discussed on page 302 in connection with facies fossils.

5. As the Standard Column grows more nearly complete we increasingly find places where sedimentation was continuous from one system into the next. In these, lacking abrupt change in the rocks and often faced with fossils intermediate between those characteristic of the two systems, we must make an arbitrary decision. These boundary problems, which can also occur within the systems, emphasize the fact that in the last analysis the Standard Column is only a convention, an arbitrary man-made set of subdivisions of what we may someday be able to prove is a continuous record of sedimentation enclosing samples of continuously evolving life on earth. It is extremely useful, and will remain so as long as geologists all over the world agree to draw the boundaries at the same places according to the same criteria. Maintaining this cooperation is a major purpose of the International Geological Congresses held every fourth year in different parts of the world.

6. In view of the continuing effort to perfect the Standard Column, it is not surprising that there are slight differences of opinion about some of its details. Some of these arise from local conditions. Almost everywhere, for example, the Carboniferous System is easily divisible into two parts, the upper of which contains nearly all the important coal deposits. In the United States these parts are so widespread and distinctive that they have each been given system rank —the lower called *Mississippian* and the upper *Pennsylvanian*. The U.S. Geological Survey cautiously waited almost fifty years, following the proposal of these names in 1906, before officially adopting them as full-fledged systems; whether or not they will ever be much used outside North America remains to be seen.

7. Continuing exploration and study have turned up sequences of strata that are superior to those in the original type area for some of the systems—superior in the sense that they are more nearly complete and more fossiliferous, and also sufficiently studied and known that geologists all over the world can make sound comparisons with them. Most geologists would probably agree, for example, that at the present time the best Cambrian sequence is near Oslo, the best Devonian in Belgium, and the best Triassic in the eastern Alps. On the other hand, the finest known examples of Permian, Jurassic, and Cretaceous rocks are still those in or close to the areas in which they were first defined.

The Geologic Time Scale and Fossil Record of Life

Although the Standard Geologic Column represents a pile of sedimentary rocks, it is also the record of a definite amount of time—the sum of all the separate times required for the deposition of the strata, and the pauses between them, in each of the systems. Each of these lengths of time is a geologic *period*, with the same name as the corresponding system. Thus the Jurassic Period is the part of geologic time during which the Jurassic System accumulated. As indicated in Figure 291, the periods are grouped into *eras* and may be subdivided into *epochs*, the whole scheme constituting the *Geologic Time Scale*. Only rocks and their material inclusions, such as fossils, belong in a system; events, such as submergence of the land, are assigned to a period or other appropriate unit of time.

Note that the Time Scale is derived from the Standard Column which, although it tells us nothing about duration, has proved to be the most useful tool we have for placing widely separated geologic events in their proper order.

William Smith at first used fossils simply as distinctive ingredients of the rocks, almost as one might give attention to unusual pebbles. The fossils proved more useful than pebbles, however, because in any thick sequence of strata the same assemblage never recurs. (Minor exceptions to this may occur locally in similar beds deposited within a relatively short time span.)

At the heart of this generalization is the fact that each known species of animal or plant has a fairly definite *geologic range* in time; it first appears in the geologic record in strata of a particular geologic age and, if no longer living, disappears at a younger horizon. A century and a half of collecting and comparing has established the ranges of thousands of fossil species. This accumulated knowledge is the chief tool in using fossils for long-range correlation; beds containing horse teeth are always above those with dinosaur bones and the two fossils are never mixed, so we know the relative ages of strata containing either, no matter where they occur.

The geologic ranges of some major groups of fossil and living plants and animals are plotted vertically in Figure 292. The horizontal lines are spaced to agree with present estimates of the actual durations of the periods, based upon measurements of radioactivity (page 310). The pinching and swelling of the columns represents in a very general way the changing relative abundance and diversification of the different groups as estimated from known fossil remains.

Not all the groups are of the same rank; they have been selected partly to illustrate variety in geologic ranges. Note, for example, that trilobites are found only in Paleozoic (Cambrian through Permian) strata. The genus *Glossopleura*, however, is limited to mid-Cambrian time, which is ground for considering at least part of the Bright Angel shale to be Middle Cambrian. This age assignment is supported by the presence of the brachiopod genera *Lingulella* and *Paterina* whose ranges overlap that of *Glossopleura*. Among younger brachiopods the productids are limited to Carboniferous and Permian time, and *Derbya* and *Meekella* are known only from the late Carboniferous (Pennsylvanian) and Permian. These fossils alone do not place the Kaibab limestone very closely in the Time Scale, but the presence of the nautiloid *Domatoceras* helps a little by eliminating a late Permian age. With Cambrian at the bottom and Permian at the top, it is now apparent why we called the horizontal beds in the Grand Canyon the "Paleozoic Strata" (page 249 and Fig. 238).

Some groups of fossils are more useful than others for dating rocks. The fusulines, for example, underwent rapid evolution and were rather widespread during Pennsylvanian and Permian time; more than 30 varieties are known from Permian rocks alone. Thus the presence of any fusuline marks a rock as Carboniferous or Permian, and an assemblage of species often allows very accurate correlation of rocks containing these particular fossils. Sponges, on the other hand, are much less useful because as a group they exhibit relatively little diversification

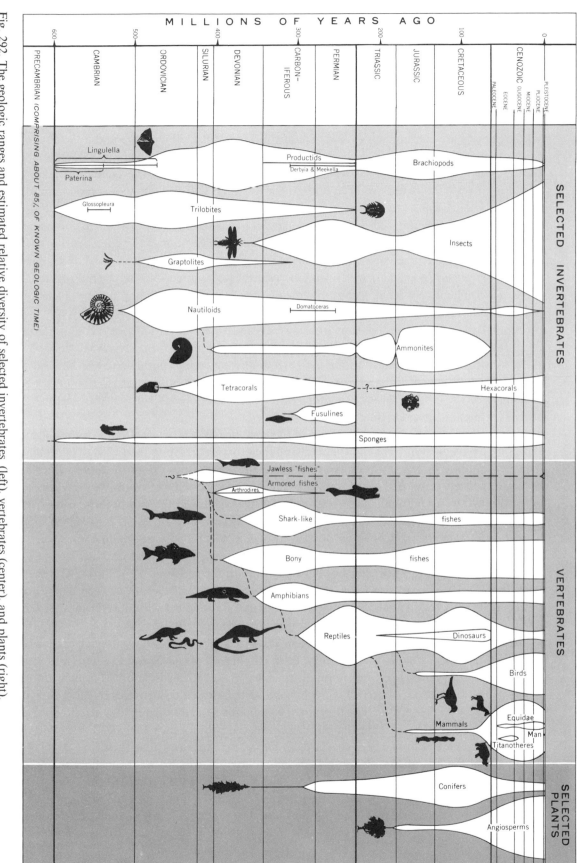

MILLIONS OF YEARS AGO

PRECAMBRIAN (COMPRISING ABOUT 85% OF KNOWN GEOLOGIC TIME)

CAMBRIAN | ORDOVICIAN | SILURIAN | DEVONIAN | CARBON-IFEROUS | PERMIAN | TRIASSIC | JURASSIC | CRETACEOUS | CENOZOIC | PLEISTOCENE

PALEOCENE · EOCENE · OLIGOCENE · MIOCENE · PLIOCENE

SELECTED INVERTEBRATES

Lingulella
Paterina
Productids
Derbyia & Meekella
Brachiopods

Glossopleura
Trilobites

Insects

Graptolites

Nautiloids
Domatoceras

Ammonites

Tetracorals
?
Hexacorals

Fusulines

Sponges

VERTEBRATES

Jawless "fishes"
Armored fishes
Arthrodires
Shark-like   fishes
Bony   fishes
Amphibians
Reptiles
Dinosaurs
Birds
Mammals   Equidae
                 Man
Titanotheres

SELECTED PLANTS

Conifers

Angiosperms

Fig. 292. The geologic ranges and estimated relative diversity of selected invertebrates (left), vertebrates (center), and plants (right).

in proportion to their long range; one sponge-like group, however, is a world-wide guide to early Cambrian strata.

The vertebrates, and two representatives of the plant kingdom, are included to round out this abbreviated view of life through time. Note that the arthrodires (primitive armored fish) occur only in Devonian rocks. This is one of several reasons for assigning the Temple Butte limestone to the Devonian, which in turn gives us another reason for believing that the disconformity beneath it represents a greater time gap than the one above; the fossils show that two whole periods are missing immediately below the Temple Butte limestone and none above.

Note that the mammals have been abundant and varied for only the past 100 million years, during which many distinctive groups, such as the Titanotheres, have come and gone; that modern horses (Equidae) have been around for less than half the range of the mammals; and that man is so recent as to be almost unplottable on this chart.

We observe today that sediments accumulate in a wide variety of environments ranging from dunes, floodplains, and deltas to estuaries, coral reefs, and the floor of the open ocean—to name but a few. Each of these environments tends to account for a distinctive type of sediment, a *sedimentary facies* (page 273) which may be referred to as "reef facies," "sandy facies," etc. Each environment also tends to have its own *biofacies*, a distinctive population of plants and animals which is contemporaneous with different populations inhabiting the others.

Similarly, in ancient sedimentary rocks we do not expect to find identical fossils entombed in all the sediments of a given age. Just as bones of desert animals buried on the floodplain of the lower Colorado River near Yuma, Arizona, today are contemporaneous with shells derived from the rich marine life inhabiting the Gulf of California, so footprints in the nonmarine Supai red sands of the eastern Grand Canyon can be contemporaneous with marine shells in limestones formed at the same time in southern Nevada.

Thus we may also distinguish biofacies of the past, or *facies fossils*, meaning assemblages that represent a particular ancient environment and therefore a correspondingly limited distribution. The reef-building algae and corals and their associates, for example, constitute a biofacies; every detail of their occurrence as fossils indicates that the requirements of warm, shallow, clear, and aerated seawater which limit these organisms to special sites in today's oceans, also governed their fossil counterparts in the past. (A case in point is described on pages 318–321.)

But environments of deposition do not always exert such a strong influence. At the other extreme are forms of marine life that float or swim near the surface and are thus likely to have very wide distribution. Their shells or hard parts fall to the sea floor regardless of the environment there and may be incorporated in the bottom sediment regardless of its type. These are the opposite of facies fossils; they are obviously more useful for world-wide correlations, but tell us less about the environment of accumulation of the enclosing rocks.

As hinted on an earlier page, the environmental influence on fossil assemblages is one factor which limits their precision in correlation. In the Cambrian of the Grand Canyon, for example, *Lingulella* and *Paterina* are found only in the silty and shaly beds, most of which are in the Bright Angel formation. Note in Figure 269, where the known occurrences of these two fossils are indicated by L's and P's, that they are distributed vertically throughout the uninterrupted shale of the eastern Canyon but that in the central and western part *Lingulella* is limited to the two shale levels, 300 to 400 feet apart. We interpret this to mean that these two brachiopods were most at home in the sandy muds that accumulated a little farther from shore than did the coarse sands of the Tapeats formation. As the shoreline advanced eastward their environment shifted, and they moved with it. After they reached the region that is now the eastern part of the Canyon, an increase in the supply of mud produced a tongue of shale reaching far to the west, and again they moved with their facies, with the result that in the western Canyon *Lingulella*, at least, *re*appeared 400 feet above—and long after—its first occurrence. Clearly these fossils alone cannot be used to make accurate correlations within these several hundred feet of strata. If, instead of being in the Grand Canyon where the strata are completely exposed, these rocks were in the eastern United States or England, where only scattered outcrops are available, such a facies influence would probably go undetected and it would be possible to make the mistake of concluding that the two westernmost occurrences of *Lingulella* were of the same age; the error would be supported by the fact that both groups of *Lingulella* occur in a thin shale between two limestones.

Similarly, if in response to shifting environments, a facies fauna arrived in one place some time after it left another, we might, in the absence of other information, mistakenly conclude that the two similar fossil assemblages were contemporaneous and erroneously assign the rocks to the same place in the Geologic Column. Errors of this kind are likely to be reduced by the presence of fossils from swimming or floating animals (such as *Domatoceras*) that are relatively insensitive to facies.

One of the most important biofacial problems in geology is that of establishing adequate correlations between the marine and nonmarine faunal successions. It is evident in Figure 292 that the amount and complexity of life on land has increased with the passage of time. In Mesozoic and Cenozoic (i.e., Triassic to present) rocks, particularly, it has been possible to work out successions of reptiles and mammals that are as important in correlating land-laid sediments as the classic marine faunas are in dealing with the more widespread marine sediments. But there are as yet too few places where the two successions can be related to each other—places where we can be sure that this assemblage of bones and teeth is contemporaneous with that assemblage of shells. This gives special importance to places where, for example, bone-bearing river deposits of any age can be found interfingering with shell-bearing marine strata.

The Duration of Geologic Time

Let us turn now from the ordering of geologic events to the question of their duration.

Until recently the earth's history was thought to be rather short. Many Christians believed that the earth was created in 4004 B.C., and most people envisioned the time available as so short that mountains and valleys, if they were not original earth features, could be explained only as the products of great convulsions, most of which must have taken place just beyond the reach of man's memory and records. When, however, in the eighteenth century, men began to realize that valleys are the work of the streams that flow in them, that many mountains are composed of rocks derived from older sediments (necessarily accumulated in low places), and that most of these sediments accumulated extremely slowly, the immensity of geologic time gradually became apparent.

Serious efforts to measure the length of geologic time began in the late 1800's. Lord Kelvin, assuming that the earth's high internal temperatures are relics of an originally molten state, used observed subsurface temperatures and rates of surface heat loss to reach the conclusion that the earth was between 20 and 40 million years old. This plausible method became obsolete a few years later when through the discovery of radioactivity it was found that heat is generated in the earth's crust and that Kelvin's basic assumption was thus invalid. An independent approach, to the age of the oceans, was first made by Joly, who divided the total amount of sodium in their waters by the amount annually delivered to them by rivers. Both of these are measurable quantities, within reasonable limits, and he obtained an answer of about 90 million years. But it is only a hypothesis that the oceans were initially salt-free, and the assumption that today's rivers are typical of all time is certainly unjustified; there have been many long periods in the geologic past when the continents were much lower and more extensively flooded than at present, and during these runoff must have been greatly reduced. Also, of course, there are huge bodies of rock salt on the lands (see page 372); these represent sodium withdrawn from the oceans, and even if allowance is made for those that are known, there are probably others as yet undiscovered. These factors and others—such as the return to the oceans of sodium from old marine deposits, and direct additions from submarine volcanism—suggest that Joly's figure is too small by some unknown amount.

Many attempts were also made to determine the time represented by given thicknesses of sedimentary rock by using estimates of their rate of accumulation. It was reasoned that if, in a given area, an average of one inch of sediment X is now deposited each year, then 100 feet of X probably took 1,200 years to accumulate. But observed rates of sedimentation range from almost unmeasurably small fractions of an inch per century to many feet per hour and make it almost impossible to estimate the average for any large deposit. Furthermore, as pointed out on page 70, the very existence of stratification proves that the rate of accumulation fluctuates, and the surfaces between successive strata probably often record pauses longer than the time required to deposit any of the beds. Unfortunately most sediments do not contain reliable clues to how fast they were deposited—or to the duration of the intervals between layers. Perhaps the most interesting byproduct of these efforts has been a number of estimates of the maximum thicknesses of the systems; the grand total for the whole Geologic Column, using maxima from all over the world, is about 400,000 feet or 80 vertical miles—and still growing.

There are a few sequences of strata—notably some Pleistocene lake deposits in middle-latitude North America and Europe, some Eocene shales in Wyoming, and some Permian evaporites in Texas—whose thin cyclic laminations are believed to represent seasonal variations in accumulation. An example of such annual layers, known as varves (page 70), from the deposits of a Pleistocene lake is shown in Figure 293. The light layers are believed to have been deposited in the more agitated waters of summer, and the dark ones to have accumulated under quieter conditions when the lake surface and inflowing streams were frozen in winter. On this

Fig. 293. Seven years' worth of Pleistocene varves, near Seattle, Wash. The total thickness shown is about 5 inches. (*Photo courtesy of J. Hoover Mackin.*)

basis 7 years are represented by the 5-inch thickness shown here. Variations from year to year often permit sequences of varves to be matched from one deposit to another. However, like the analogous tree rings, these sequences are extremely limited from the geological point of view because they represent such short time spans—rarely more than a few hundred years at any one locality—and such special, localized conditions. Also, at some localities there is un-certainty about the annual origin of the layers.

Radioactivity has provided geology with its first and only means of measuring the duration of long periods of geologic time in years. The techniques were all originated in the present century and are being continually refined, as new methods and instruments are developed in the Atomic Age.

A *radioactive* element is one which spontaneously radiates energy, commonly in the form of *alpha particles* (helium nuclei) or *beta particles* (electrons); both are generally accompanied by *gamma rays* (high frequency X-rays). These penetrating emissions come from the nucleus of the atom and are independent of any chemical combination into which the element may enter, because chemical activity involves only the concentric shells of orbiting electrons outside the nucleus.

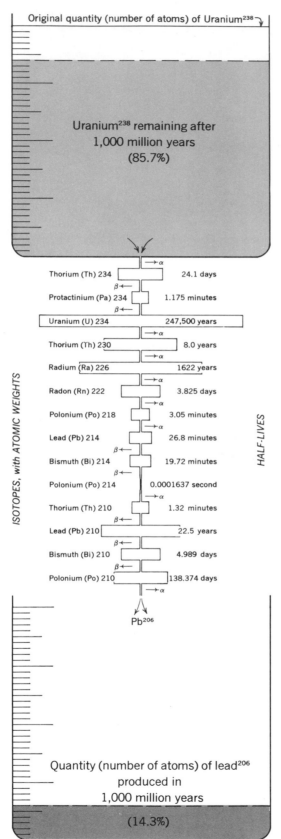

Original quantity (number of atoms) of Uranium²³⁸

Uranium²³⁸ remaining after
1,000 million years
(85.7%)

ISOTOPES, with ATOMIC WEIGHTS

HALF-LIVES

	→α	
Thorium (Th) 234		24.1 days
β←		
Protactinium (Pa) 234		1.175 minutes
β←		
Uranium (U) 234		247,500 years
→α		
Thorium (Th) 230		8.0 years
→α		
Radium (Ra) 226		1622 years
→α		
Radon (Rn) 222		3.825 days
→α		
Polonium (Po) 218		3.05 minutes
→α		
Lead (Pb) 214		26.8 minutes
β←		
Bismuth (Bi) 214		19.72 minutes
β←		
Polonium (Po) 214		0.0001637 second
→α		
Thorium (Th) 210		1.32 minutes
β←		
Lead (Pb) 210		22.5 years
β←		
Bismuth (Bi) 210		4.989 days
β←		
Polonium (Po) 210		138.374 days
→α		

Pb²⁰⁶

Quantity (number of atoms) of lead²⁰⁶
produced in
1,000 million years

(14.3%)

22
GEOLOGIC
TIME

But energy cannot be produced out of nothing. In this case the price is the loss from the nucleus of the alpha or beta particles whose escape alter its atomic weight or charge, or both, thus transforming it into a different element. The first transformation of uranium of atomic weight 238 (i.e., U^{238}), produces a second or daughter element that is also radioactive, this in turn produces still another, and finally, at the end of 15 such steps, a stable (nonradioactive) descendent is reached in the form of lead with atomic weight 206 (Pb^{206}). The process resulting in these changes is called *radioactive disintegration* or *decay*.

Although it is impossible to predict when any particular atom of a radioactive element will disintegrate, there are such enormous numbers of them that statistical treatment gives very reliable results, and therefore the fraction of them that will disintegrate in a given length of time can be quite accurately determined. The rate of decay is commonly expressed as the *half-life*—the time required for half of the nuclei to disintegrate. The situation is analogous to that in a human population numbering hundreds of millions; it is impossible to say when any individual will die, but after keeping records for a few years we can determine the death *rate* rather accurately and can therefore predict how long it will be until only half the individuals in the present population are still living. Carbon¹⁴, for example, has a half-life of about 5,600 years;

Fig. 294
Diagrammatic representation of U^{238} disintegration series, visualized as the slow draining of changing atoms from one container to another in hour-glass fashion. The relative quantities of the intermediate daughter elements are proportional to their half-lives, indicated at the right, but the range of these values is so enormous that their spread has been reduced by a factor of 10 billion to make possible the graphic suggestion of quantities used here.

any number of C^{14} atoms, protected from outside gains or losses, will be reduced by half in this period. In the next 5,600 years half of the remainder will disintegrate. Thus with the passage of each 5,600 years the C^{14} atoms that remain will become successively one-half, one-fourth, one-eighth, one-sixteenth, and so on, of the original number.

Each daughter element in a radioactive disintegration series has its own half-life. Some half-lives are measured in thousandths of a second, some in minutes, others in days, in years, or in millions of years (Fig. 294). When U^{238} is newly incorporated in a mineral (such as pitchblende), the atoms of the first daughter element (thorium of atomic weight 234, or Th^{234}) accumulate until as many of them are disintegrating as are forming. At this point a secular equilibrium is reached in which, on the average, each atom added to the Th^{234} is balanced by one that disintegrates. This state fixes the quantity of this daughter that can exist in this system at a definite fraction of the number of atoms of the parent. This fraction is determined solely by the decay rate, or half-life, of the daughter. The same applies to each successive descendent, except the stable end member Pb^{206}. Because of the great differences in decay rates, the relative amounts of the various daughter elements are far from equal; those that are produced rapidly and decay slowly occur in large quantities, those born of elements with long half-lives but which themselves disintegrate rapidly exist only in minute amounts. This is shown diagrammatically, with enormous reduction of the differences, in Figure 294. Note that since the decay rates are fixed, the relative number of atoms in each of the daughter elements is also fixed once equilibrium, the steady state, has been established throughout the series (which requires about a million years). The actual amount of each depends almost entirely on the quantity of the original parent; a large amount of uranium sets up a big chain, a small amount a little one. Regardless of size, however, with the passage of time the uranium, being unreplenished, is gradually depleted while the quantity of lead, which does not decay, grows larger. During this process the change in quantity of the intermediate elements is negligible for most purposes because U^{238} disintegrates so very slowly (90% remains after about 685 million years).

It is this increase in the ratio of lead to uranium as time passes that is of greatest interest to geologists. All minerals that contain U^{238} also contain U^{235} and frequently Th^{232}, each of which stands at the head of a similar but not identical disintegration series. Thus, in this group of elements, we have the following to work with:

Radioactive Element	Stable End-product	Half-life in Years
Thorium^{232}	Lead^{208}	13,900,000,000
Uranium^{238}	Lead^{206}	4,510,000,000
Uranium^{235}	Lead^{207}	713,000,000

The two varieties of uranium and the three of lead are *isotopes*, which differ from each other only in atomic weight; although indistinguishable by ordinary chemical methods they can be separated by delicate instruments sensitive to the slight differences in mass of their nuclei. Note that not only will the three ratios Pb^{208}/Th^{232}, Pb^{206}/U^{238}, and Pb^{207}/U^{235} increase with time, but also the ratio Pb^{207}/Pb^{206} because Pb^{207} is produced about six times as fast as Pb^{206} (note their half-lives; the relative amounts of the parents, U^{235} and U^{238}, seem to be constant wherever uranium occurs).

Geologic ages are generally calculated from several such ratios, which provides a kind of internal check on the results. The importance of precautions of this kind stems from the failure of most natural specimens to meet all the prerequisites for a simple and direct age determination. These prerequisites, the ideal conditions, are:

1. The mineral must have contained none of the disintegration products of uranium or thorium, especially lead, when it originally crystallized.

2. There must have been no gain or loss of any member of the series from or to the surrounding rock during the life of the mineral.

3. The decay rate must have been constant throughout the life of the mineral, regardless

not only of its chemical state, but also of all extremes of temperature and pressure to which it has been subjected in the earth's crust.

4. The decay rate, or half-life, of the parent uranium or thorium must be accurately known.

5. Our measurements of the amounts of the different elements and isotopes in the mineral today must be accurate.

The first two of these conditions are probably never fully met. The third is probably always true, and the uncertainties arising from the fourth and fifth are determinable and relatively small. The ages we obtain are therefore only as good as our ability to make corrections.

Concerning the first two conditions: It would simplify matters if we could be absolutely sure that all the disintegration products of uranium and thorium always escape from their parents while the latter are molten or in solution, but are trapped in the crystal lattice when the parent elements become part of a solid mineral. In many cases this seems to have been essentially true, but in some (those involving recrystallization during metamorphism, for example) there is uncertainty about when the system became closed and the ratios began having time significance (i.e., when the "clock" started running) and therefore about exactly what geologic event is being dated. Usually some common lead is also present, but fortunately it contains a distinctive isotope of nonradiogenic origin (Pb^{204}) and can thus be distinguished from the accumulating radiogenic lead.

The problem of leakage during the life of the solid crystal is more difficult. One of the daughter elements of U^{238}, for example, is radon (Rn^{222}), a gas that can escape from the crystal lattice, especially at high temperatures. Such loss penalizes the production of succeeding daughter elements, which results in less lead at the end and therefore too small an age by the Pb^{206}/U^{238} ratio. (This leakage can be measured in the laboratory, but the conditions do not duplicate all those to which the mineral has been subjected.) However, since Rn^{222} leakage decreases the production of Pb^{206} but not of Pb^{207} (in whose series it does not occur) the age as determined from the Pb^{207}/Pb^{206} ratio is too great. Thus when for the same specimen the age given by this ratio is considerably higher than the one given by Pb^{206}/U^{238}, radon leakage may have taken place. The principal check we have in this method is the internal consistency of the ages based on the three lead ratios: Pb^{206}/U^{238}, Pb^{207}/U^{235}, and Pb^{207}/Pb^{206}.

Concerning the third and fourth factors: Decay rates, or half-lives, have been measured many times by different people in different laboratories and are considered well established for most of the elements; any errors arising from these figures are probably insignificant compared to other uncertainties. Furthermore, these rates seem to be truly constant; no extremes of temperature or pressure which man has yet been able to impose on the radioactive elements has produced any significant change in their decay rates.

Three other methods of dating are based on the following radioactive transformations:

Radioactive Isotope (Natural Abundance)	Stable Isotope	Half-life in Years
Rubidium^{87} (27.8%)	Strontium^{87}	50,000,000,000
Potassium^{40} (.012%)	Argon^{40}	1,470,000,000
Carbon^{14}	Nitrogen^{14}	5,600

In other words, about one-fourth (27.8%) of the atoms of natural rubidium are radioactive and by emission of an electron become atoms of strontium, the rate of decay being extremely slow. About one atom in 8,400 of potassium is radioactive; the nucleus captures an electron from its own innermost shell and becomes argon, the accompanying adjustments in the orbiting electrons giving rise to feeble X-rays. The strontium and argon methods have certain advantages: they can be used on more common minerals (mica, for example) than can the lead-uranium technique, and there are no long chains of daughter elements with their associated complications. Offsetting these somewhat are the disadvantages of very weak radiation, very small quantities, and the fact that argon is a gas and is not retained equally well by all minerals.

Carbon[14], the "sweep secondhand" of the radioactive clock, is continuously produced through the bombardment of nitrogen in the atmosphere by cosmic rays from outer space. As a result, a small and apparently uniform fraction of the carbon compounds in air and water, and therefore in all living things, is radioactive C^{14} (about one atom in 10,000 billion). When an organism dies it stops exchanging air and water with its surroundings and the residual C^{14} decays with a half-life of about 5,600 years, the amount remaining at any time being a measure of the interval since death. This makes it possible to date organic substances such as wood (including charcoal), bone, seeds, and shells back to about 40,000 years ago, beyond which there is so little C^{14} left that reliable results are difficult to obtain. The validity of the ages obtained by this method depends on the assumption that the total C^{14} available to plants and animals has not changed during the past 40,000 years. This assumption seems to be safe for about 5,000 years before about 1870, an interval within which there are numerous archeological dates that can be used as checks. Earlier dates are consistent but not rigorously provable. Those from about 1870 to 1954 show the effect of "dead" carbon added to the atmosphere by burning coal and oil, and those after 1954 have to be corrected for the sharp rise in C^{14} resulting from atomic-bomb tests.

With this variety of radioactive elements, most of which occur in several different minerals, it is often possible to obtain two or more independent ages for the same rock. These often fail to agree closely, which is another indication that the conditions enumerated above are rarely all met in any one specimen, but the discrepancies sometimes provide hints as to what happens to different elements and minerals deep in the crust.

The geological application of radiometric ages often involves difficulties of another sort. The ages we obtain almost certainly represent the time since the mineral crystallized. This information is directly useful in dating the time of solidification of igneous rocks, and in some cases for dating episodes of recrystallization in metamorphic rocks, but it tells us nothing about when a mineral became part of a sedimentary deposit. In principle we can date sediments that contain fresh interbedded lavas or tuffs, but in practice the most useful (fossiliferous) sediments are marine, and their interbedded volcanics, if any, are usually too greatly altered by chemical reaction with the seawater to be datable. For such reasons there are as yet only a few rocks whose geologic ages are accurately known that also meet the stringent requirements for accurate radiometric dating. Because of this, the scale of years along the left margin of Figure 292 is doubly uncertain, especially for pre-Cretaceous rocks.

On the positive side, although the uncertainty in radiometric age determinations is commonly plus-or-minus 10% or more, the ages usually agree well with estimates based on purely geological relationships. In western North America, for example, there are several complex batholiths whose geologic ages, determined from their relations to fossiliferous sedimentary rocks, range from probably late Jurassic through definitely mid-Cretaceous; the corresponding radiometric ages are consistent with this time span, the oldest being at least 150 million years and the youngest around 100 million. Radiometric ages have been especially useful in dealing with Precambrian rocks; in the absence of usable fossils, such "absolute" age determinations offer the only method we have of correlating rocks and events throughout the world.

Unless some of the basic assumptions about the constancy of decay rates are wrong, radiometric ages have proved the earth to be much older than most people had previously thought possible. The oldest rocks so far measured have ages close to 3.5 billion years—and most of these are geologically younger than their surroundings. On the other hand, if the crust were older than about 5.5 billion years there should be no detectable U^{235} left today. These facts, and some theoretical considerations, point to an age of *about* 4.5 billion years for the crust and an upper limit of perhaps 6 or 7 billion years for the origin of the earth as a whole.

According to the best current correlations, it would appear that the Paleozoic Era (Cambrian through Permian periods) began about 600 million years ago, the Meozoic Era (Triassic through Cretaceous periods) about 225 million years ago and the Cenozoic Era (Tertiary and younger rocks) about 65 to 70 million years ago. These three figures are worth remembering for the perspective they give to geologic histories presented only in the comparative terms of the Geologic Time Scale. From them we realize that the entire usable fossil record of life on earth, the complete Standard Column, represents only the last 15% of the probable history of the crust. The events recorded in the Paleozoic strata of the Grand Canyon, beginning with the Tapeats sandstone and ending with the Kaibab limestone, encompass about 375 million years—all included in Step Two of our first analysis. The obviously longer and more eventful part of the story, covered by Steps Three to Ten, extends from about 600 million to about 1,700 million years ago and is revealed only in the deep parts of the Canyon where most visitors never go. (The approximate 1,700-million-year date comes from radiometric determinations on granite and pegmatite intruded into, and therefore younger than, the Vishnu schist.)

V Case Histories

Examples of how a knowledge of geological principles may be used to decipher the record of past events and thus to read earth's autobiography.

23 PROCESS AND HISTORY AS GEOLOGIC TOOLS

In Part I we examined the nature and origin of the rocky materials which compose the accessible part of the earth's crust. In Parts II and III we investigated the processes, operating both within and upon the crust, that modify the internal structure and the surface shape of these materials. In Part IV we put these processes into the long perspective of geologic time, first by using the evidence of sequence of events that is built into the rocks, and then by relating them to the time scale based on radioactivity.

Parts I to IV constitute an introduction to the essence of geology—prejudiced in this case toward large, outdoor features that can be recognized in photographs. But to stop here would be like learning to fly an airplane and then never going anywhere. In Part V, therefore, we will apply these elementary principles of geology by examining a number of geologic features with the purpose of deciphering their geologic histories. The features to be examined range from small to large, obscure to well known, simple to complex. The selection has been influenced, of course, by the requirement that in each case some of the important evidence for the reconstruction of the history be reasonably discernible in photographs. It will be found, also, that in almost every example there are at least some details of the account which are attended by unsolved problems. Uncertainties and all, these histories are samples of the fruits of the geologist's labor, of what he has learned to see in the landscape that is not apparent to the casual observer, of what at the outset we termed geological insight.

To some the phrase "geologic history" may imply that there is a choice and that we have elected to take a backward look—that we are dwelling on the past when we could and perhaps should be concerned with current events. Note, however, that because of the great length of geologic time and the slowness of most geologic processes, very few events in geology can be classified as current. The effects of an occasional flood, volcanic eruption, earthquake, or landslide can be witnessed and treated as news. But we have to go well back into human prehistory to encompass any really important geologic change or to determine the full significance of almost any geologic process. If we want to study or "watch" geology in action, therefore, we cannot avoid a long look back in time. Indeed, such looks are one of the best ways we have of recognizing what is going on around us today. Awareness of geologic history gives us a running start toward understanding the modern scene. It will also help to prepare us for speculation on what this scene implies regarding processes and causes deeper in the earth (Part VI).

Fig. 295. General view of the Cascade Range, looking north-northwest from 15,500 feet.
Mt. Jefferson in left foreground; left to right in the distance, St. Helens and Hood; just to
the left and to the right of Hood, and barely discernible on the horizon, Rainier and Adams.

24   VOLCANISM

Crater Lake and the Evidence for Mount Mazama

The Cascade Range consists chiefly of a long wide ridge about 100 miles inland from the coast of Oregon and Washington. Throughout most of its length its dense forest cover further distinguishes it from the lower grassy plateaus on the east and cultivated valleys on the west. As may be seen in the view above (Fig. 295), here and there along the crest of this broad dark upland rise the lofty volcanic cones for which it is famous—Glacier Peak, Rainier, Adams, Baker, and St. Helens in Washington, Hood and Jefferson in Oregon, and Shasta and Lassen in California. All of these reach above 10,000 feet and Rainier and Shasta exceed 14,000 feet. In addition to these conspicuous giants there are many hundreds of small volcanic cones and vents, some of them associated with unweathered lavas probably not more than a few hundred years old. Indeed, Shasta blew out ash in 1786, St. Helens erupted several times between 1831 and 1854, Baker was active four times between 1843 and 1870, Lassen erupted in 1914–1917, and Rainier is still hot enough to emit a little steam from its snow-filled summit crater. Clearly, most of the height of the central and southern Cascades is the simple result of piling up volcanic rocks on an older land surface; perhaps, geologically, the process is not yet finished.

Cradled in the crest of the Cascades of southern Oregon is Crater Lake, whose deep blue waters nestle like a sunken jewel in the forested slopes that surround it. The lake is 5 to 6 miles in diameter and the crater rim rises 600 to 2,000 feet above its surface. The deepest part of the crater is almost 4,000 feet below the highest point in its rim.

Both the present land surface and the many layers of lava and ash exposed in the inner wall of the crater slope uniformly outward from the rim. That there was once a volcanic cone rising above the site of the present lake is thus beyond doubt. Indeed, a dark lava flow that once poured down a steep valley on its outer slope is now visible in cross-section at the far right in Figure 296. Let us look further into the evidence bearing on the height and shape of this former volcanic cone and consider how and when it disappeared.

The size of the base is comparable to that of the larger Cascade cones of today, approximately intermediate between Shasta and Rainier. If we restore the upslope parts of the layers over the lake, using dips comparable to those that now exist in Shasta and Rainier, a cone at least 12,000 and possibly 15,000 feet high is formed. This ancestral volcano, whose lower flanks we now see around the lake, has been named Mt. Mazama.

There is considerable evidence that Mt. Mazama supported glaciers. Two U-shaped valleys are truncated at the rim on the southeast side of the lake (nearer side of photograph in Fig. 297, and far left side of Fig. 168). These cross-profiles are very different from the youthful canyons that streams alone would produce on the flanks of a steep volcano (cf. Fig. 66). There are also terminal moraines in Munson Valley, at the extreme left in Figure 297, and glacial deposits occur with and below the flat-surfaced accumulation of pumiceous ash in the main valleys in the foreground. Clearly, Mazama not only was high enough to have glaciers but these had cut characteristic deep valleys on its flanks before the last explosive (ash-producing) eruption. Furthermore, at more than a dozen places inside the rim glacially polished and striated lava can be seen, some of it overlain by till. The lowest of these is but a few hundred feet above the lake. Glacial deposits are reported to occur in several places beneath the valley-filling lava flow on the right in Figure 296. Taken together, these features prove that Mazama supported glaciers during most of its later history—while hundreds of feet of volcanic rock were being added to its slopes. The maximum extent of the ice is suggested by the widespread occurrence of glacial deposits up to 5 or more miles—some even 17 miles—from the lake; when these were deposited Mazama must have been almost enveloped in ice. On the other hand, some beds of the last ash to be erupted seem to lie on top of the till and to be undisturbed in several valley bottoms close to the rim; if this is true the ash probably fell on the ground rather than on top of ice, which would indicate that at least some of the glaciers had melted back up the slopes before the last pumice fell.

Fig. 296. Part of the western rim of Crater Lake; Wizard Island on the left. There are two vertical dikes in the cliff, just left of center, and Llao Rock at the right is the cross-section of a dark lava flow filling a glaciated valley on the slopes of Mt. Mazama.

Fig. 297. *Upper:* General view of Crater Lake, looking northwest. The U-shaped notch in the crater rim, just left of center, owes its shape to glaciation of this valley by a tongue of ice descending the now-vanished upper slopes of Mt. Mazama. *Lower:* Drawing based on the scene above but restoring, in ghostly fashion, the summit of Mt. Mazama.

These observations enable us to reconstruct the stage given in the drawing at the left—just before the final eruption and disappearance of Mazama.

The volume of the restored portion of Mt. Mazama is about 17 cubic miles, which is thus the amount of rock to be accounted for in the disappearance of the mountain. This huge volume of rock must have either blown up or been engulfed, or both. There is precedent for all three possibilities in the histories of other volcanoes whose activities have been observed in more recent times. The best argument for explosion would be the recognition of adequate amounts of debris blanketing the nearby landscape. The majority of geologists who have studied the deposits of ash, pumice, and other fragmental material that spread as much as 600 miles northeast of the lake (note the indication of a prevailing wind) estimate their volume at not more than 12 cubic miles in their present frothy and broken state. Probably at least half of this represents new molten material whose eruption carried away the Mazama summit. If we assume that the remaining half represents pieces of Mazama, and make allowance for the fact that their volume in today's broken state is greater than it was as relatively solid mountain, then we have found something less than 6 cubic miles, or about one-third, of the missing cone. Might this be because a great deal more has been removed by erosion? Some say yes, but most think not; trees killed and charred by the last ash fall have C^{14} ages of about 6,600 years, so it was probably not longer ago than this that the summit of Mazama vanished. The narrow gully in the flat-floored valley in the center of the view at the left is probably a sample of the relatively small amount of erosion since that time. If so, allowance for post-eruption erosion still leaves the major part of Mazama unaccounted for.

It is difficult to escape the conclusion that a large part of Mazama collapsed into the crater because of withdrawal of magma from below—a phenomenon that has been observed many times, on a smaller scale, at Kilauea on Hawaii. So, even though we are not sure exactly how or why such withdrawal takes place, the lack of adequate evidence for any alternative leads most geologists to believe that most of Mt. Mazama, probably at least two-thirds of it, collapsed into its own vent.

In any case, the latest chapter in the volcanic history of Mt. Mazama was the eruption of two perfect cinder cones. One, Wizard Island (at the left in Fig. 296), is exposed, and by tree-ring counts must be at least 800 years old. The other, known as Merriam Cone, is 1,320 feet high and a mile across at the base (almost exactly the shape and dimensions of Parícutin) but is entirely submerged near the north margin of the lake.

Fig. 298. General view of the southern Guadalupe Mountains on the New Mexico-Texas boundary, 95 miles east of El Paso. Looking northeastward. Note facies changes from left to right.

25 INTERPRETATION OF SEDIMENTARY ROCKS

The Permian of the Guadalupe Mountains Area: *A Study in Sedimentary Facies*

The variations from place to place in sedimentary layers deposited at the same time can be both exasperating and illuminating. On the one hand they often make it difficult to know which strata in one place are exactly contemporaneous with a particular sequence in another place, hindering the interpretation of the geologic history by complicating the accurate assembly of its scattered parts. On the other hand, when such facies changes can be worked out, they often add enormously to our understanding of past conditions. Here is an example.

The view above (Fig. 298) is northeastward toward the Guadalupe Mountains, which lie astride the Texas-New Mexico border. El Capitan, the promontory at the right end of the white cliff, is 95 miles east of El Paso, and the Carlsbad Caverns are 35 miles farther on, in the distance at the left.

The strata displayed in the prominent cliff at the top of the mountains also form most of the low hills to right and left of the cloud shadow in the foreground. Since they are nearly horizontal in both places there must be one or more faults between them. The principal one is evident along the base of the steep slope, especially at the right; this fault and others are more prominent in Figure 300. The height of this west face of the Guadalupe Mountains is thus primarily the result of differential uplift totalling 2,000 to 4,000 feet along the north-south fault zone, while its shape is the result of weathering and erosion that have caused the escarpment to retreat.

The southeast front of the mountains, which begins behind El Capitan in Figure 298, is of a different origin. In Figure 299 El Capitan is at the left and this front extends off to the right. Here, and at many other places along this front, the strata in the canyon walls can be followed without a break into the lowland on the south; note, for example, that there is no disruption of the bedding as the resistant cliffy limestones near the center of this view are traced into the thin sandy layers in the smooth slopes at the right. At the foot of the mountain slope, where a fault might be expected, there is instead an abrupt facies change at the place where the massive limestones become small ledges that extend as short fingers into the sandstones.

The most notable geological feature of the southern Guadalupe Mountains is the limitation of the massive limestones, so prominent in El Capitan, to a relatively narrow belt along the crest of the southeastern escarpment. This belt is visible as the white cliffs that extend into the background of Figure 298 and the right distance of Figure 300, and may be seen in the walls of many of the canyons along this front. To the south, as we have seen, the massive limestone is replaced by south-dipping ledges that interfinger with weak fine-grained sandstones that extend from the base of the slope out into the valley. To the north the prominent limestone changes into more-extensive, thin, ledge-making beds; this is very clear if the large cliff in Figure 298 is followed to the far left. The same thing happens to the cliffs in the background.

In short, if we were to examine the top 1,000 feet of strata at successive points from the lowlands on the south (right in Fig. 298), up the canyons and across the upland toward the north (far left in Fig. 298), we would find that the same layers that are dark sandstone at the south become ledgy limestones in the canyons, massive limestones at the crest, and thinner beds of limestone north of the crest. Still farther north the thinning limestone beds interfinger with layers of gypsum, salt, red shale, and sandstone (shown in Fig. 233).

Each of these contrasting sedimentary facies has its own distinct fossil assemblage. The fine-grained dark sandstones in the south contain more than 220 species representing a marine fauna normal for water 1,000 to 1,500 feet deep. The massive limestones along and near the crest contain more than 250 species, less than one-third of which are also found in the dark sandstones; a few varieties of reef-building sponges and algae are very abundant in this assemblage. The thin limestones to the north contain a shallow-water marine fauna of 86 species, only 4% of

Fig. 299. Part of the southeastern escarpment of the Guadalupe Mountains. Note abrupt change in facies from resistant limestone in the mountains to weak sandstones in the lowlands.

which also occur in the massive limestones; farther northward these are reduced to a few hardy forms capable of living in abnormally salty water. Yet, despite the sharp differences between these groups of fossils, they all belong to the same part of the Permian Period—in fact they occur, as we have seen, in a demonstrably contemporaneous set of beds. These are facies faunas, corresponding to the contrasting environments in which the different sedimentary facies accumulated.

Using both the evidence provided by the fossils and the character of the rocks themselves the Permian scene may be reconstructed (Fig. 300). The massive limestone is the core of a Permian reef, one of several which partly outline an ancient marine basin in this part of Texas and New Mexico. Behind the reef, to the north, was a shallow back-reef lagoon and beyond that mud flats and stagnant ponds where evaporation and oxidation were dominant. Here the salt, gypsum, and red beds accumulated. On the seaward (southeast) side of the reef the pounding surf helped produce an accumulation of broken reef material, a sort of submarine talus of reef debris similar to that found in association with many modern reefs. At the bottom of this fore-reef slope, which forms the southeast front of the mountains, the debris abruptly interfingers with the sandstones of the basin floor. The sands, being of wholly different composition, must have come from another direction in the basin.

The restoration in the back part of the diagram includes the essential facts but with some simplification of detail. Here we can see the relation of the massive limestones, whose near end is so prominently displayed north from El Capitan, to the old reef. The rough, steep southeast front (to the right of El Capitan in Fig. 299) is simply the topographic expression of the superior resistance to erosion of the reef and its now-cemented seaward debris.

The Carlsbad Caverns are the work of groundwater, which, during much later geologic time, dissolved some of the limestones near the far end of the reef.

Fig. 300. The Guadalupe Mountains from the south. Deeply eroded fault scarp at the left, partly exhumed
reef at the right. Restoration in drawing depicts conditions during part of Permian time.

26 FAULTING

The Keystone Thrust Fault

About 20 miles west of Las Vegas, Nevada, lies the north-trending mountain front shown in Figure 301. All the exposed rocks are Paleozoic and Mesozoic strata. Since they all dip to the west (left), we should expect to find them successively younger in that direction.

The most conspicuous unit is the resistant light-colored sandstone that forms the main part of the rough cliff near the center of the view. This is the Aztec sandstone, 1,500 to 2,000 feet thick and devoid of fossils. To the west (left) the Aztec is overlain by from 8,000 to more than 22,000 feet of dark strata consisting mostly of limestones and dolomites with somewhat steeper westward dips. The lowest of these rocks, just above the Aztec sandstone, contain Cambrian fossils and the successively higher strata belong, in order, to the succeeding Paleozoic systems up through the Permian (out of sight to the left). In other words, on top of the Aztec sandstone there is a nearly complete Paleozoic sequence whose total thickness is two to five times the thickness of the corresponding strata in the Grand Canyon.

In the other direction, it can be seen that the Aztec sandstone is underlain by weaker strata that produce smoother slopes along the lowermost few hundred feet of the big cliff. These are the upper part of a sequence more than 1,500 feet thick that includes several thin limestones containing abundant Triassic fossils. Still farther to the right (beyond this view) it can be seen that these are underlain, in proper sequence, by Permian strata that include the Kaibab limestone.

In short, in the near half of this scene the oldest strata rest upon the Aztec sandstone and everything is in order above them, and the youngest fossiliferous strata lie immediately below the Aztec and everything is in order below them. In terms of the Standard Column, the bottom half of the pile overlies the middle. Either the top or the bottom of the Aztec sandstone is not a depositional contact, and the geologic age we assign to this unfossiliferous rock will depend importantly on which it is. If the Aztec is in sedimentary sequence with the rocks above it, it must be Cambrian or older, but if it is in sequence with those beneath it, it must be Triassic or younger. The choice represents a difference of about 350 million years.

Fig. 301. General view, looking north-northwestward, of the Spring Mountains in southern Nevada. The sharp boundary between light and dark rocks, along the top of the 2,000-foot cliff at the left, marks the trace of the Keystone thrust fault.

Fig. 302
Upper: Detail along the sole of the Keystone thrust; cross-bedded Jurassic(?) sandstone below, brecciated Cambrian dolomite above. *Lower:* Same, showing blocks of the underlying sandstone incorporated in the brecciated lower part of the overriding plate.

The answer is easily found in the details along the two contacts. As might be guessed from Figure 301, there is no fault at the top of the weak Triassic beds; the Aztec sandstone must, then, be younger than these strata. Because it lies below Cretaceous rocks elsewhere (outside this view), it is referred to as Jurassic(?), the query denoting the uncertainty arising from the indirect evidence for its age.

This means that along the upper contact Cambrian rocks rest on Jurassic(?) strata—a position that, in the local stratigraphic pile, is two to four miles above where Cambrian would be expected. In many places along this contact the upper few feet of the underlying sandstone is ground almost to a powder. Everywhere the lower part of the overlying Cambrian dolomite is thoroughly brecciated; the resulting jumble of large and small fragments is clearly discernible in both parts of Figure 302, even though percolating groundwaters have cemented them into a coherent mass. Here and there chunks of the underlying sandstone are incorporated in this breccia: notice the pieces of striped sandstone embedded in the dark carbonate rock in the lower part of Figure 302. Also, this contact with its zone of crushed rock does not really parallel the bedding as it seems to do in the general view; toward the west (left) the dip of the crush zone steepens and it truncates successively older beds downward in the Aztec. It is clear that this is a fault along which the dark older rocks have slid into a position on top of the younger sandstone, and since the fault surface descends to the west it is probably from that general direction that these overriding rocks came.

We conclude then that a great gently sloping fracture must have developed in the outer crust and that the Paleozoic rocks rode up this, as up a ramp, crossing successively younger strata until they arrived on top of rocks at least as young as Jurassic(?). In thus surmounting strata with a total thickness of two to four miles they must have moved several tens of miles horizontally.

Note that a knowledge of the local stratigraphy—the normal sequence of strata, and, through their fossils, where they fit into the Standard Column and Time Scale—is essential to full comprehension of this magnificent thrust fault.

Figure 303, at the right, shows a closer view of the Keystone thrust fault near where it disappears from the right background of Figure 301. The horizontally banded ridge in the middle distance, comprising the middle and lower parts of the Paleozoic sequence, is the same one that was at the far right in Figure 301. Here again the Paleozoic strata lie immediately above the thrust (D); on the ground the fault can easily be traced from the area of Figure 301 into this one.

However, whereas in the first view all the rock directly below the thrust plane was Aztec sandstone of nearly uniform thickness that lay almost parallel to the fault surface, here the rock below the thrust is of several kinds, with discordant dips. In the foreground are two tilted blocks separated by a nearly vertical fault (C), which, in the background, also offsets the thrust fault (D) and its overlying Paleozoic rocks. In each block the strata dip to the northeast. The lower part of each block consists of Aztec sandstone (16), identical to that in the first scene, and the upper part is again brecciated early Paleozoic carbonate rocks (7 ±). The contacts between these (A), where dark old rocks overlie light young ones, are thus segments of the Keystone thrust involving the same rocks in the same relationship seen in Figure 301.

The rather startling interpretation to which these observations point is that the Keystone thrust has overridden itself. The simplest explanation would be that after the Paleozoic strata had moved eastward over the Aztec sandstone (to the right along the surface marked A in the drawing) the thrust plate and overridden rocks were broken into blocks by steep faults, such as B and C. These blocks were then tilted eastward, perhaps through persistence of forces similar to those that caused the eastward thrusting. Erosion subdued the relief, and then, with renewal of thrusting, the upper plate rode across the tilted blocks (somewhat as though it were moving to the right across the surface in Figure 241). The upper plate seems to have been moving over the land surface at this stage, because stream gravels derived from distinctive rocks in the overriding plate are now found in gullies cut into the underlying rocks, some of which, furthermore, have apparently been sheared by the advance of the overriding thrust plate. (For a much smaller present-day analogue of this process, see Figure 379.)

A second matter of considerable interest is the history of movement on the steep fault C. Only renewed movement on this fault could have offset the Paleozoic strata in the background after movement on the main Keystone thrust surface (D) had ceased. The exact direction of movement on C is not known. If we assume that it was essentially vertical, then the apparent displacement (see page 104), in the foreground, measured by the offset of A, is approximately 2,600 feet, and the apparent displacement where the fault goes out of sight over the distant ridge is about 900 feet, measured by the offset of D. Here, the simplest explanation would require at least four events: (1) thrusting, (2) movement on the steep fault, resulting in tilting of the blocks, (3) renewal of thrusting, resulting in the blocks being overridden, and (4) renewal of movement on the steep fault, resulting in offset of the thrust plate and subjacent strata in the background. The larger apparent displacement in the foreground would thus record both episodes of movement on the steep fault, and the smaller displacement in the background would record only the second. Note, however, that although the conclusion that there were at least two episodes of movement on C is absolutely inescapable, we cannot be sure of the actual directions or amounts.

Fig. 303. Looking northwestward at La Madre Mountain, a spur of the Spring Mountains west of Las Vegas, Nevada. The major structural features are delineated in the drawing.

27 UNCONFORMITIES

Two Unconformities near Turtle Bay

In the foreground of the view below (Fig. 304) we are looking north along the strike of dark sediments that dip steeply to the west (left). Here and there they contain fossil ammonites of Cretaceous age. The sequence is very thick and extends well beyond this scene to both left and right.

These Cretaceous strata are conspicuously truncated by a second set of beds, dark in its lower part and almost white above. These strike northeast and dip gently toward the northwest (upper left). Abundant marine fossils, especially in the lower beds, show these rocks to be of upper Miocene age.

The geometry of this discordant relationship could be the result either of movement on a thrust fault like the one just discussed, or of an angular unconformity. The geologic histories implicit in these alternatives are drastically different, so it is important to make the correct interpretation. In this connection note that the surface of discordance follows precisely the bottom of the lowest Miocene bed. This would be almost unavoidable if the Miocene had in fact been deposited upon this surface. A fault, on the other hand, would have to have been produced after the Miocene rocks were deposited, and that it should then so accurately find and follow the base of this series is highly unlikely. Besides, the rocks show no signs of breaking or crushing along this contact. We conclude, therefore, that the discordance is an unconformity.

The scene in the photograph at the right (Fig. 305) clarified in the drawing below, is the same area viewed southwestward; the dark flat-topped mesas in the right middle distance of the first scene are in the right foreground of this one. These are highly fossiliferous marine Pliocene sediments; similar patches scattered over a wide area prove that they are remnants of a once extensive and still nearly horizontal sedimentary blanket. Since they rest on both Miocene and older strata, they could not have been deposited until erosion had removed part of the Miocene to form a land surface that exposed tilted Cretaceous as well.

The analysis of Figure 112 was based on another view of these Miocene-Pliocene relationships—and outlined only half of the crustal unrest recorded here.

Fig. 304. Steeply dipping Cretaceous strata (foreground) unconformably overlain by lighter colored Miocene beds north of Turtle Bay, Baja California.

Fig. 305. Dark Cretaceous strata (left) unconformably overlain by light Miocene beds, and both unconformably overlain by horizontal Pliocene (mesas in right foreground). Looking southwest to Turtle Bay (upper left), Baja California.

28 LANDSLIDES

Landslide in San Antonio Canyon

The upper view at the right is downstream in a mountain canyon. An old road winds along near the canyon floor and a newer and straighter one is cut into its right wall. In the center is a mound about 250 feet high which almost obstructs the canyon; the stream and older road skirt around it at the far left. Although the stream flows on a bed of bouldery gravels everywhere else, in its narrow course around this mound it has cut a trench 50 feet deep in solid granitic rock.

The picture below (Fig. 306) was taken along the winding road at the left and shows that the mound itself is composed of a jumble of angular blocks of fresh crystalline rock, some of them more than 15 feet across. There is no matrix of gravel and sand, no stratification, and almost no variety in the rock types represented. It cannot be a stream deposit. But it rests on one, visible in the right foreground of the same view, that is composed of rusty stratified sands and gravels whose pebbles, of many different lithologic types, are worn by abrasion and softened by weathering.

The brecciated rock of the mound, the hummocky shape of its upper surface, and the obvious scar high at the right in the steep canyon wall show that the mound is a landslide resting on older stream gravels. Originally its toe must have reached the opposite (left) side of the canyon. This created a dam, behind which newer stream gravels accumulated to a depth of about 200 feet, thus raising and widening the canyon floor. (The drawing represents this stage.) At first most of the stream water probably percolated through the slide and the gravels beneath it, but eventually the floor was high enough to send spring runoff through the notch around the toe. Along this route the water was not only confined in a narrow channel, but owing to the raised floor from which it started, its gradient was steepened to twice the 350 feet per mile it has above and below the obstruction. These factors help to account for the deep trench it has cut through both the slide and the underlying bedrock.

Fig. 306. Steep exposure near the toe of the slide in San Antonio Canyon; coarse angular slide debris resting on old stream gravels (lower right).

328

Fig. 307. *Upper:* Looking southwestward down San Antonio Canyon in the San Gabriel
Mountains, California. *Lower:* The same scene as it probably appeared just before
surface flow was re-established around the toe of the slide.

Fig. 308. Upslope view, southward, over the lobe of dark marble breccia spread beyond the mouth of Blackhawk Canyon on the north flank of the San Bernardino Mountains in southern California.

Limestone Debris Beyond the Mouth of Blackhawk Canyon

The view above is southward over a wrinked lobe of dark rock spread out on the floor of the Mojave Desert. In Figure 309 we are looking at the same feature in the opposite direction, northward from a position over the foothills in the background of the first scene. The loosely aggregated angular fragments of which the lobe is composed are shown in Figure 310; very few of these are as much as 10 inches across, the average is about 1 inch. There are two or three large blocks of quartzite breccia near the north end, one of them more than 25 feet across, but otherwise the fragments consist almost entirely of altered gray crystalline limestone (marble). A silt or clay matrix is conspicuously absent. The material is spread in a layer that is 30 or more feet thick near the base of the mountains; the outer margin (foreground, Fig. 308) is about 50 feet high and the thickness here is probably at least 100 feet.

This thin sheet of thoroughly shattered marble, with almost no other rocks in it, is 1½ to 2 miles wide, almost 5 miles long, and lies on a desert floor whose average slope away from the mountains is 2½°. Its upper surface is conspicuously hummocky, in marked contrast with the distributary pattern characteristic of adjacent slopes and found on the alluvium now covering much of the central upper part of the lobe (Figs. 308 and 309). There is some tendency for the hummocks to form transverse ridges that are slightly concave downslope; between many of them are closed depressions, some of which contain small playas. The edges of the sheet, especially close to the mountains, are marked by sandy ridges (Fig. 309) that rise slightly higher than the surface of the sheet itself.

All geologists who have examined this hummocky lobe agree that it represents a large prehistoric rockfall or debris flow of some type—commonly referred to as the Blackhawk Slide—and that the source rock is the limestone forming the first crest in the background, more than 1,500 feet above the head of the slide. Because some of the very low hills at the right in Figure 309 are remnants of slightly older sheets of similar kind, it seems clear that the top of this mountain-front has collapsed more than once, sending tongues of debris out over the desert floor.

Despite its relatively good preservation several features of the Blackhawk Slide are puzzling: How did such a thin sheet travel so far over so gentle a slope—and without becoming contaminated? Why are the surface wrinkles concave downslope instead of convex downslope as is normal in debris flows (Fig. 121), rock glaciers (Fig. 118), and many lavas (Fig. 62)? What is the meaning of the marginal sandy ridges? One might guess that the mass moved as a very fluid mudflow, perhaps with the consistency of soupy concrete mix, but this seems impossible in the absence of a clay-rich matrix to supply the necessary lubrication. Some have suggested a rockfall similar to those observed at Elm, Switzerland in 1881 or at Frank, Canada in 1903 (page 126), but the volume of the Blackhawk mass is enormously greater than that of the debris from either of these. Could a rockfall have given such a mass the mobility and momentum to travel 5 miles on a 2½° slope? Perhaps, if the debris was traveling at high velocity and was lubricated by a gas or air. Since there is no known basis for postulating an explosion, the most likely source of the necessary speed would seem to be a near free fall. A drop from 1,500 feet might impart a velocity of 150 to 200 miles per hour, and good mobility might ensue if enough air were trapped and compressed in the process to serve as a lubricant. Many of the details of such an explanation are very difficult to imagine; to some geologists it raises about as many problems as it solves.

The unusual properties of this debris lobe seem to demand an unusual mechanism, but exactly what it was is still an unsolved problem.

Fig. 309
The Blackhawk Canyon debris lobe looking downstream (north). Note marginal ridges and alluviated central area.

Fig. 310
Vertical exposure of porous marble breccia in the Blackhawk Canyon debris lobe. Hammer handle is 16 inches long.

29 EROSION

Topographic Inversion at Table Mountain

Table Mountain, a winding flat-topped ridge that can be traced for more than 40 miles across the foothill belt at the western base of the Sierra Nevada in central California, is shown at the right. Its barren summit is the top of a layer of dark lava less than 200 feet thick and, as now exposed, 200 to 2,000 feet wide. Along its length it descends toward the southwest (the direction of this view) about 100 feet per mile—a slope of 1°.

At first glance, Table Mountain appears to be the sinuous remnant of a sheet of lava, most of which has been eroded away. But if this is true, why did the erosion not leave other mesas and buttes scattered about, instead of only one twisting ridge?

Let us look more closely. Table Mountain rises above a landscape of low relief that has been developed across steeply dipping metamorphic rocks containing some plutonic intrusions. The lava of Table Mountain is separated from this crystalline basement by 100 to 300 feet of sedimentary layers. The lowest of these are thin patches of gold-bearing river gravel which may be as old as Eocene, but the rest is younger stratified volcanic detritus consisting chiefly of well-sorted volcanic sandstones and tuffs that contain lenses of clay and siltstone and, near the top, tongues of coarse volcanic conglomerate. Scant plant remains in some of these beds and potassium-argon ages from associated volcanics indicate an early Pliocene, or perhaps late Miocene, age. The composition of the volcanic sediments is strikingly uniform (andesitic) and similar to that of widespread lavas higher in the Sierra. Such volcanic detritus occurs in irregular patches throughout several thousand square miles of the Sierran west slope and evidently once formed a continuous sheet that is now represented by erosional remnants or is found preserved under such protective covers as the Table Mountain lava. Followed eastward up the Sierran slope, the volcanic detritus thickens to several thousand feet, implying that it was spread westward by streams that drained the higher country during and after the eruption of the corresponding andesitic lavas there. The well-sorted and rounded character of the major part of the deposit favors such an origin; the lenses of clay and silt would represent occasional slack-water or pond deposits, and the tongues of coarser material are probably mudflows that came down some of the channels.

In other words, just prior to the eruption of the Table Mountain lava the landscape probably consisted of a composite alluvial blanket of volcanic debris with a rather flat constructional surface, like a great bajada, grooved here and there by the shallow channels of intermittent streams. Buried under this blanket were older gravels containing alluvial (= placer) gold washed down from bedrock outcrops higher on the slope, and protruding through it were scattered hills and ridges of crystalline rock too high to be completely covered by the few hundred feet of gravels and volcanic debris.

From this we infer that one of the late eruptions produced a lava stream which traveled many miles down the slope, most of it probably confined to one of the stream channels. (This stage is suggested in the background of the drawing.) With cessation of volcanism and subsequent uplift and westward tilting of the Sierra Nevada, all the volcanic cover was subjected to erosion of increasing vigor. The blanket of volcanic detritus, being relatively incoherent, was mostly stripped off, re-exposing large areas of the old surface on crystalline rocks. But the long narrow lava flow, being relatively thick and resistant, still stands in relief on this surface. In many places it protects an underlying sample of the detrital volcanics, and, in a few, deposits of placer gold.

Because what was originally a channel has thus become a ridge, this is an example of reversal, or *inversion*, of topography.

Fig. 311. *Upper:* View southwestward along Table Mountain, west of Jamestown in the Sierra Nevada foothills, California. *Lower:* Drawing emphasizing the important geological features and, in the background, restoring the scene to the conditions that probably existed just after the lava flowed down the channel.

Marine Terraces on the Palos Verdes Hills

The wave-cut bench, characteristic of exposed windward coasts where waves pound the shore in the narrow vertical range between low and high tides, is a crude but useful record of sea level. It is crude because it is a seaward-sloping surface, and because its position in relation to mean sea level is partly determined by such factors as the resistance of the rocks to abrasion and chemical solution, and the average size of the waves that attack it. However, even with these variables, many benches provide a conspicuous and fairly long-lived indication of the relative level of sea and land during the period in which they formed, one which is probably accurate within a vertical range of ten to twenty feet. Ancient wave-cut benches, above or below present sea level, are called marine terraces (page 194).

A general view of one of the world's best-known sequences of emerged marine terraces is shown in Figure 312. Some significant features associated with these old shorelines are illustrated in Figure 313, which shows a vertical cut across the lowest terrace. The lower two-thirds of the exposure consists of strata dipping about 20° to the left: these are silty marine sandstones containing a rich lower Pleistocene fossil fauna. The upper third is a horizontal blanket of unstratified colluvium, a nonmarine layer of mixed sediments produced mainly by creep of the weathered zone on the steeper slopes above. Between the two is a thin layer of white shells, now about 70 feet above sea level. Here they are chiefly thick-shelled clams; in other places they include sea-urchin spines, several varieties of boring mollusks, and abalone shells—all characteristic of beaches and tide pools and denoting near-shore shallow-water conditions and an upper Pleistocene age. Comparable deposits have been found on 9 of the 13 recognized terraces, including the next to highest one. The angular unconformity between lower and upper marine Pleistocene exposed in this scene and in other road cuts nearby demonstrates not only that the terraces are *cut* features, but also that there was deformation after lower Pleistocene time, after which all the terraces were cut during upper Pleistocene time. The colluvium is characteristic; it obscures the details over most of the Palos Verdes Hills and seems a reasonable result of the processes of weathering and downslope movement acting on a stepped slope draped with loose beach deposits.

What is the complete meaning of this series of marine terraces, each representing a former wave-cut bench, now distributed through a range from less than 100 to about 1,300 feet above sea level? Some of the lower ones might be the result of eustatic (page 179) changes of sea level, but the higher ones are not sufficiently widespread to suggest such an origin. On the other hand, if they are even partly the result of local deformation of the crust, then the terraces themselves, especially the earlier-formed ones, should be warped in conformity with the local uplift that raised them above the sea level at which they were cut. There is some evidence that this has happened; the Palos Verdes Hills terraces are very slightly warped—even the next to lowest is deformed near the left margin of the view in Figure 312.

In what order were the terraces cut? The fossils from the different terrace deposits all represent living species and are no help in distinguishing among them. One of the presumably youngest terraces was found by C[14] to be more than 30,000 years old—too old for accurate dating by this method. We can reason that they were probably cut in sequence, either from highest to lowest or vice versa—as the land either rose out of the sea or sank into it. If the land were sinking the material cut from each successively higher terrace would be spread over the terraces below it and tend to obscure them. Furthermore, to reach their present position above sea level the terraces would have had to be lifted out of the sea and the wave action during this process might be expected to destroy or modify them still further. It is most likely, therefore, that they

Fig. 312. Pleistocene marine terraces on the seaward slope of the Palos Verdes Hills south of Los Angeles, California. Looking southeastward. (*Photo by J. S. Shelton and R. C. Frampton.*)

Fig. 313. Angular unconformity between lower and upper Pleistocene marine deposits, topped by about 8 feet of colluvium. South side of Second Street, San Pedro, California.

were cut while the land was rising. This would make the highest terrace the oldest and the lowest the youngest, a conclusion supported by the fact, illustrated in the right foreground of Figure 314, that a part of a given terrace level may be missing because it has been consumed by local widening of the next lower one (which therefore must be younger).

Knowing the general tendency of both sea level and land levels to change with time, and having here a remarkable record of such changes, it is only natural to wonder whether the part that shows above water represents the whole story. More than a hundred submarine profiles, carried to depths of 500 to 700 feet at various points along the coast from Santa Barbara to San Diego as well as around near-shore islands, have revealed at least 5 recognizable terraces on the sea floor down to a depth of about 500 feet. These are notably better developed on the seaward (windward) sides of the islands than on their protected landward sides, indicating that the prevailing winds of the past were also from the west. Coring, diving, and submarine photography have provided evidence that these terraces, too, are *cut* into Pliocene and older sediments; in all essentials they are comparable to those exposed on the Palos Verdes Hills. Being relatively protected from erosion, the submarine terraces can be traced over greater distances than those on the land; their depths reveal warping (relative uplift toward the shore) whose effect is greatest on the deepest terrace and diminishes with the successively shallower ones, indicating that the lowest is probably the oldest. It is difficult to prove whether they were cut

Fig. 314. Looking north over the lower Palos Verdes Hills terraces. The prominent terrace in the center of the view has been separated from its continuation (under buildings in lower right corner) by local widening of the lower terrace at the highway level.

by a single halting rise of sea level or by successive lowerings of diminishing magnitude. Two kinds of evidence suggest that they are probably late Pleistocene: (1) Several cores taken on a terrace under about 350 feet of water off the Palos Verdes Hills contain dead shallow-water algae with C^{14} ages of less than 25,000 years. (2) Some of the drowned river mouths along the nearby coast (Fig. 153), which were deepened when sea level was lower, are cut into rocks younger than middle Pleistocene.

If all the interpretations up to this point are correct, a neat problem is set before us. It looks as though the emerged terraces were cut from top to bottom, before the submerged terraces were cut from bottom to top. Since the Palos Verdes Hills seem to occupy the crest of the up-warp shown by the more extensively mapped submarine terraces, a plausible explanation would be that they were uplifted by this warping, perhaps in the early part of late Pleistocene time, and then the eustaticly lowered sea level of the last part of the Ice Age cut the submarine terraces during its return to the present level. This return must, then, have been more rapid than any continuing warping, such as that which deformed the submerged terraces.

One may well ask: Why are the terraces so uniformly stepped, especially if some were produced by crustal upwarping and some were produced by eustatic changes in sea level? Do both of these processes consist of a series of pauses, during which wave action is effective, that alternate with periods when the movement is relatively steady for distances of 100 feet or so, during which the waves cannot cut a well-defined bench? The submarine terraces may be stepped because of fluctuations in sea level arising from climatic changes affecting the volume of the ice; there is abundant independent evidence to support this idea. Other things being equal, climatic cycles of long duration that substantially change the volume of ice on land will produce corresponding swings in sea level. The maximum lowering from this cause, during the last episode of Pleistocene glaciation, is estimated to have been about 400 feet, based both on old shorelines and on calculations of the volume of the ice. It is thus probable that the submerged terraces off Palos Verdes Hills were cut during one or more such times of lowered sea level, the pauses during which the benches were cut representing intervals of relative climatic stability.

A possible explanation of the emerged terraces follows from combining the effects of a cyclically rising and falling sea level with relatively steady uplift of the land. If the two rise together for a period, wave cutting will be concentrated at one level. Then, if sea level falls while the land continues to rise the relative position of land and sea may change so rapidly that there is little chance of the waves leaving their mark. Under this hypothesis the spacing between cut terraces will be greater than the swing of sea level. For example, if sea level rises and falls through 50 feet at the same rate that the land is rising, the terraces will be 100 feet apart even without allowing for a pause at the low end of the sea-level swing.

31 PLEISTOCENE GLACIATION

The Scablands of East Central Washington

Few regions anywhere in the world possess a more distinctive geologic history more grandly displayed than the lowlands of southeastern Washington. Most of the area is included in the drawing at the right (Fig. 315), which looks northward from the Oregon line. The foreground portrays the scene as it is today but in the more mountainous terrane in the back the ice has been restored as it probably appeared late in the Pleistocene epoch. South of the ice sheet lie 10,000 square miles of relatively flat country with an average elevation of about 1,500 feet, surrounded by mountains—notably the Cascades on the west and the Rockies on the east, as well as many smaller mountains and hills on the south. The depression thus enclosed, about 120 miles across, slopes downward toward the southwest. The drainage is collected by the Columbia River system around the north and west margins (partly covered by ice in this scene) and by the Snake River along the south, the two joining for a southwest departure through the Horse Heaven Hills near the southwest corner.

This is the Columbia Basin, the northwest part of a great sea of almost undeformed lavas that extends into Oregon and southern Idaho and is known as the Columbia Plateaus. The total area of the Plateaus is about ten times that of the portion considered here. But in this sample are recorded two remarkable geological stories. The first concerns the accumulation of these *Columbia River basalts*, one of the largest outpourings of lava known on any continent. The second concerns the unique way the surface of part of this lava plateau was sculptured by floods during the melting of the Pleistocene ice. Let us consider them in the order in which they took place.

Some of the finest natural vertical exposures of the Columbia River basalts are found in the walls of upper Grand Coulee. The lower view at the right (Fig. 316), showing the upper 500 to 600 feet of the east wall, is typical. Individual flows differ noticeably in thickness and in the pattern of their columnar joints, probably because they came from sources at different distances and arrived with slightly different temperatures and viscosities. At the tops of some flows there are weathered zones or even ancient soils, some with fossil tree stumps in place, and in a number of places there are thin layers of tuffaceous sediment between flows. The last are most common near the margins of the basin and include both alluvium washed out onto the lavas from the surrounding highlands and shallow-lake deposits formed when the advancing lava flows temporarily obstructed drainage. Clearly there were some long pauses between successive eruptions.

In a few places, also near the margins, exposures of the granitic and metamorphic rocks beneath the lavas give some indication of the total thickness of the flows and the nature of the landscape buried beneath them. At Steamboat Rock, near the north end of Grand Coulee (Fig. 317), a buried ridge of granite at least 600 feet high is prominently exposed beneath several hundred feet of lavas in the coulee walls. Steptoe Butte (Fig. 319), south of Spokane, is a hill of older rock which the lavas, visible in the left foreground, surrounded but never covered. The greatest exposed thickness is 100 miles farther south in the mile-deep Snake River gorge, where over 4,000 feet of lava remain despite some loss by erosion; there are places in the lower walls of the gorge where 2,000-foot hills on the granitic pre-lava surface may be seen beneath the basalts. Apparently, near the margins of the basin, the lavas spread over and buried a landscape of considerable relief, perhaps a continuation of the surrounding mountains as they were at that time. The floor under the lava in the deep central part of the basin may be flatter and the total thickness of the lava much greater—one can do little better than guess. But even using a conservative estimate of half a mile for the average thickness in the Columbia Basin, the volume of the lava in this area alone is around 5,000 cubic miles—fifty times the volume of the San

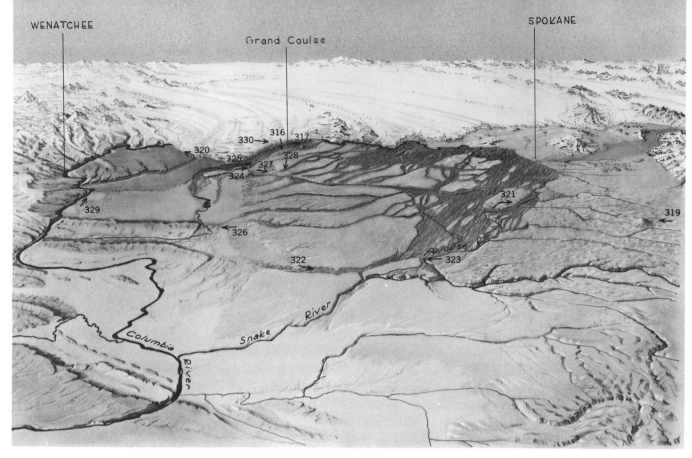

Fig. 315. View northward over eastern Washington emphasizing the scablands topography on the Columbia River basalts. In the background the Pleistocene ice front is shown in about the position it probably had at the time of the hypothetical Spokane Flood. Numbered arrows indicate, by figure numbers, the locations and directions of accompanying photographs.

Fig. 316. Columbia River basalt flows exposed in the upper 500 feet of the east wall of Grand Coulee.

Franciscan volcanic field, part of which appears in Figure 5. The total volume of the Columbia River basalts, including the extensions in Oregon and Idaho, has been estimated at 35,000 cubic miles.

Huge basaltic outpourings of this type are characteristically almost devoid of volcanoes in the popular sense; there are no great cones like Shasta and Rainier (which, though nearby, are younger and of different composition from the basalts). There is also very little ash in proportion to lava, indicating that there was not much explosive activity. But there are a great many dikes; these cut some flows but connect with others, for which they obviously served as feeders. These observations mean that these basalts came from *fissure eruptions*, the lava having welled up through and spread out from cracks that were at least several miles long. Probably because there was little loss of heat through explosive activity, and the outpouring was rapid in terms of volume per hour or day, some of the giant flows were more than 300 feet thick and some possessed sufficient mobility to spread at least 60 miles from their sources. As pointed out above, there were sometimes long pauses between eruptions, during which weathering and sedimentation took place. Much of the time the scene probably resembled that in Figure 318 (except for the man-made cultivation patterns). This is the margin of a recent black lava flow, erupted from a fissure out of sight to the left. It has spread over older lavas which have weathered long enough to develop a soil. Note the sharp contrast between the flat lava plain and the mountains against which the flows lap out in the distance.

Volcanic accumulations of this type are known as flood or plateau lavas or, because they are always dominantly composed of basalt, as *plateau basalts*. Plateau basalts account for the largest known volumes of volcanic rocks on the continents and may also be widespread on the floors of some ocean basins. Any theory regarding processes within the earth's crust must take account of the composition and occurrence of these masses; the volume of some exceeds that of a large mountain range. It is not surprising that the crust has sagged under the Columbia Basin; more than 5,000 cubic miles of rock have been removed from below and added to the load on top.

Over 100 species of trees, shrubs, and other plants are represented in the fossil leaves and other remains found in the sediments sandwiched between the lavas around the margins of the Columbia Basin. These include pines, conifers, and sequoias; a great variety of deciduous trees such as oak, elm, poplar, willow, maple, magnolia, ginkgo, birch, alder, laurel, and chestnut; and a number of shrubs, grasses, pondweeds, and mosses. This assemblage not only indicates a climate more humid than is found in the region today (partly because the moisture-trapping Cascades had not yet been erected to cast their rain shadow across eastern Washington) but it also serves to date the major part of the eruptions as Miocene. This age is also indicated by tongues of the basalt which, in the lower Columbia gorge, interfinger with Miocene marine sediments (Astoria formation).

Even the course of the Columbia River seems to have been affected by the lavas. From its origins in Canada the Columbia has a rather straight southward course for more than 130 miles until, at the junction with the Spokane River, it encounters the edge of the Columbia River basalts. At that point it turns abruptly westward and for a like distance follows the outer margin of the lavas around the Big Bend to Wenatchee. Throughout this loop its course lies almost exactly along the lava contact, a circumstance that could hardly be accidental; it seems almost certain that the spreading basalts crowded the river into this position, which, before the sagging of the Basin, was the low crease between the sea of spreading lavas along its left bank and the surrounding mountains on its right. At Wenatchee it abandons this route and cuts across the southwest corner of the lava plain, probably deflected by local eastward flowing lavas and great alluvial fans whose source was in the Wenatchee Mountains and growing Cascades to the west.

A remarkable blanket of loess (see page 197) mantles most of the basalt surface in the Columbia Basin. It consists chiefly of unstratified buff-colored silt a few tens of feet thick.

Fig. 317. Part of a granitic ridge buried by basalt flows and then partly exhumed
by the cutting of upper Grand Coulee.

Fig. 318. Edge of Quaternary lava flow on Snake River plain. Looking northwest, 10 miles west of
Idaho Falls.

Toward the northeast it thickens to around 100 feet, becomes a little finer grained, and acquires the characteristic "stormy sea" topography visible around Steptoe Butte in Figure 319. Most of the rainfall (7 to 20 inches annually) is absorbed by this pervious deposit and the near lack of runoff results in smooth slopes conspicuously free from fresh rills and gullies. The same conditions permit farming on the slopes by methods that would invite disastrous soil erosion under most other circumstances. The loess is popularly known as the Palouse "soil" because of its typical development in the area of easternmost Washington drained by the Palouse River system. Because of the success with which wheat crops can be harvested in alternate years by dry-farming methods, it has proved to be one of the state's major agricultural assets.

Geologically, the Palouse loess is somewhat puzzling. It is clearly an eolian blanket, probably more than 50 cubic miles in volume, deposited by winds blowing across the basin from the southwest. In some places there are buried soil profiles within it, represented by B horizons and layers of caliche, so there must have been pauses during its accumulation. Fossils have so far been of little use in dating the loess. It must, of course, be younger than the Miocene basalts it covers, and older than the late Pleistocene meltwater channels that have been cut into it (see below). In at least one place some of the loess seems also to be interbedded with late Pleistocene glacial drift; quite probably none of the loess is older than Pleistocene.

The Palouse topography (Fig. 319) also poses a problem; the loess is almost certainly too fine-grained to have behaved like accumulating dune sand. If, as seems probable, the even level of its crests is inherited from the original top of the deposit, then it was originally thin and is not yet deeply eroded. But if erosion is just beginning to assert itself, why is the pattern of ridges and depressions so intricate and why is so much of the surface in slopes? The permeability of the loess may be an important factor, but can it determine the *pattern* of this "stormy sea" topography? The puzzle invites imaginative speculations. For example, many observers have noted that, as is clearly visible in Figure 319, most of the hills are steeper on the northeast (right) than on the southwest; this may be related to the fact that patches of snow linger longer on slopes facing away from the sun. Such a possibility suggests other complex origins. Perhaps, for example, the loess has been laid (more than once?) over a blanket of drifted snow which, thus insulated, disappeared so slowly and gently that the loess came to rest on the basalt with initial variations in thickness that reflect the snow surface in reverse (loess hills on the sites of snow hollows?). Or perhaps the explanation lies in the unusually large number of agents available here—wind in combination with snow, rain, and meltwater—and their interplay over hundreds of thousands of years.

Turning to the unusual topography that has been cut into the surface of the loess-covered lavas we encounter a geological story apparently unlike any other in the world. The essential elements are these:

During the Pleistocene the mountains surrounding the Columbia Basin were glaciated. U-shaped valleys and associated moraines show that tongues of ice moved toward the lava plain from several directions; northwest of the lava plateau, Lake Chelan—50 miles long, 1 mile wide, and up to 5,000 feet deep—occupies a valley that is a spectacular example of this glacial erosion (Fig. 315, upper left). However, because of the low altitude of the plain and the protective moat afforded by the Columbia River valley, significant advances of the ice out over the lavas could take place only where a number of glaciers converged to produce a large tongue. This happened along the northern rim of the Basin, in the Okanogan Valley and north of Spokane, resulting in two lobes that each reached about 30 miles onto the lavas. The scene at the right (Fig. 320) demonstrates the sharp contrast between the hummocky boulder-strewn moraine left by the Okanogan Lobe, which moved in from the right, and the flat unglaciated lava plain beyond the glacial limit to the left. (Other views of this boundary are shown in Figs. 160, 216, 217.) Except for these and a few lesser invasions along its northern margin the lava surface was never under ice.

Fig. 319. Steptoe Butte, a knob of basement rock showing through the Columbia River basalts
(visible in left foreground) and their blanket of Palouse loess.

Fig. 320. The arcuate outer margin of the terminal moraine left by the Okanogan lobe of
Pleistocene ice. West of Moses Coulee, looking northwest.

The remainder of the basin surface, south to the Snake River, is streaked with a gigantic system of abandoned channels trending southwestward down the dip slope toward the low point of the basin. The widest are more than 20 miles across and the deepest cut 900 feet below the surface. Along most of them the Palouse loess has been stripped off and the basalt surface scoured on an enormous scale to produce *scablands*.

An example of one type of scabland channel crosses the scene in Figure 321. In the foreground and background is typical Palouse topography—smooth and rounded. Along the two-mile-wide channel some of the basalt flows have been partly ripped off, leaving a complex pattern of interlaced smaller channels between residual island-like buttes. At the left is the end of a long narrow lake, and another smaller one lies to right of center. Both occupy long depressions scooped out of the basalt by the denuding current. Some of these rock basins are more than 100 feet deep and 10 miles long (Fig. 322).

It has been estimated that about 2,800 square miles of the Columbia Basin, about one-sixth of its total area, have been thus made into scablands. The most interesting feature of these channels is the evidence that their origin was almost certainly dominated by one or more gigantic floods of short duration rather than erosion at normal rates by rivers of normal size. One of the strongest arguments for this is the overall pattern of the channels (Fig. 315). From a number of sources along the north and northeast margin of the lava they divide and reunite in a braided network that from place to place comprises anywhere from 3 to 11 old watercourses all draining approximately southwestward. As may be seen in Figure 323, in some places the current split or shifted, leaving a Palouse loess island with a sharp prow pointing upstream. In other places the water spilled over the low divide from one channel into another. Huge potholes (Fig. 324) and rock basins (Fig. 322) occur not only along the channel bottoms but also on some of the divides at places where they were crossed by spill-over currents. In short, the

Fig. 321. A typical scabland channel along which the Palouse loess and some layers of basalt have been stripped away. Looking northeastward, between Rock Lake and Ewan.

Fig. 322. View eastward along Washtucna Coulee, believed to be part of the former course of the Palouse River, which was excavated by a branch of the scabland-cutting drainage into a closed depression 100 feet deep and more than 10 miles long.

Fig. 323. Upstream point of an "island" of Palouse loess, on either side of which pass scoured scabland channels. Looking west near the mouth of the Palouse River.

channels were filled to overflowing in many places. Where this has happened one thinks of looking for evidence of sediment that could have choked the channel, but there is a remarkable absence of such deposits except in the lowest part of the Basin, where waters were ponded. The potholes and rock basins of the scablands are largely free of sediment that might have been left by any waning current; the basins have not been filled in, the long profile has not been smoothed out, the watercourses did not reach a graded condition. Evidently the currents stopped rather abruptly.

Further evidence of extraordinarily large floods may be seen in the relics of former cataracts. The best known is Dry Falls in the southern part of Grand Coulee, part of which is visible in Figure 325. This great wall is 400 feet high and almost 3 miles wide—two and a half times as high and almost five times as wide as Niagara Falls today. Some of the cascades, like the one shown in Figure 326, were 10 miles wide and 10 miles long. Notice in this, as in other views of the channel floors, that the lava flows have not been worn and shaped by abrasion, as is common along normal rivers; instead they have either been peeled off whole or left as steep-sided remnants. The planes of weakness between flows and the closely spaced vertical jointing help to explain this, but it takes quite a current to remove thousands of solid columns of basalt without at the same time rounding off what it leaves. This too points to an enormous flood of short duration rather than long-continued erosion of normal proportions.

The few deposits directly related to the scablands floods fit well with this interpretation. Granitic boulders as large as 20 feet across, almost certainly derived from the granitic ridge near Steamboat Rock (Fig. 317), have been found 50 miles downstream from this part of Grand Coulee. Huge gravel bars 100 to 150 feet high occur in the expected places (Fig. 327)—at the mouths of tributaries, just below obstructions, or in stretches where slack water should occur. These show downstream foreset bedding of the long deltaic type (i.e., not in horizontal sets) and sometimes include blocks of basalt more than 10 feet in diameter. It is notable that with one possible exception these bars are *not* terraced by later and lesser currents. In a number of places floodwaters from a main channel backwatered into inactive tributaries and left deposits of sand and gravel with foreset bedding dipping *up*stream.

Fig. 324. Scouring and potholes in channel east of Dry Falls, south of Coulee City. Looking east-northeast.

Fig. 325. A part of the Dry Falls precipice, Grand Coulee, looking upstream. Note the scouring
of the basalt surface and potholes (right foreground).

Fig. 326. A portion of the Drumheller Channels, 100 square miles of cataract floor over which
a major part of the scabland drainage once flowed. Looking northwest.

Fig. 327. Giant gravel bar in Crab Creek (center, inside curve of railroad).
Looking east toward Wilson Creek.

Many of the gravel bars bear giant ripples on their surfaces (Fig. 328)—asymmetric wave-like undulations up to 20 feet high and 300 feet from crest to crest. Similar giant ripples occur on bars along both the Columbia (Fig. 329) and lower Snake rivers, indicating that these streams shared in the torrent of flood waters at one time. If we knew more about the quantitative relationships of ripple amplitude and wavelength of ripples to grain size and the depth and velocity of the generating current, we could better estimate the kind of flow necessary to produce these overgrown examples. At present we can only say intuitively that since the dimensions of the ripples and the coarseness of the sediment are enormous, so must have been the current that produced them.

Granting the strong probability that only colossal floods could produce an interlaced network of simultaneously occupied channels whose currents peeled off whole layers of jointed lava, scooped out miles-long basins in solid rock, spilled over divides, formed huge cataracts and cascades, transported 20-foot boulders for 50 miles, left gigantic bouldery gravel bars bearing elephantine ripples, dumped backwater deposits upstream into local tributaries, and subsided so rapidly that these features show virtually no effects of subsequent normal-scale erosion or sedimentation, the question is: can we find reasonable circumstances and mechanisms to account for them? The best answer seems to be this:

Whenever any part of the fluctuating ice front advanced onto the north margin of the lavas, it must have blocked the drainage in the west-flowing Spokane-Columbia moat. This water would then have spilled over onto the lavas at one or more points behind the ice dam to join with meltwater streams coming directly from the ice lobes. These currents must have cut initial channels along several courses whose different locations recorded the shifting positions and

Fig. 328. Giant ripples on the surface of a gravel bar at Wilson Creek. Looking south.

Fig. 329. Giant ripples on the inside of a bend in the Columbia River, near Trinidad. Looking upstream.

heights of the ice fronts and dams. (Note that in this unusual case the drainage pattern was *not* established through headward growth; the water was abruptly introduced at the *upper* end of the scablands drainage system.) In particular the east side of the Okanogan Lobe must have blocked the Columbia River at the head of Grand Coulee, diverting its flow southward to help in the cutting of that great trench (Fig. 330). Probably at the same time meltwater from the south front of the lobe collected to cut Moses Coulee 10 miles farther west (Fig. 315). Glacial striae on basalt within the upper end of Grand Coulee prove that the ice advanced after the Coulee was partly cut, and the hanging valley in Steamboat Rock (Fig. 330) suggests that a smaller discharge crossed this area in a different direction before it was completed. Both substantiate the idea that the position of the ice front fluctuated.

At about the time of the maximum ice advance, the Spokane Lobe, by filling the valley containing Pend Oreille Lake (50 miles northeast of Spokane), blocked the westward flow of the Clark Fork River, which reaches more than 250 miles back into the northern Rockies. The result was an enormous five-pronged Pleistocene lake centered near Missoula, Montana, which filled Bitterroot, Blackfoot, Flathead, and numerous other valleys. Shorelines, beach deposits, and deltas show that its surface was about 4,150 feet above sea level, 1,000 feet above Missoula, and 2,000 feet above Spokane. This great body of water, glacial Lake Missoula, had a total area of about 2,990 square miles, a maximum depth of 2,000 feet, and an estimated volume of over 500 cubic miles, held in by ice at the west and north and by mountains on the east and south.

Ice dams are not permanent. The break-up of this one left abundant evidence of the flood it released. At various places along the Clark Fork in western Montana, leading to the outlet of glacial Lake Missoula, there are gravel bars bearing giant ripples up to 50 feet high and as much as 500 feet from crest to crest. Where the flow was constricted by narrows, the steep valley walls are swept conspicuously clean up to more than 1,000 feet above the valley floor. These gravel bars and clean valley walls are taken as evidence that the lake emptied suddenly when

Fig. 330. Looking northeastward across upper Grand Coulee and Steamboat Rock.

the ice dam gave way. Calculations suggest that the discharge may have reached a maximum of over 9 cubic miles per hour—1,900 times the average flow of the Colorado River in the Grand Canyon and well over 100 times flood stage on the lower Mississippi. Such a discharge probably lasted only a few days; at half the estimated maximum discharge the available water would have drained out in three days. The flood would have stopped almost as abruptly as it started.

When the water was first released there must still have been ice to the north and west, so the only place the waters of glacial Lake Missoula could go was along the ice margin toward Spokane. Here, depending on the position of the ice front toward the west, the flood either spilled directly out onto the lavas or started down the Spokane-Columbia moat until, either blocked by further ice or finding the capacity of this trench inadequate (or both), it spilled over at one or more points west of Spokane. Perhaps this whole sequence of events took place more than once. In any case, this presumed Spokane Flood—probably produced by sudden emptying of glacial Lake Missoula—provides a possible mechanism for supplying a huge amount of water, abruptly turned on and abruptly turned off, to supplement the normal interplay of ice and river in accounting for the unique features of the scablands.

Geology was not a science until the legendary Noachian flood and six-day creation were replaced by explanations derived from careful study of rocks. The doctrine that past events should be explainable through no more than reasonable extensions of observable processes has played a vital role in substituting the plausible for the preposterous and the feasible for the fanciful in geology. Many geologists, mindful of this important truth, have been reluctant to accept such a catastrophic mechanism to explain the scablands and have tried to find other explanations based on combinations of more ordinary processes. This has stimulated healthy controversy but does not seem to have seriously weakened the Spokane Flood hypothesis. The sheer magnitude of the whole scabland complex and the many ways in which it exceeds the bounds of normal stream erosion and deposition seem to justify, if indeed they do not demand, an outsize agent operating under extraordinary circumstances.

Why are similar scablands not known elsewhere? A fairly strong answer is to be found in the fortuitous combination of circumstances necessary to supply all the essentials: a large area of closely jointed and gently tilted lava flows so situated among mountains that a fluctuating ice front could both divert a major river onto their upper edge and suddenly release huge volumes of impounded water across their high margin. No less important is the role of the Columbia River trench, which, since its last ice blockade was broken, has collected the drainage that would otherwise have moved out onto the lavas from the surrounding mountains, thus leaving the scabland channels beheaded and almost dry. The channels now receive only the runoff from precipitation that falls directly on them—an amount so scanty that they have undergone very little change since they were cut. The Columbia trench around the upper boundary of the scablands is thus a principal reason for their preservation.

Fig. 331. View from 35,000 feet above sea level, looking northward along the border between California and southern Nevada. The nearest playa is 8 miles long.

Pleistocene Lakes of the Great Basin

A typical landscape in the semi-arid southwestern United States is shown in the scene above (Fig. 331). The light patches are playas, the nearest one being 8 miles long and almost 2 miles wide. Between these and the dark mountains are long smooth gray alluvial slopes, or bajadas. (A closer view of these relationships as they occur at the south end of Death Valley, 40 miles west of this scene, was shown in Fig. 148). Topography of this kind prevails over a large area lying west of the Colorado Plateaus and extending from Oregon on the northwest to Arizona and Mexico on the south.

About 185,000 square miles of this area—centered in Nevada and including adjacent parts of California, Oregon, and Utah—is without drainage to the sea and is therefore known as the *Great Basin*. All the runoff within the Great Basin collects in a few shallow salty lakes, such as Pyramid Lake in Nevada, Mono Lake in California, Great Salt Lake in Utah, and in scores of playas like those shown above. From these it rapidly returns to the atmosphere by evaporation. Hydrographically the Great Basin is essentially a giant complex of potential intermontane puddles, most of which are dry most of the time.

The Great Basin is also a land of extreme contrasts. Many of the mountain ranges reach elevations of 10,000 to 13,000 feet, high enough to collect sufficient rain and heavy winter snow to support extensive forests, even in the southern part of the region. Most of the depressions between mountains, on the other hand, are well below 5,000 feet and tend to be arid and relatively barren of vegetation, especially in the south. Among them are Death Valley and its neighbors, which include the hottest, driest, and lowest places in the United States.

But the very conditions that account for the parched and desolate scenes in some of these basins today are also partly responsible for the preservation of subtle topographic features that, in a more temperate climate, would have been obscured by vegetation and modified or removed by creep. Consider, for example, the scene below (Fig. 332) which shows part of the east side of Searles Lake basin, 35 miles southwest of Death Valley. The alluvial slope built from the mountains at the left toward the basin at the right is trenched by the channels, or "washes," along which runoff now intermittently flows. Crossing these channels is a system of delicate terraces which, throughout a belt up to two miles wide, faithfully follow the contours of the slope, regardless of whether it is composed of hard rock, as in the left foreground, or unconsolidated alluvium, as in the middle distance. These terraces consist of the same kinds of pebbles that compose the fans, but along the terraces the pebbles are noticeably more rounded and better sorted as to size. When other kinds of evidence, such as the lake deposits under the adjacent playa, are added to these features it is clear that the pebbles were produced by waves reworking the alluvium of the fan and that the terraces represent different levels of a former lake. The lake was obviously younger than the fans and probably older than some of the channels cut through them. The highest well-marked beach is at least 640 feet above the playa floor, which is a minimum figure for the maximum depth of the water.

There is much additional evidence for the former existence of a lake. Rocky outcrops along some of the terraces are coated with a more or less spongy calcareous deposit, called *tufa*, which is characteristic of the present shores of Pyramid Lake, Mono Lake, and the Salton Sea. It seems to be produced, often with the help of algae, from warm waters rich in dissolved salts. Tufa deposits are unknown along ocean shores, but are common around some hot springs and in shallow closed basins whose waters can escape only by evaporation and therefore leave behind increasingly saline solutions.

Fig. 332. View southward along the east side of Searles Lake basin, showing many shorelines left by ancient Searles Lake.

Fig. 333. Looking northward along the west shore of ancient Lake Bonneville,
just west of Wendover, Utah.

Throughout the Great Basin such shoreline features are generally more conspicuous on the east than on the west sides of the individual depressions, implying that during their formation the prevailing winds were from the west—as they are today. But in the larger basins even the upwind shores are quite distinct, as shown by the scene in Figure 333, 120 miles west of Salt Lake City. Here there are more than a dozen well-defined beach lines left by ancient Lake Bonneville, a huge body of water that was ancestral to Great Salt Lake in northwestern Utah. Note that at all levels these shores delineate a point of sand, or spit, projecting to the right into the lake (foreground). The only place where the shores are conspicuously destroyed by later erosion is on the near side of the spit.

In addition to the sequences of beaches, many basins also include appropriately situated bars, deltas, and caves that help to complete the picture of ancient shoreline processes. A prominent gravel bar, somewhat dissected by later erosion, is shown in Figure 334.

Fig. 334. View southward along the gravel bar left near Afton Canyon by wave action along
the east shore of ancient Lake Manix, Mojave Desert, California.

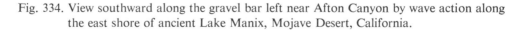

When these prehistoric lakes were first studied in detail, almost 100 years ago, it was realized that the story they have to tell is not necessarily as simple as it looks. For one thing, although the level of a lake with sustained outflow is fixed by the elevation of its outlet, a lake that does not receive enough water to overflow will change in level according to the balance between its gains and losses. Note, for example, the irregularly declining level of Great Salt Lake, which, even in the short time of 112 years, has fluctuated through a range of 20 feet (Fig. 335). Not only will such a lake cut many faint shorelines instead of one bold one, but the younger marks it leaves may be either above or below older ones.

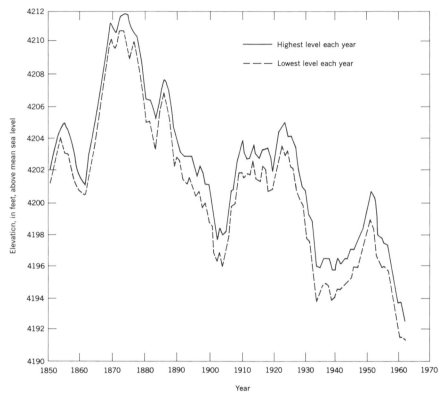

Fig. 335
Yearly water levels of Great Salt Lake during the period 1851–1963. (*From U.S. Geological Survey*, Water Resources Review, Annual Summary—Water Year 1963.)

Some of these considerations apply to the view in Figure 336, which shows Lake Bonneville shorelines just north of Salt Lake City, on the east side of the lake. (These are comparable to those shown in Figure 133, some 25 miles to the south, as well as to the west-shore features visible in Figure 333.) The well-developed uppermost bench, more than 1,000 feet above the present lake, strongly suggests that the lake stood at this level for a long time and was unable to rise any higher—both of which are conditions that would naturally follow if the lake had had an outlet at this elevation. Leveling and tracing of the shorelines show that there was indeed such an outlet, through Red Rock Pass, whence the outflow joined the Snake River near Pocatello, 150 miles to the north. The route is marked by appropriate scouring, and downstream from Pocatello there are scabland channels, alcoves, dry falls, and huge gravel bars containing boulders up to 20 feet across, indicating that giant floods coursed down and overflowed the Snake River. Some of the effects are similar to those in the channeled scablands of Washington. The floor of Red Rock Pass is now about 350 feet below the highest lake level, indicating that as it was deepened by the discharge it probably also determined some of the less prominent levels occurring this far below the top terrace. (It is unlikely that these faint shorelines (Fig. 336) are older than the prominent one immediately above them, as they might be if they were cut during

Fig. 336. Shorelines of ancient Lake Bonneville on a spur five miles north of Salt Lake City, looking south-southeast. The highest bench is approximately 1,000 feet above Great Salt Lake, which is out of the view to the right.

the period when the lake was rising to the higher stabilized level. Debris from the cutting of such a prominent bench, spread by offshore currents during the relatively long stable period it represents, would almost surely obscure weaker pre-existing terraces immediately below it.)

Even in today's most arid part of the Great Basin there is widespread evidence not only of former lakes but also of connecting rivers. The view in Figure 337 shows part of one of the many gorges cut by such rivers in the southern part of the Great Basin. Only a powerful stream working rather rapidly could produce such steep walls, yet these canyons are now either dry or contain only an intermittent trickle. Small fish and snails with marine affinities, remarkably adapted to living in pools of warm saline water, now live in Death Valley and a number of other isolated basins. This may mean that the basins were not only once joined together but also were connected to the sea via the Colorado River. On the other hand, such shore birds as sea gulls and pelicans undoubtedly frequented the inland lakes in ancient times, as they do today, and it may be that eggs of marine organisms arrived in mud clinging to these visitors.

Fig. 337
Abandoned stream gorge through the Arrow Canyon Range in southern Nevada, looking downstream. Man standing on dry riverbed gives scale.

One system of connected basins started in the Owens Valley along the east base of the Sierra Nevada and then turned eastward into the Great Basin. The last three depressions along this chain are present-day Searles Lake (Fig. 332), Panamint Valley, and Death Valley. The view in Figure 338 is along the east side of Panamint Valley. In the lower right corner is an edge of the dried mud surface of the playa on the basin floor. In the background is the steep flank of the Panamint Range, behind which, 25 miles distant and 1,200 feet lower, is Death Valley. Here the Panamint mountain front presents a dominantly tectonic landform against which erosion is just beginning to assert itself. This condition is almost certainly the result of active uplift of the mountains relative to the basin; as evidence note, for example, the numerous small fault scarps that cross not only the light surface of the old fan remnants in the foreground but also the modern alluvium at the far right, clearly demonstrating that there has been movement here within the past few hundred years. The upper light slopes in the background are banded by a succession of shorelines, the highest of which is about 1,100 feet above the playa. This figure gives approximately the maximum depth of water in the former lake, uncorrected for later sedimentation on the basin floor or displacement on the faults. The slopes on which the shorelines are preserved are the surfaces of three large remnants of alluvium that appear to be what is left of a mass of fan-like deposits that have been cut through and largely removed by canyon runoff. A fourth and much smaller remnant farther to the right retains three of the highest shorelines.

At first glance one might assume that the canyons that cut through the old fans are younger than the shorelines and the lake. But if this is so, one wonders why the delicate shorelines survived both the time and the runoff required to cut the canyons, and why the lake was shrinking while it was receiving such runoff. It seems more likely that the destruction of the old fans was largely accomplished *before* the shorelines were cut—perhaps by the very runoff whose accumulation helped to produce the lake—and that only minor changes have taken place since. Indeed, the interplay of erosion, faulting, and shore processes in shaping this scene might be quite complex. It has not been adequately studied.

Fig. 338. A portion of the west base of the Panamint Range, California, showing shorelines on faulted and eroded remnants of alluvium.

Fig. 339. General view northward in central Death Valley. The flat part of the floor is here about
5 miles wide and 250 feet below sea level. Mormon Point is the far end of the sloping
ridge which projects into the Valley from the right foreground (see Fig. 340).

The next depression eastward presents the scene in Figure 339, which is a view northward along the lowest segment of the Death Valley trough. The flat and locally salt-encrusted floor is here about 5 miles wide and the portion shown is 45 miles long, all of it more than 200 feet below sea level. Although this is the lowest point in the Great Basin (and in North America), it did not serve as a sump for the whole province because so many of the individual basins did not overflow. Death Valley was, however, the end of two noteworthy systems of lakes and rivers. One was the Owens Valley–Searles Lake–Panamint Valley chain mentioned above. The other originated far southwest in the San Bernardino Mountains and involved four or five lakes before reaching Death Valley; today this course is followed by the Mojave River, which now rarely gets as far as the second or third basin before drying up (at the surface, at least).

Shorelines in Death Valley are neither abundant nor conspicuous. Figure 340 includes some of the best preserved, which are on the low end of Mormon Point, the ridge projecting into the valley from the right foreground in the last view. The highest of these shorelines are at least 300 feet above sea level, indicating a lake about 600 feet deep, more than 100 miles long, and 1,300 feet lower than Panamint Valley, the next basin upstream. As in Panamint Valley, the terraces are best preserved on remnants of gentle alluvial slopes whose positions have helped them survive later erosion. Stone artifacts and other evidences of human occupation have been found on and below some of these terraces, implying that prehistoric man may have lived here during the waning stages of the lake.

When, despite huge quantities of inpouring sediment (note the alluvial fans in Fig. 339), a small area of the crust remains a basin, lower than all its surroundings, there must be a reason. The condition is even more anomalous than that of the hill which persists despite weathering and erosion. Other things being equal, raising a basin floor by sedimentation takes place faster than lowering a mountain by erosion, because (a) a great deal of rock can be removed from a mountain without lowering its summit at all, and (b) the volume of the sediment thus produced is greater than that of the solid rock from which it was derived.

Both persistent mountains and persistent basins are usually to be explained by movement of the earth's crust. Evidence of relative downfaulting of the Death Valley depression can be seen almost everywhere around its margins. On the near slopes just beyond the road in Figure 340 are several low scarps in the alluvium that trend across the drainage lines and must have been caused by faulting. Similar ones can be seen in Panamint Valley (Fig. 338). Another may be seen on the far side of the fan in the center foreground of Figure 339. The aerial view in Figure 341 includes a prominent fault on the west side of the valley along which the depression has subsided about 10 feet relative to the mountains (in the background). Note that this fault splits into two near the center of the view, and that renewed movement on the nearer branch has taken place so recently that the resulting very low scarp has not yet been fully erased by the active wash in the center. Clearly the displacement has not all occurred at one time. The downfaulting shown in Figure 341 has also lowered the base level of the fan, with resulting rejuvenation of its drainage pattern. Beyond the fault scarp the drainage lines are now incised into the older (lighter colored) surface of the bajada.

Fig. 340. Shorelines preserved on faulted and eroded remnants of alluvium
at Mormon Point in Death Valley.

Fig. 341. Recent fault scarp across lower end of Hanaupah Canyon alluvial fan
on west wide of Death Valley. View is upslope.

Other evidence of active deformation in Death Valley, shown by warped shorelines and
modern measurements with tiltmeters, has already been mentioned (page 156).

The fact that shoreline features are rather well preserved throughout the Great Basin, even
when developed on loose sand and gravel, has led all who have studied them to conclude that
they are geologically quite recent. But the likely hypothesis that they are in some way connected
with the Pleistocene Ice Age, is more difficult to prove and refine than one might think. The
scene below (Fig. 342) includes one of the few places where clearly defined lacustrine and glacial
features occur together. The view is upstream at the mouth of a glaciated canyon issuing from
the east base of the Sierra Nevada. Its well-developed U-shaped profile is continued a short
distance out of the mountains between low, curving morainal embankments—similar to but
smaller than those of Green Creek (Fig. 207), eight miles away. On the near side of the larger
moraine is a series of closely spaced horizontal shorelines reaching to about 650 feet above the
present level of Mono Lake, whose shore is now a mile away. The relationship proves that there
was at least one high stage of the prehistoric forerunner of Mono Lake that came after the

Fig. 342. Westward (upstream) view of the mouth of Lundy Canyon at the east base of the
Sierra Nevada near Mono Lake, California. One lateral moraine is striped by lacustrine
shorelines in the foreground and both are cut by a fault along the range front.

maximum advance of the valley glacier that built the moraine. The final stage of the large lake must therefore have been Pleistocene or younger. (So, too, must be a small amount of relative uplift of the Sierra Nevada; note the dark fault scarp, at least 30 feet high, crossing the valley floor and cutting the moraines just above the center of the view. This break is in line with the mountain front and proves postglacial displacement.)

Did the prehistoric lakes rise as a result of the same precipitation that produced the snow that fed the glaciers? Or were they a result of the later melting of that snow and ice, or of post-glacial rainfall, or both? Would the warmth required to induce melting also produce offsetting evaporative losses? Some evidence bearing on these questions comes from C^{14} dates, which will be discussed below.

When all the evidence from beaches, bars, deltas, spits, connecting gorges, caves, and tufa is assembled, it can be shown that well over 100 of the depressions in the Great Basin were once occupied by lakes. Most of those that are known are shown on the map in Figure 343. This compilation is crude in the sense that it represents no precise point in time and shows neither the fluctuations in size of the individual lakes nor the evanescent connections between them. The map simply depicts their approximate maximum extent, regardless of whether they were strictly contemporaneous. A few inland lakes and brackish-water bodies outside the Great Basin are also shown (for example, the one in the San Joaquin Valley in central California, and the Salton Sea at the Mexican border—described on pages 376–381) because they probably existed at about the same time.

The many preserved shorelines in the Great Basin, considered as a widespread set of originally horizontal reference lines, like natural contours engraved on the landscape, might seem to afford an unusual opportunity to measure later warping and faulting throughout the region. Except where destroyed by wasting and erosion, any single shoreline that is no longer continuous and level at one elevation must have been broken or deformed by movements of the crust. This idea is at least as old as Gilbert's 1890 study of Lake Bonneville, but has never been thoroughly developed on a regional scale. To do so would require skillful surveying of a very large area, combined with careful field work, in order to be sure that closely spaced shorelines were not confused with each other as they were followed over the great distances involved.

Despite such difficulties, Gilbert found that in the eastern half of Utah's Great Salt Lake basin the shorelines left by Lake Bonneville are no longer level. The highest, representing an area almost equal to that of Lake Michigan today, ranges through a vertical distance of about 150 feet. Another shoreline, about 375 feet lower, differs as much as 100 feet from place to place. For each, the highest elevations are found on a hill that was formerly an island in the lake. Gilbert drew contours on these elevations and found that the remnants now define gentle domes, the one for the upper shoreline being steeper than the one for the lower. Clearly the land has been warped upward in the vicinity of the ancient lake, and the amount of uplift is larger for the higher—and probably older—lake level. (More recent measurements, extending Gilbert's observations, have shown that the domes are almost perfectly centered over the vanished lake, being highest where the lake was deepest, and that the maximum uplift of the higher shoreline is at least 210 feet.) The amount and distribution of this warping led Gilbert to eliminate faulting as an important factor and to conclude that it was probably caused by the slow recovery of the crust from the position to which it had been depressed by the weight of the water (estimated at 10,000 billion tons) in the 1,000-foot-deep lake. This is not unreasonable in view of the fact that Lake Mead, behind Hoover Dam, depressed its shores 6½ inches during its first 15 years— which is a very small fraction of the probable life of Lake Bonneville.

The sediments that accumulate in a lake contain evidence regarding its history that can be found nowhere else. The few Great Basin lake floors and playas that have been sampled by core drilling have yielded some interesting information. Some of the lake deposits are composed entirely of fine silts that become sandy and conglomeratic near the borders. This is exactly what one would expect from examination of the sediments on their surfaces today—playa muds grading into alluvial fan material around the borders. But 50% or more of some accumulations consists of evaporites such as salt, gypsum, and a variety of other water-soluble minerals—chiefly chlorides, carbonates, borates, and sulfates of sodium, calcium, potassium, and magnesium—ultimately derived, of course, from the weathering of rocks.

We reason that the mud layers were produced during times of active runoff from the surrounding mountains and therefore represent periods of relatively wet climate. On the other hand, the layers of salts could have formed only from waters that had reached saturation and were relatively clear, so that these layers imply the low runoff and high evaporation of a dry climate. But this simple relationship does not completely explain all the mixed and interbedded associations of salts and sediments found in the cores. For example, Searles Lake (which is actually a wet playa and contains the largest known deposit of water-soluble salts in California) is a member of the chain of lakes that formerly reached from the east slope of the Sierra Nevada to Death Valley. This chain collected water from a 130-mile segment of the mountain front and probably maintained its flow even during some periods when evaporation far exceeded precipitation near its lower end. Under these conditions the upper lakes could serve as settling basins for the suspended sediment load and, because of evaporation losses, the outflow of each would be both clearer and brinier than its inflow. In the absence of local runoff, this could result in relatively high concentrations of salts toward the lower end of the chain. The combination of such conditions with fluctuations in local runoff can account for a great variety of deposits.

Every Great Basin depression that has been studied in detail gives evidence of having been filled with water at least twice. Originally this interpretation was based on such geological evidence as two distinct ages of shore features or two sets of lake deposits separated by erosion surfaces and tongues of subaerial alluvium. More recently, C^{14} age determinations have been made on the organic content of mud layers, on shells, and on the tufa and other carbonate deposits. By this method it has been learned that the flood gravels that accumulated where Lake Bonneville outflow joined the Snake River (near Pocatello) were deposited about 30,000 to 40,000 years ago. For Searles Lake, Lake Bonneville, and the predecessor of Pyramid Lake, numerous determinations indicate that within the interval from 10,000 to about 50,000 years ago there were two periods of high water levels and mud accumulation, between which the lakes nearly or completely dried up. It should be noted that this 50,000 years produced only the uppermost 200 feet or less of the deposits in Searles Lake basin; the many hundreds of feet of lake beds beneath these undoubtedly record earlier fluctuations, beyond the reach of C^{14} dating.

The last advance of the Pleistocene ice sheet along the west shore of Lake Michigan pushed over trees whose C^{14} content places the event at about 11,800 years ago. If the C^{14} determinations in the Great Basin are comparable, then the Great Basin lakes existed in the late Pleistocene, at least, and this conclusion is independent of whether or not C^{14} gives *true* age values; it requires only that they be consistent.

Fig. 343. Map of California, Nevada, and parts of adjacent states, comparing present-day water bodies (black) with approximate maximum extents of the late Pleistocene lakes (gray). DV = Death Valley, LB = Lake Bonneville, LL = Lake Lahontan, LM = Lake Manix, ML = Mono Lake, PL = Pyramid Lake, RP = Red Rock Pass, SL = Searles Lake, SS = Salton Sea. (*Base map ©1960 by Jeppesen and Co., Denver, Colo. All rights reserved.*)

The Long Beach Oil Field

The search for petroleum, life-blood of the twentieth-century machine age, occupies more geologists than any other branch of the profession. Surface studies in all parts of the world, enriched by information from hundreds of thousands of wells, have produced a general understanding of many aspects of the origin and accumulation of oil, but no method for accurately predicting where it will be found.

Oil occurs almost exclusively in sedimentary rocks of marine origin. From this and its chemical composition it is clear that petroleum, a complex and somewhat variable mixture of hydrocarbons is formed by slow underground processes operating on organic material derived from animals or plants that accumulate with sediments, especially muddy marine ones. Whatever their origin, the minute globules of petroleum are slightly lighter than water and thus tend to rise through the interstitial waters of the enclosing sediments. If the pore spaces are sufficiently large and interconnected (i.e., if the permeability is high enough) this slow upward movement of countless droplets will continue until they either escape at the surface in an oil seep, or are prevented from doing so by some underground obstruction. Such obstructions are known as *oil traps* and must consist, essentially, of an impervious layer or zone that, owing to folding, faulting, or to up-dip facies changes, is concave from below and accessible to the migrating oil. Trapped beneath this cap, the oil gradually displaces much of the pore water to form a *reservoir*, or "*pool*," bounded below by water and above by natural gas or the impervious rock. The oil, gas, and water occupy the interstices of the reservoir rock (usually sandstone, but sometimes porous limestone or fractured rocks of other kinds) and are kept from escaping upward by the shape of the overlying impervious roof.

Fig. 344. View northwestward over Signal Hill and the Long Beach oil field as it appeared in 1930. (*Photo by Spence.*)

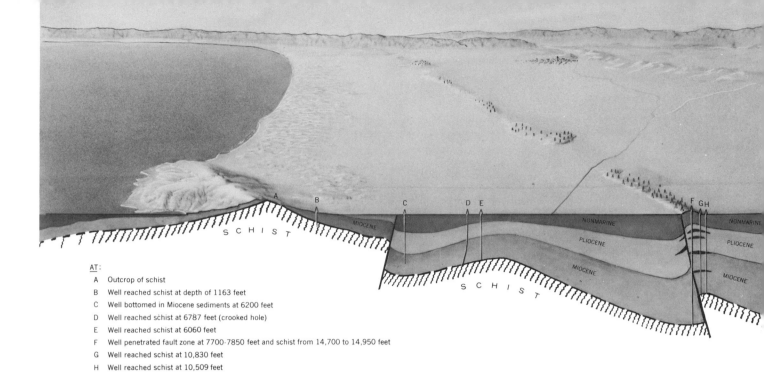

AT:
A   Outcrop of schist
B   Well reached schist at depth of 1163 feet
C   Well bottomed in Miocene sediments at 6200 feet
D   Well reached schist at 6787 feet (crooked hole)
E   Well reached schist at 6060 feet
F   Well penetrated fault zone at 7700-7850 feet and schist from 14,700 to 14,950 feet
G   Well reached schist at 10,830 feet
H   Well reached schist at 10,509 feet

Fig. 345. Diagrammatic view northward over region west of Los Angeles. The front of the block
    shows a vertical cut from Palos Verdes Hills (left) through Signal Hill (right). Constructed
    principally from information obtained from wells. Black lenses symbolize oil pools.

The search for oil is a search for traps, but the only known way of learning whether or not a suspected trap actually contains oil is to drill into it.

Anticlines and domes are the most obvious oil traps and were the first to be recognized. They still are basically responsible for most of the world's known reservoirs despite the later recognition of several other types of traps and almost endless subtle and complicated variations and combinations. Let us consider a well-known example.

The Long Beach oil field lies about 20 miles south of Los Angeles, California. In the photograph, taken in 1930, before the forest of wooden derricks had begun to give way to more modern equipment, the long narrow shape of the field is revealed in the pattern of wells. Beneath them is a low ridge whose high point, Signal Hill, rises about 300 feet above its surroundings. This is the surface expression of an anticline so recently active that early Pleistocene sediments are involved in the arching.

As shown in the drawing, deep drilling has revealed the probable explanation for the fold. A steep fault in the schist that forms the sloping crystalline floor beneath the sediments has produced a sharp step. The sedimentary blanket has necessarily accommodated itself to this step in an asymmetrical anticline whose steep side corresponds to that of the step. The vertical displacement in the schist is about 4,000 feet but diminishes upward to less than 400 feet near the surface. Offsets in oil zones suggest that there has also been right lateral slip on the fault. It would appear that this fault, and several smaller faults in the schist that do not reach the surface, have been active since Miocene time; either accumulated displacements are proportional to age, or the more compressible sediments have absorbed some of the offset. Both have probably played a part in the upwardly diminishing displacements. This is one of the very few sedimentary anticlines in the world whose relations to underlying crystalline rocks are even this well known.

Since its discovery in 1921 the Long Beach field has yielded nearly a billion barrels of oil. Because there are traps and producing zones at many depths down to 10,000 feet, the cumulative production per surface acre, now over half a million barrels, is one of the highest in the world.

Bingham Canyon Mining District

The enormous man-made pit in the scene at the right, 20 miles southwest of Salt Lake City, is the central operation in one of the richest mining districts in the world. Following its discovery in 1863, the area was mined by underground methods for more than 40 years, the chief products being lead, silver, and gold. In 1906 this open pit was begun near the head of Bingham Canyon and since that time copper has been the principal metal obtained. Total production from the district, through 1962, includes more than 16 billion pounds of copper, 4 billion pounds of lead, 1,500 million pounds of zinc, 250 million pounds of molybdenum, 200 million ounces of silver, and 11 million ounces of gold. The total combined value of these metals plus several others produced as by-products is around $5,000,000,000. What is the geological explanation for this extraordinary concentration of mineral wealth?

Ore is rock from which a useful metal can be commercially extracted; the definition is economic, not geologic. Most of the ore in the Bingham Canyon district occurs within and around an irregular stock of granitic rock, about a mile across. From this stock numerous sills and dikes extend into surrounding thick Pennsylvanian sandstones containing interbedded limestone lenses. The granite, which has a somewhat ragged texture and is extensively altered, bleached, and mineralized, yields about 0.76% metallic copper and 0.015% molybdenum from small grains of sulphide minerals that are fairly uniformly disseminated through the rock. The concentration tends to be highest near the center and to decline slightly toward the margins. Such very low grade, disseminated ore is usable only if handled on a large scale. In Bingham Canyon the rock is blasted free (the annual consumption of blasting powder is over 8,000,000 pounds) and then loaded by power shovels onto standard-gauge ore trains which, as may be seen in this view, travel along the stepped levels. In 1962 over 29,000,000 tons of ore were thus mined; in most years an equal or greater quantity of waste rock is removed. The ore goes to nearby mills where it is crushed and the metal sulphides concentrated before being sent to a smelter for more costly processing in which the metal is chemically separated from the minerals. Some of the rock that is too low grade for milling is spread in great dumps (right foreground of Fig. 346). From artificial ponds on their surfaces acid waters seep through the dumps, leaching out additional copper, which is precipitated from the waters before they are pumped back to repeat the process. Well over a thousand tons of copper are recovered each month by this method.

What such low-grade, disseminated ores lack in richness they more than make up in reliability, for there is little danger that they will unexpectedly pinch out or be cut off by a fault, as can so easily happen in smaller and richer deposits.

Surrounding the stock is a zone up to about a mile wide in which the principal ore minerals are lead and zinc suphides, although minerals containing copper and silver are also important. These occur in veins which appear to follow fractures in the older rocks and in irregular masses embedded in some of the limestone lenses. Some of the veins are connected with the stock and many of the larger irregular masses have formed where a vein intersects a body of limestone. Often it is clear that the minerals were carried in solutions that spread from the vein-filled fracture, dissolving the limestone and replacing it with a variety of new minerals. These veins and replacement deposits are mined by underground methods since the ore bodies are small and it is desirable to handle as little as possible of the barren wall rock.

Among the many ways in which ore deposits can form, the three found here—dissemination, filling of fissures, and replacement of limestone—are all types that are important the world over. And all three illustrate why most ore deposits are ascribed to igneous sources. Not only is the ore centered in and zoned around an igneous intrusion; mineralization clearly took place in the surrounding rocks after they had formed, partly along cracks from which it spread into rocks susceptible to wholesale chemical changes. The nearly universal association of ore deposits with igneous intrusions is strong evidence that some of the latter are accompanied by

Fig. 346. Looking southward up Bingham Canyon (left foreground) in the Oquirrh Range, south of Great Salt Lake. The open pit, on whose far wall alone there are 8 ore trains of at least 16 cars each, is well over 1,000 feet deep in the place where a hill rose 1,400 feet above the canyon in 1906.

hot, chemically active fluids, either liquid or gaseous, which under favorable conditions can introduce abnormal quantities of unusual minerals into ordinary rocks.

Unexpected support for this interpretation has recently come from a mile-deep well drilled next to the Salton Sea (page 376). It was located near some active hot springs and recent volcanoes with the purpose of tapping natural steam for power. After a few months of testing this purpose was overshadowed, geologically, by the discovery that the well had reached briny waters with temperatures well above 500°F—the highest ever encountered in a well. Still more remarkable, after flowing through a large 275-foot-long horizontal pipe for three months these solutions had coated the inside of the pipe with an estimated 5 to 8 tons of dark deposit containing about 20% copper and lesser but astonishingly large amounts of silver and a host of other unusual elements normally found concentrated only in ore deposits. It would appear that here, for the first time, man may be able to study ore-depositing solutions at work instead of drawing inferences millions of years after they have cooled off or dried up and their host rocks have been exposed by erosion.

The question of why a few intrusives are accompanied by active mineralization, while the vast majority seem not to be, remains a mystery.

367

33 LOCAL DEPRESSIONS IN THE CRUST

Meteor Crater and Zuni Salt Lake

Man's accelerating exploration of space has led to growing interest in the geology of the moon's surface. This interest has, in turn, stimulated more intensive study of craters on earth whose origins may shed some light on the causes of those on the moon. What origins can be deduced for our terrestrial examples?

Since craters are steep-sided, closed depressions, the search for the origins of craters is a search for ways in which nature might have disposed of the rock that once occupied them. Either it went down or it went up, and so a simple genetic classification of craters comprises two groups: (a) those formed by subsidence or collapse, and (b) those formed by explosion. Volcanic activity may produce either type, or a combination of the two (page 314); groundwater solution has produced many of the former (Fig. 233), and there are quite a few whose origin is uncertain.

Figure 347 is a view of Meteor Crater, 20 miles west of Winslow, Arizona. What could have created this huge bowl—4,000 feet across and 600 feet deep—in the almost undeformed strata

Fig. 347. Meteor Crater, 18 miles west of Winslow, Arizona. Looking northwest from 4,500 feet above the plateau.

of the Colorado Plateaus? Because it is situated on a limestone plateau exhibiting sinkholes and other solution features, an origin involving groundwater solution and subsequent collapse might be considered a reasonable possibility. But let us look further.

1. It is clear in Figure 347 that the nearly horizontal strata of the plateau are sharply bent *up* at the crater rim; the highest points, despite subsequent erosion, are more than 150 feet above the surrounding plateau surface.

2. It may also be seen that the ground surrounding the crater is distinctly lumpy and, if one looks closely, dotted with huge blocks. The blocks are of the same rock that forms the main cliff inside the rim (Kaibab limestone) and the lumps are low hills of fine debris corresponding to the weaker rocks exposed in the crater. The debris in these hills exhibits a crude orderliness: the fragments at the top represent rocks exposed near the bottom of the crater and the debris deepest in the hills comes from the rocks highest in the rim. This proves not only that the debris was derived from the crater, but also that the surface rocks were thrown out first, followed in order by successively deeper layers. If the crater had been produced by gradual enlargement of a vertical pipe the debris should be better mixed. (There is no volcanic material here, although Quaternary basalts and cinder cones are present 10 to 20 miles distant.)

3. The rim and the debris hills have a distinctly lighter hue than their surroundings. This is produced by white pulverized quartz sand derived from sheared Coconino sandstone exposed in the deeper half of the depression. Faint light plumes extending northeastward from the crater (to the right in Fig. 347), probably indicate that prevailing winds have redistributed some of this distinctive material; the wind direction they imply agrees with that indicated by modern dunes and windstreaked surfaces over a large area of the Plateaus.

In addition to these features visible from the air, there are three that require hands-and-knees study and laboratory work for identification.

4. The deposits in and around the crater contain many lumps of natural *silica glass* produced by fusion of the shattered Coconino sandstone. The formation of this substance requires temperatures of at least 1,000°C; most other occurrences of it are the result of lightning striking dune or beach sand. The quantity here probably exceeds that at any other known locality.

5. Some of this shattered sandstone and glass contains the first known natural occurrences of coesite and stishovite, two rare varieties of silica. (Quartz is the common variety). Until their discovery at Meteor Crater in 1960 and 1961, coesite and stishovite were known only as artificial substances produced in high pressure laboratory apparatus. Coesite forms only at pressures above about 150 tons per square inch and at temperatures of at least several hundred degrees. Stishovite, although of the same composition, is almost twice as heavy as quartz and forms only under pressures on the order of 500 to 1,000 tons per square inch and temperatures above about 750°C. The pressures required for coesite would not be expected in the earth at depths of less than about 30 miles, or for stishovite at less than about 150 miles. To form them at the surface thus requires extraordinary compression.

6. Although drilling and gravity measurements have proved that no large piece of meteoritic iron lies buried under the crater, small to microscopic fragments are abundant; recent estimates place their total mass at about 12,000 tons.

The first three features are compelling geological reasons for believing that Meteor Crater is of the explosive type, and the last three indicate that the explosion resulted from the impact of a meteorite. By comparing this crater with the similar but smaller ones produced by nuclear explosions, whose energy release is known, it is estimated that the meteorite must have been traveling at least 21,000 miles per hour. At this velocity it would need to have been about 112 feet in diameter. However, the velocities of observed meteors are higher, generally between 25,000 and 160,000 miles per hour. Within this range many combinations are possible; e.g., if it were traveling 33,000 mph an 80-foot meteorite could probably have done the job; at 62,000 mph a 65-foot object would probably have been adequate.

One hundred thirty-three miles away, on the plateau of western New Mexico, is the shallow mile-wide crater shown in Figure 348. The low part of the depression is occupied by Zuni Salt Lake, which has no outlet and must be nearly saturated with sodium chloride most of the time, for salt has been gathered from its shores and shallow floor by Mexicans and Indians since before the opening of the West. On the south shore of the lake are two black cinder cones. This depression resembles Meteor Crater superficially, but the geologic evidence points to an entirely different origin.

In the upper wall of the depression Cretaceous sandstones are overlain by patchy thin lava flows. Above these the surrounding lowland is covered with a blanket of coarse volcanic ash in which are many fragments of sedimentary rock, among them fossiliferous sandy limestones much older than anything exposed here or nearby. This blanket thins with distance from the depression, the maximum thickness of 100 feet occurring in the ridges along the north and southwest rims.

Although solution of evaporite or limestone beds at depth is a possible explanation of the crater's origin, the cinder cones and mantle of ash thinning away from the depression indicate that it has also, more than once, been a center of volcanic activity. The steep rim almost certainly means that there has been some collapse, and probably therefore some enlargement of the depression since the eruption of the ash and its inclusions. It is reasonable that if an explosive eruption is abruptly terminated and no lava wells up in the conduit, a shallow hole will be left over the vent. This will gradually be filled by slumping ash and bedrock, and by deposits formed in any pond or lake that occupies the crater. As these become compacted there may be further subsidence and widening of the depression.

Ship Rock (Fig. 19) and many other necks like it may have had similar histories. The deformed tuff and mixed rocks in their exposed parts are about what might be expected several hundred feet beneath Zuni Salt Lake and suggest that basically this scene and Ship Rock may differ only in the depth to which erosion has uncovered the volcanic pipes.

Explosion pipes of the kind represented by Ship Rock, filled with varying proportions of ash, foreign rocks brought up from below, others that have fallen in from above, and often with dikes irregularly penetrating the whole mixture, are known as *diatremes* (from Greek roots meaning *through* and *hole*). Ship Rock and its surroundings have been so deeply eroded that we see a fairly deep segment of the pipe. Zuni Salt Lake may represent a much earlier aspect of a smaller example, corresponding more nearly to the stage restored in the drawing in Figure 19.

These two craters of unlike origin represent two of the less common ways in which small nontectonic depressions are produced on earth. Others, involving solution of limestone by groundwater (sinks) and melting of ice (kettle holes) have been touched upon on earlier pages. On the moon, however, the situation may be just the reverse. If it has always lacked an atmosphere and water, the depressions on its surface should be impact and volcanic craters, and in the near-absence of weathering and erosion should last almost indefinitely; if sinks or kettle holes are found we will have to make drastic revisions in our thinking about lunar history.

Fig. 348. *Above:* Zuni Salt Lake, 74 miles south of Gallup, New Mexico, as seen from the northwest. *Below:* The same from the south.

Structures Associated with Large Salt Deposits

Next to glacier ice, which flows readily under its own weight, natural salt is the weakest major rock in the earth's crust. Salt is produced by evaporation in lagoons and shallow arms of the sea (such as parts of the coast of Baja California) or in closed basins (like the Caspian Sea, Great Salt Lake, and Death Valley). Owing to the chemical complexity of natural waters, salt deposits are usually associated with lesser amounts of other water-soluble compounds—particularly calcium sulphate.

Buried salt has been found in many parts of the geologic column and at many localities. Notable deposits occur in the Cambrian of southern Persia, Silurian of Michigan, Pennsylvanian of Utah and Colorado, Permian of Russia and of Texas and New Mexico, Triassic of Germany, Jurassic(?) along the United States Gulf Coast, and Pliocene-Pleistocene of Utah, Nevada, and California. Many salt deposits are thousands of feet thick; the total volume of the Permian salt in western Texas and eastern New Mexico is estimated at 25,000 cubic miles. Salt is mined on all continents; in some places—for example, the Salzburg district of Austria—the mines have been worked since Pre-Roman times.

The upper scene at the right (Fig. 349) provides the unusual spectacle of salt erupting at the crest of an anticlinal ridge and flowing glacier-like in a mile-long tongue down the steep flank of the fold. Embedded in the salt are numerous blocks of other rocks, among them a sliver (nearly a mile long) of trilobite-bearing Cambrian limestone and shale, now lifted to a position above some of the Cretaceous strata forming the crest of the ridge. Since the salt bears samples of rocks from near the bottom of the stratigraphic column it probably penetrates the whole sequence, and unless there has been obscure faulting, the parent strata that supply the extrusion must be at least as old as Cambrian. The aggregate thickness of the arched and penetrated strata (Cambrian through Pliocene) is more than 20,000 feet, which is probably also a measure of the total ascent of the salt injected into this anticline from below.

Salt plugs, or *domes*, of which this is an example, are common. There are more than 150 in southern Persia. Some have erupted through the floor of the Persian Gulf; the salt in the upper parts of these has all dissolved away but the many included blocks, among them igneous rocks of uncertain origin, remain as mounds, some of which make low islands.

Owing to its easy solubility in rain and groundwater, rock salt rarely shows at the surface. The unusual exposures in Persia are made possible not only by the arid climate (less than 2 inches annual rainfall) but also by the fact that the salt is actively rising—or has stopped only recently. Even so, the surface of most exposures is darkened by an accumulation of insoluble inclusions left behind as salt is removed in solution.

A related situation in western Colorado is shown in Figure 350. The conspicuous "bowl" (Sinbad Valley) lies on a broad gentle anticline whose axis extends from upper left to lower right across this view, just beyond the canyon of Roc Creek in the center. Along this axis the land surface has collapsed in the pattern of a giant keyhole, producing a jumble of blocks and slides composed of the exposed Triassic and Jurassic strata. A well in Sinbad Valley penetrated 9,900 feet of salt, associated evaporites, and a little shale, after going through about 400 feet of less-soluble material at the surface. Enough other wells in this area have penetrated similar rocks to establish the widespread presence at depth of Pennsylvanian salt beds. From these the salt flowed into the cores of this and several parallel broad anticlines where it has more or less intruded the overlying rocks. In the humid climate of Colorado the salt is not exposed at the surface; indeed, owing to solution by groundwater and its low strength, it has not even been able to support the arch of the anticline.

The surface evidence of two salt plugs along the Texas-Louisiana coast can be seen in Figure 351. Each produces a circular mound about a mile and a half in diameter which rises very slightly above the surrounding marshes and sea level. The shape is emphasized by the

Fig. 349
Left: Kuh-i-Namak ("mountain of salt") in the Dashti district of southern Persia. The salt has pierced the summit of an anticlinal limestone ridge rising 4,000 feet above the plain and flows down both flanks in glacier-like tongues more than a mile wide and 2 to 3 miles long. This air view is toward the southwest. *Below:* Looking west near the summit; note the dark insoluble residues on the salt surface. (*Photos courtesy of Aerofilms, Ltd., England.*)

Fig. 350. View northward over Roc Creek Canyon and the oval basin of Sinbad Valley, just east of the Utah-Colorado line. Sinbad Valley and its narrow, shallow extension toward the lower right, are the result of collapse along the crest of a gentle anticline, brought about by solution of thick salt beds in its core.

associated pattern of vegetation and by the ring of oil wells that for many years have produced 7,000 to 10,000 barrels a day, chiefly from depths of about 8,000 feet, in a narrow zone around each mound. From the rocks encountered in these wells it is clear that each mound is situated over a nearly vertical cylinder of salt. At High Island less-soluble residues capping the salt reach to within about 300 feet of the surface; at Avery Island salt is found as little as 16 feet down. The surrounding Miocene and younger strata, which are nearly horizontal in this region, have been bent up around the flanks of each salt plug; in many places the detailed structure includes steep wrinkles cut by numerous small faults. These complex structures, in which much of the oil is trapped, are the result of deformation accompanying the rise of the salt from its source, probably at least 20,000 feet down.

There are more than 200 such salt domes along the Gulf Coast. Most of them have no surface expression at all and many are offshore; the majority, therefore, have been located by geophysical methods (low gravity readings over the low density salt). Some domes are mined for salt and associated sulfur as well as drilled for their peripheral oil. The average Gulf Coast dome represents the upward migration of about four cubic miles of salt from a presumably Jurassic source at an estimated four to eight miles below the surface.

The fundamental reason for the existence of salt domes is simpler than the details of their structure. In all these examples, the original salt beds lie beneath many thousands of feet of younger sediment. Since salt has a viscosity only slightly greater than that of glacier ice, and ice flows under the weight of an overburden only a few hundred feet thick, it is to be expected that salt would flow readily under several miles of overlying rock. Its mobility in response to pressure differences is increased by the fact that salt is lighter than most of the compacted sediments that overlie it. Clearly, deeply buried strata of salt are about as stable as jelly under a rug. With shifting loads at the surface, or wrinkling, salt will flow as a viscous liquid toward areas of minimum pressure. Once a local hump in the mobile layer is initiated, the weight of the surrounding overburden will tend to squeeze the fluid and lighter material toward and into this bulge as an upward intrusion.

Fig. 351. *Upper:* High Island salt dome on the Gulf Coast of Texas 30 miles northeast of Galveston. Maximum elevation, 22 feet. *Lower:* Avery Island salt dome, 9 miles southwest of New Iberia, Louisiana. Note surrounding salt marshes (near and far sides), barge canals for oil well drilling (left), and cypress swamp (right). Maximum elevation, 152 feet.

Subsidence of the Crust and the Salton Basin

Although uplifts are the most conspicuous and easily studied products of crustal deformation, it is probably true that throughout geologic time tectonic depressions have been about equally extensive and numerous. The blanket of accumulating sediments, and often water, that covers these depressions, makes investigation relatively difficult unless they have later been elevated and their infilling sedimentary rocks have been cut into by erosion.

Yet there are compelling reasons for studying basins while they are still subsiding and before they have been severely deformed. If in the long run the areas of uplift in the crust are in any sense balanced by areas of subsidence, then the record of the extent, magnitude, and timing or age of the latter is largely to be found in the sediments filling the basins. If we are to learn how they go down, even to the very small degree that we have learned how mountains go up, it is important to investigate them while they are basins and before they become something else. Furthermore, the intimate relation of known oil fields to sedimentary basins has added strong economic incentives to the study of these features in all parts of the world.

There are several methods of analyzing a basin *in vivo*, so to speak. One is to study the structural framework of the uplands surrounding it, and any exposures in its margins, as a guide to projecting the known into the unknown. The result is an educated guess about the interior of the basin, based on trends in the exposed rocks that are closest to it and therefore most likely to have been affected by the same deformation that produced it. Another method is based upon the fact that the acceleration of gravity varies slightly according to the density and proximity of near-surface rocks. Sensitive instruments can detect these differences and thus indicate the positions of any buried masses of light rock (such as salt) or heavy rock (such as volcanics) that are embedded in the accumulating sediments. If there are no such irregularities in the basin fill, the pull of gravity will be slightly less over deep parts of the basin than over shallow parts because of differences in the distance to the more dense (usually crystalline) rocks of its floor. In this way gravity measurements can be used to gain some idea of the shape of the floor of the basin. Estimates of depth from the same readings involve assumptions about the densities of the fill at different levels and are usually approximate at best.

Supplementary information on both shape and depth is often obtained from seismic surveys. Explosive charges are set off and delicate electronic instruments, arranged in a pattern on the surface, record the time intervals between the initial impulse and the arrival at each instrument of waves reflected from contrasting strata and the crystalline basement rocks. The reflection from the floor is usually easy to distinguish from the others, and, since the travel times of the shock waves are accurately known, the determination of the basin's depth will be as accurate as the velocities assumed for the passage of the waves through the various sediments penetrated. The same technique can also be used to determine the dip and strike of gently deformed and distinctive contacts at depth—such as a limestone layer overlain by shale.

Of course the best information comes from wells drilled to the floor of the basin and systematically sampled by means of cores or collections of the cuttings, but because of the expense, very few purely exploratory wells are drilled. Under favorable conditions, however, the information obtained from one hole can be extended to a large area by means of gravity and seismic methods. The best-known basins are those in which enough oil has been found to stimulate (and pay for) extensive drilling.

Although dozens of North American basins are better developed and better known, the Salton Basin of southeastern California is examined here because some of the evidence for its subsidence is at the surface and can be photographed. The drawing (Fig. 352) at the right looks southward and shows that the Salton Basin lies in the north end of the depression occupied by the Gulf of California. The low central part of the basin is occupied by the Salton Sea, whose surface is approximately 225 to 230 feet below sea level, the exact figure depending on the

Fig. 352. General view southward over the Salton Basin to the head of the Gulf of California.

fluctuating balance between inflow (mostly from irrigation) and evaporation. North of the Salton Sea is the Coachella Valley and south of it the Imperial Valley, both known for their highly productive irrigated land. The whole topographic depression, the Salton Basin (about 1,700 square miles of which are below sea level) is separated from the gulf by the landward part of the Colorado River delta. The river enters the trough from the east and has built a broad low alluvial cone that is 65 miles wide above sea level and reaches across to the mountainous wall on the west. As a result, the lowest pass between the Gulf of California and the Salton Sea is near the west side and about 40 feet above sea level. This alluvial barrier keeps the sea from pouring into the depressed Salton Basin.

There are a number of reasons for believing that the Salton Basin has been an actively sinking part of the crust during late Cenozoic and recent time. The youngest widespread marine strata in the trough are fossiliferous sandy shales of probable early Pliocene age, recording an invasion of the gulf that reached all the way to the north end. Abundant oyster shells and related fossils in some of these beds indicate water not more than a few hundred feet deep. Today these strata are folded and faulted around the margins of the basin and appear in wells in the central part at depths of more than 10,000 feet. This deep position of a shallow-water deposit proves subsidence, and the fact that the sediments overlying it are mostly nonmarine shows that, during most of the time since the early Pliocene, sedimentation has been rapid enough to keep most of the basin filled to or above sea level and thus to keep the sea out.   *377*

The amount of subsidence is shown even more impressively by the crystalline rocks. Figure 353 is a view southward along the west margin of the basin (upper right in Fig. 352). At the right is part of one of the large areas of low-relief peneplain-like topography that adjoin the Salton Basin on the west and northeast. The average elevation of these erosional flats on crystalline rocks is about 4,000 feet. Note how abruptly they end at the margin of the depression. Evidence from seismic surveys indicates that the crystalline floor of the depression is at least 10,000 to 15,000 feet below sea level—perhaps more than 20,000 feet in some places—making the total relief on the crystalline rocks at least three to four miles. It is highly probable that most of this was produced by displacements of the crust—displacements comparable in magnitude to the present height of the Rocky Mountains or Sierra Nevada and twice the uplift of the Kaibab limestone at the rim of the Grand Canyon. From the point of view of crustal movement the Salton Basin is a respectable negative mountain range.

Fig. 353. The abrupt southwestern margin of the Salton Basin, as seen looking south near the California-Mexico boundary. The edge of Laguna Salada (Fig. 352) appears at the upper left.

Near the margins of the basin there is visible evidence of recent subsidence. In several places old alluvial slopes such as those shown in Figure 354 (west of the Salton Sea, looking south) are crossed by groups of faults along which the land surface has been stepped down toward the basin. In some places (Fig. 355) thick remnants of alluvial fans, originally built into the basin from surrounding highlands, are now being eroded. Their basinward ends terminate abruptly instead of merging with the basin floor as they must once have done. This is a direct result of continued displacements of the kind shown in Figure 354 and faintly visible at the right in Figure 355. The excessive steepness of the basin margins is partly responsible for the rock slide shown in Figure 356. Such oversteepened slopes invite destruction by weathering and erosion and can only be created and maintained by faulting. The presence of hot springs in the basin, the many earthquake epicenters that have been located there, and the measurable displacements that have taken place since 1940 (page 415) indicate that the depression is still tectonically active.

It is often difficult to tell how much of the depth of a depression has been caused by uplift of its surroundings and how much by local sinking. Here, although it is very likely that the mountainous margins have been somewhat elevated, the thick sequence of nonmarine sediments lying thousands of feet below the level of nearby seawater shows that a large part of the disturbance of the crust has been real subsidence.

The history of the Salton Sea and its predecessors includes some puzzling details. The present water body was created in 1905–1907 when for the better part of two years the Colorado River

Fig. 354. Faults crossing old alluvium west of Truckhaven on the west side
of the Salton Basin. Looking south.

Fig. 355. Two large remnants of alluvium (lighter color) along the west side of the Salton Basin.
The Salton Sea is out of view at the right.

took over an inadequately controlled irrigation ditch and most of its water poured into the dry depression until the lake surface reached −198 feet, about 30 feet higher than it now stands. After the breach at the head of the ditch was finally closed, evaporation lowered the lake about three feet a year until, in 1923, it stood at −250 feet. During this time the acreage under irrigation was rapidly increasing, and since 1923 the drainage from this land has balanced or exceeded the evaporation loss (which is potentially about 6 feet a year), so that the water level has been slowly and irregularly rising. It will not rise indefinitely, however, because the lake is so shallow and its shores are so flat that as the level rises the surface area increases about twice as fast as the volume. Because the total loss by evaporation is determined primarily by the surface area, the evaporation loss will eventually counterbalance any reasonable and likely inflow, thus bringing the lake level to a relatively stable state.

Downstream from Yuma (Fig. 352) the Colorado River flows down a broad deltaic hump of its own building. Under such conditions it is normal for a stream to shift its course from time to time as sedimentation chokes a distributary channel or as flood waters undercut or overflow a bank. Before protective levees were erected by man the river probably poured into the Salton Sink whenever its course shifted to the north (right) side of its delta. Records show that it flowed into the basin about half a dozen times between 1842 and 1905.

Old shorelines have been reported up to about 160 feet above sea level in some parts of the Salton Basin. The most prominent is at about +44 feet and is nearly continuous around the basin; a segment along the west side is shown in Figure 357. The vanished body of water delineated by this mark is known as Lake Cahuilla or Lake LeConte. The beaches, bars, spits, calcareous deposits, and fossils associated with this shore show that the lake maintained a constant level longer than any other Recent flooding of the depression, and both its state of preservation and C[14] dates indicate that it disappeared only a few hundred years ago. Under the circumstances this might seem to indicate that Lake Cahuilla was an arm of the sea and that the shore has been uplifted. But the shells imbedded in the shore deposits are not marine types and the shoreline is almost undeformed. Constant level in a lake, on the other hand, is very unlikely unless there is a fixed outlet and sufficient inflow to keep the level up to the overflow point.

These observations lead to some interesting questions. Where did the overflow at +44 feet take place and what maintained the level? If the outlet extended across the delta of the lower Colorado, there is no clear indication of its location today; could it have been obliterated by sedimentation during the past few hundred years? How long could a channel with a stable level have been maintained in such unconsolidated deposits? The higher shorelines pose even greater problems of the same kind. Do they represent many different water levels or displaced fragments of one older shoreline? If the former, what were the dams? If the latter, was the water an arm of the sea with the present levels representing uplift, or was it a lake at some unknown level with the possibility that some of the remnants now mark subsidence? How would the fact that sea level was lower by several hundred feet during parts of the Pleistocene affect these alternatives? Were any of these water bodies contemporaneous with any of the Pleistocene lakes in the Great Basin (Fig. 343)?

It may be that hard rock barriers have come and gone several times between the Gulf of California and the Salton Basin. If so, the various shorelines are further indirect evidence of tectonic activity in this trough.

Fig. 356. Rock slide on steep west margin of Salton Basin. Some of the blocks on the surface of the slide are as large as houses.

Fig. 357. Lake Cahuilla shoreline at approximately +44 feet, west of Salton Sea. Below the water level the rocks are darkened by a coating of calcareous tufa. Looking southwest.

34 LOCAL UPLIFT OF THE CRUST

The Black Hills

The rocky materials of the crust are best displayed in mountains, permitting direct study of greater thicknesses of sediment and larger masses of igneous and metamorphic rock than are ordinarily exposed in lowlands. Furthermore, in the structure of mountains, we see some of the finest records of diastrophism, the deformation of the crust, whose explanation remains one of the principal problems of geology. Let us consider the effects of diastrophism by examining a relatively uncomplicated example.

Topographically, the Black Hills rise as an isolated group of low mountains astride the Wyoming-South Dakota boundary. In plan the high ground is crudely oval, approximately 50 miles wide by 100 miles long, with the length oriented northwestward. In profile the hills rise 2,000 to 4,000 feet above the surrounding Great Plains, high enough to be mantled with dark forest vegetation, whose contrast with the surrounding grasslands is responsible for their name.

Structurally, as shown in Figure 358, the Black Hills are a blister in the Great Plains, a flat-topped dome of old rocks with twice the length and breadth given above, surrounded by concentric belts of successively younger strata whose original continuation over the top of the dome has largely been removed by erosion. In the east half of the central Black Hills the sedimentary cover is entirely gone and erosion now cuts into the relatively weak metamorphic and igneous rocks of the basement complex. That these crystalline basement rocks are Precambrian is well established on two counts: fossiliferous Upper Cambrian marine sediments lie directly on them with depositional (i.e., nonconformable) contact, and pegmatites which cut across (and are therefore younger than) the other elements of the basement complex, have radioactive ages of about 1,600 million years—which means that the complex was formed at least a billion years before the Cambrian.

The sedimentary blanket warped up in this dome is 1½ miles thick and includes beds ranging from Upper Cambrian to Upper Cretaceous in age. As may be seen in the illustration, the most prominent units are (1) the Carboniferous limestones that cover the west half of the almost flat central part of the dome and produce a conspicuous infacing cliff enclosing the area where erosion has cut through to the basement, (2) an outer ridge produced by the massive Cretaceous Dakota sandstone, and (3) the lowland belt between these, which was called "The Race Track" by the Indians but is now known as Red Valley because it is underlain by weak Triassic red beds and gypsum. Additional but less conspicuous concentric ridges and lowland belts reflect the same general structure outside these more prominent ones (cf. Fig. 90).

Two features of this example of mountain building deserve special note. First, the time of the first and principal uplift is rather well established. The strata involved in the deformation include beds as young as Late Cretaceous, while Oligocene fresh-water deposits lie unconformably across the eroded edges of the upturned layers and locally rest directly on the Precambrian (light-colored southeast margin of dome in this view). Thus in the interval between Late Cretaceous and Oligocene, erosion managed to remove enough of the blanket of sediments to expose the crystalline basement in at least a few places. Relative to their immediate surroundings the Black Hills must, therefore, have been uplifted during this interval by at least the thickness of the rocks removed, or about 7,500 feet, which is a minimum measure of their structural relief. Continuing erosion since Oligocene time has increased the exposed area of pre-Oligocene rocks.

Second, the deformation here seems to have consisted of a simple, vertical, almost piston-like rise of less than 20,000 square miles of crystalline basement rocks. Conspicuously absent are the complex patterns of major faults and evidences of horizontal compressive forces, such as belts of parallel folds and low-angle thrust faults, that are typical of most mountain masses—including the Rockies, less than 120 miles away to the west. If igneous activity played a significant role the evidence is still concealed; as already pointed out the pegmatites, and therefore probably also the plutonic rocks (for example, the granite of Mt. Rushmore in which the faces of four presidents have been carved) were intruded at least 1,600 million years ago—long before even the first sediments were laid down. If the granite is related to any mountain-making process, it was a much earlier one.

Might the doming of the Black Hills have been produced by later igneous intrusion? There are at least ten small shallow intrusions scattered about the north half of the uplift: the Devils Tower is probably the best known. These intrusions penetrate strata as young as Late Cretaceous, yet by early Tertiary time were sufficiently exposed to shed pebbles into Oligocene conglomerates and even to be overlapped by these deposits. Thus both the principal uplift of the Black Hills and the intrusions in the north must have taken place during the same time interval. This interval was probably at least 40 million years long and it has not been possible to relate the uplift and intrusions more accurately in time. Because the intrusions are fine-grained porphyries displaying such features as columnar structure and, in some cases, laccolithic form, they must be of shallow origin and were therefore probably side effects, only indirectly related to the main deformation.

We see here a simple demonstration of the truism that the real cause of uplift is unknown and operates out of sight within or below the older crystalline rocks of the crust.

Fig. 358. High-altitude oblique view northwestward over the Black Hills, drawn with
emphasis on significant details of the topography and geologic structure.
Big Horn Mountains in left background.

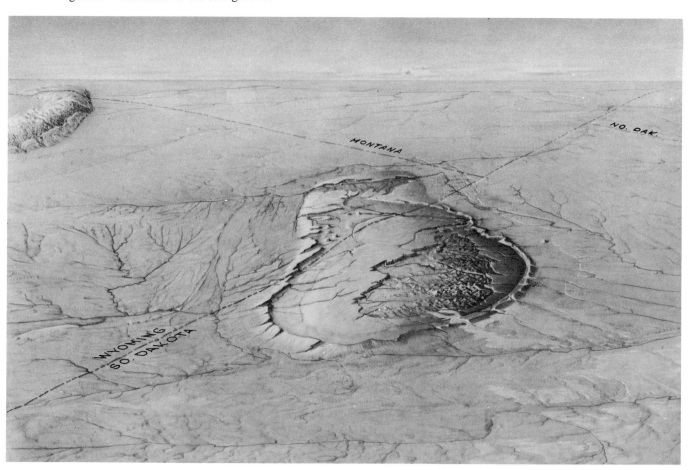

Fig. 359. General view of the crest of the southern Sierra Nevada.
Looking northwest from about 18,000 feet.

The Sierra Nevada and Yosemite Valley

The Sierra Nevada (Spanish: snowy, jagged range) of eastern California is the largest single mountain range in the conterminous United States. It reaches well above timberline and the steep rocky slopes of its summit region include many cirques, small lakes, and other evidences of Pleistocene glaciation (see, for example, Figs. 170, 207, 208). The view above (Fig. 359), which is northwestward across the region near Mt. Whitney, shows the abrupt and fairly straight eastern escarpment and, in the upper left, gives a suggestion of the much gentler westward slope. The range is about 350 miles long and 55 to 80 miles wide.

The origin of a mountain range may be thought of in terms of two separate but related stories; one concerning the origin of its rocks and the other the origin of its shape. The first story is recorded in the rocks themselves and concerns the evolution of the materials out of which the mountain mass is made, beginning with the earliest discernible rock-producing event. If the mass includes Precambrian rocks the time span covered by this account may be very long. The second story concerns the mountains as a topographic feature—when and how they acquired their present height and form. This involves the opposed effects of tectonic and leveling processes and is principally recorded in large- and small-scale features of the shape of a mountain and in the sediments derived from it by erosion. Although the time covered by the geomorphic history is usually relatively short it is often possible to recognize several stages within it.

Having emphasized the story of the rocks themselves in our examination of the Grand Canyon (pages 248–259) we will treat this aspect of Sierra Nevadan history very briefly and concentrate on the second aspect—the origin and history of the present shape of this mountain range.

The Sierra Nevada is composed of granite and related plutonic rocks in which are embedded roof pendents of metasedimentary and metavolcanic rocks ranging in age from fairly late Jurassic back at least to Ordovician. Resting nonconformably on eroded surfaces across these rocks are erosional remnants of Late Cretaceous and younger sediments, and tuffs and lavas of Miocene and Pliocene age. The granite is thus younger than Late Jurassic and older than Late Cretaceous. Careful mapping has disclosed that the Sierra Nevada batholith consists of a number of similar but distinguishable masses with radiometric ages ranging from about 135 to about 85 million years—implying that the batholith was not produced all at once, and that the process may have occupied much of the rather long interval from latest Jurassic to about mid-Cretaceous time. By late Cretaceous time it had been partly unroofed and since then has, in different places and at different times, been receiving blankets of sediment or volcanics, or undergoing further erosion. Wells drilled in the San Joaquin and Sacramento valleys west of the range have encountered the batholith at increasing depths westward, showing that the granite exposed in the Sierra Nevada is only part of the whole mass. On such evidence the granite is extended beneath the valleys on the front of the drawing in Figure 360.

A first clue to the origin of the present-day Sierra Nevada is given by its general shape, clearly shown in the drawing (Fig. 360). The obvious idea that it is part of an enormous block tilted gently downward toward the west is supported by several kinds of evidence. For example, there are unmistakable signs of recent faulting along the base of the steep eastern slope. The

Fig. 360. Perspective diagram of central California emphasizing the relations, from left to right, of the San Andreas fault zone, Coast Ranges, Central Valley, Sierra Nevada block, and western Basin and Range Province. The front of the diagram is a vertical cut to show the general structure; note the vertical exaggeration.

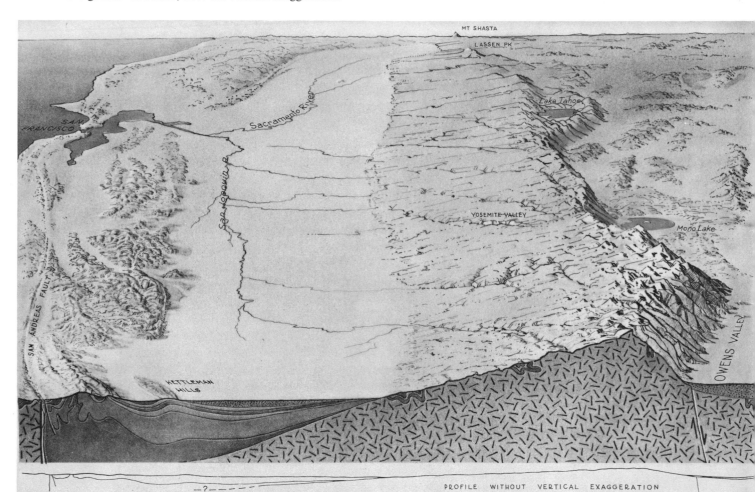

Fig. 361. *Upper:* Looking southwest over Crater Mountain to the east scarp of the Sierra Nevada
from a point about 15 miles south of Bishop. *Lower:* Closer view along the fault
in the left middle distance of the scene above. Note the evidence of its effect on
groundwater. The cinder cone on the fault line is Red Mountain.

upper scene at the left (Fig. 361) shows the traces of three of these faults cutting across volcanic rocks and alluvial fans and impeding groundwater movement in the manner so common along faults in the Southwest (e.g., Fig. 237). Along most segments of the faults the relative movement has been up on the west (mountain) side; this was shown in Figure 342 and is apparent at the far left in Figure 361, *upper*. At one place (Fig. 361, *lower*) a small volcano seems to have taken advantage of a specific fracture, as though the break had provided the final passage for the explosive escape of the molten rock. Thirty miles farther south the alluvial fans are cut by a 20-foot escarpment (up on the west) that was produced in the course of a major earthquake in 1872. Although there are some places where warping rather than faulting may be the mechanism, the very existence of the precipitous eastern escarpment proves that uplift of the mountains relative to the valley has been taking place actively for a long time.

The western slope, better known because of its gold, water, timber, and recreation areas, also shows evidence of westward tilting. The view in Figure 362, looking upslope toward the crest, shows two of its most characteristic features: an extensive and in some places almost flat upland surface, and the deep westward-draining canyons that groove this surface. There is nothing in the nature of the rocks themselves to account for this surface (as there is, for example, in the Colorado Plateaus); it has developed on a variety of plutonic and metamorphic rocks whose small differences in resistance to erosion produce only minor effects on the landscape. Since the streams flowing down the slope today are obviously trenching and destroying the upland, it must have developed under different conditions. Those conditions must have included less tilt and lower elevation, for extensive low-relief topography, independent of underlying rock types (e.g., Fig. 146), is never *produced* in mountains, even though it may later be uplifted to form them. Such an earlier condition may be approximated by undoing, in imagination, the faulting along the east base of the range.

Fig. 362. Looking eastward up the relatively flat western slope of the Sierra Nevada incised by the Middle Fork of the American River.

Evidently, after standing at low elevation long enough to acquire a fairly flat erosional surface, this huge block of igneous and metamorphic rocks has been tilted westward until its rising east edge is an alpine crest and the runoff down its west slope is cutting canyons into the old surface. (The complete geologic history would, of course, begin with the origin of the limestones, shales, and other sediments whose later metamorphism to marble and schist and intrusion by granitic plutons produced the crystalline complex of which the block is composed. The subsidence necessary to accomplish this metamorphism and the uplift and erosion required to bring its products to view imply crustal activity, and probably ancestral mountains, stretching back in time far beyond the events here deduced.)

A hint that the westward tilting probably took place in successive stages is contained in the details of some west-slope drainage lines. The Merced River, for example, is one of the several trunk streams which rise in the highest and wettest part of the range and flow down this slope. If we imagine ourselves facing upstream at a point on the lower Merced beyond the effects of Pleistocene glaciation, the long profile of Bear Creek, a typical tributary from the south, has the form shown by the solid line in Figure 363. The two convexities in its lower course suggest that the Merced River once flowed in a broad valley at the level B, to which the long profile of the tributary became completely adjusted. Then the Merced deepened its valley: first to D, a level to which only the segment ACD of the tributary had become adjusted before a second rejuvenation lowered the Merced to E, forcing the tributary to develop the even steeper segment CE. The number and preservation of the segments composing the complex profile ACE differ in the various tributaries of the Merced. In some streams, where erosion has been rapid, the final segment CE has consumed much or all of the earlier AC; Many smaller streams have short CE segments and are very steep at their lower ends.

If this interpretation is correct, then by plotting the points corresponding to B and D for a large number of streams tributary to the Merced River it should be possible to estimate the positions of earlier, higher long profiles of the Merced itself. In some places such restored pro-

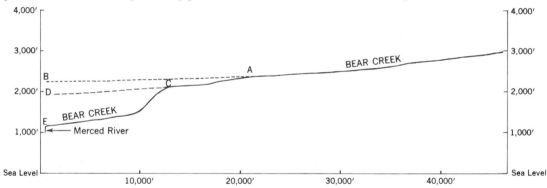

Fig. 363. Long profile of Bear Creek, a tributary entering Merced River from the south about 25 river miles below Yosemite Valley. Vertical exaggeration is approximately four times. (*Modified from Wahrhaftig, Calif. Div. Mines and Geol.*, Bull. 182, 1962.)

files are as much as 2,000 feet above its present (unglaciated) floor. The conclusion most geologists draw is that the westward tilting of the Sierra block has not taken place at a uniform rate. Major episodes of tilting accelerated the flow of the main streams that ran down the dip of the slope but had little direct effect on the tributaries that flowed more or less along its strike. The trunk streams, thus rejuvenated, eroded their valleys with increased vigor while the tributaries were merely crowded a little against their western banks and experienced the rejuvenation only indirectly as their mouths were lowered by the deepening main streams. No doubt, too, the rising east edge of the block collected more rain and snow, which increased the discharge in the main streams. The erosional effects to be expected from more water flowing down a steeper slope are obvious.

In the view of the western Sierra slope in Figure 364 some of the effects of westward tilting are combined with those of glaciation. In the background is the characteristic upland landscape of low relief with shallow valleys and rounded hills. In the foreground is Yosemite Valley, the deeply glaciated part of the Merced River valley upstream from Bridalveil Fall. Yosemite Valley is cut 2,500 to 3,500 feet below the older upland surface: the contrast between the two is overwhelming. None of the drainage lines on the upland even approaches normal adjustment to the course of the Merced River on the floor of this giant trench. The Yosemite Falls, just above the center of the view, are one of the spectacular results; for ten miles the creek which feeds the falls wanders across the upland surface at an average gradient of 300 feet per mile; a mile from the brink it steepens to twice this in a short segment that may represent the time when it was adjusting to the first deepening of the Merced; then, at the brink, it plunges almost 2,500 feet in the cascades and falls for which Yosemite Valley is famous. Such strongly convex-up profiles always imply some special cause.

Here, the obvious cause is deepening of the Merced River valley, with which the tributaries could not keep pace. But how much of this deepening was accomplished by the preglacial Merced River and how much is the result of glacial scour? The long profiles of its tributaries indicate that the deepest preglacial erosion by the Merced in this area resulted in a canyon that was narrower than the present Yosemite Valley and whose bottom was about 600 feet above the present valley floor. The present floor, however, is merely the top surface of loose sediments filling a deeper canyon in granitic rocks. The fill has been explored by seismic methods, which reveal that it is up to 2,000 feet thick, and that if it were removed there would be a lake 1,000 feet deep in the foreground of this scene and two shallower ones farther downstream. These basins in solid granite could only be the work of powerful glacial scour. The deepest one occurs just below the point where two major ice streams, the steep Tenaya and Merced glaciers, once joined. The moving ice was at least 5,000 feet thick at this point and evidently scooped out its bedrock floor somewhat in the same way that a river may scoop out its bed to form a plunge pool at the foot of a steep declivity.

The youngest major terminal moraine on the floor of Yosemite Valley is just out of sight around the bend at the upper left in Figure 364, yet the valley is almost as markedly glaciated below this point as it is above it. (The lowest evidence of ice is 10 miles farther downstream, near El Portal.) This proves that the glacial reshaping was not primarily the work of the last ice tongue to occupy the trench. Also, among the deposits left by the ice, some are deeply weathered and some are fresh, even though they contain the same granitic debris. Such observations demonstrate that the canyon of the Merced not only was subjected to at least two periods of glaciation, but that tens of thousands of years probably separated them—long enough to allow deep weathering of granite.

On these grounds we may logically assume that the fill beneath the floor of Yosemite Valley consists of two or more glacial deposits of different ages, interspersed with sands and gravels deposited by streams and with lake deposits that accumulated in the scoured basins and perhaps in later and shallower lakes formed behind morainal dams.

In short, ice occupied an existing canyon two or more times, deepened and widened it, and then—abandoning the basins it had excavated and leaving the glacial trough blocked with end moraines—left it to be filled with sediment to its present level. Although the wide floor and steep walls of the present valley are the work of glaciation, at least two-thirds of its present depth is the work of the Merced River in preglacial time—work done in response to uplift and westward tilting.

Glaciation on the upland was negligible compared to that in the valley. The distribution of polish, striations, erratics, and morainal deposits shows that some of the summits in the background of Figure 364 were never under ice. The thin patches of ice that existed between the summits accomplished very little erosion, and the scattered debris they left has had but little effect on the main features of the topography: the area is almost halfway down the west slope and too low to have accumulated much ice of its own. The upland thus retains an essentially preglacial landscape, sliced through by a tongue of ice born in higher regions, somewhat as the Grand Canyon area is a relatively unsculptured semiarid plateau sliced through by a river that originates elsewhere.

Many puzzles remain in even this late and small fraction of Sierran history. For example: When did the major episodes of uplift and tilting take place and what altitudes were reached during each?

Erosion surfaces are generally more difficult to date than rocks, and except close to the modern shores of the oceans it is often impossible to be sure about past elevations. But there is indirect evidence bearing on both. Obviously, any erosion surface must be younger than all the rocks across which it is cut and older than the oldest sediment laid upon it. Since the batholithic rocks are as young as early to mid-Cretaceous, and are overlapped by Upper Cretaceous marine sediments, the oldest erosion surface was well developed and must have been near sea level late in Cretaceous time. As we shall see, so much has happened since then that it is unlikely that recognizable remnants of this surface now remain in the higher parts of the range. Indeed, near the southern part of the block (front of diagram in Fig. 360) this buried Cretaceous surface, as located in a few wells, is apparently a little steeper than the exposed western slope of the mountains, whereas the reverse was probably true when it was first formed (assuming that it was cut by west-flowing streams). This relationship implies that the exposed part of the crystalline block has been cut well below its Cretaceous profile and that probably all of its surface is therefore younger than the surface beneath the sediments in the San Joaquin Valley.

North of Yosemite Valley (Fig. 360) there are patches and stringers of Eocene sands and gravels that lie on the granite and are capped with volcanic rocks of Miocene and Pliocene age. The Eocene gravels have received considerable attention because of the placer (alluvial) gold they contain. They are preserved chiefly along the courses of old rivers whose valleys were subsequently filled with the volcanics that now protect them. The sample of Eocene topography thus preserved is small and incomplete, but it includes several major stream courses that can be used as a basis for speculation on the tilting. One Eocene river has been traced for about 110 miles: the upper and lower thirds of this trend toward the southwest, like the modern drainage, and now have gradients of 80 to more than 150 feet per mile; the middle third flowed northwestward and now has an average gradient of less than 30 feet per mile. If we assume that the middle third lies parallel to the axis of rotation (i.e., along the strike), so that its gradient was not affected by the tilting, it should provide a sample of the corresponding original gradient. If so, then the 40 to 60 feet per mile by which the adjacent segments are now steeper is a measure of the post-Eocene tilting. Over a distance of 75 miles, which is approximately the present width of the range in this region, such a change would add about 4,000 feet to the elevation of the crest provided the southwest edge of the block was not depressed in the process. Similar crude estimates have been made from at least three other streams, but because of the many assumptions involved they do little to strengthen the already uncertain result.

"Post-Eocene" is a lot of time—about 40 million years by present estimates. That there has been at least some tilting in the past 10 million years is implied in the history of the Miocene-Pliocene volcanics. As pointed out on page 332, they include many volcanic conglomerates and other stream deposits; during their accumulation the westward slope must have been so gentle that deposition of sediment was the dominant condition. Now, however, this one-time blanket of reworked volcanic debris is being stripped from the mountains (Fig. 365). The change from sedimentation to erosion must have taken place after early Pliocene and was almost certainly caused by some combination of uplift and westward tilting.

Fig. 364. *Upper:* Air view obliquely down Yosemite Valley. Glacier Point near left; Yosemite Falls,
above the center, drain a typical hanging valley (page 216). *Lower:* Drawing emphasizing
the morphology of the scene above by removing the vegetation patterns.

Fig. 365. Remnant of layered andesitic volcanic debris of probable Miocene and Pliocene age resting on granite just north of Silver Lake (background). View is to the south.

Another approach to the dating of any large uplift is through study of the sediments it has shed. The erosion stimulated by each increase in the height or tilt of the Sierra block must have produced new deposits along the west base of the range. But here the method is difficult to apply because we are trying to distinguish small pulses which are relatively close together in geologic time. Critical importance thus resides in the details of the sediments; we need to be able not only to recognize, but also to date and correlate throughout the region, thin units of (1) fine-grained sediments derived from well-weathered low-standing Sierran rocks and (2) coarser, more rapidly deposited detritus spread by rejuvenated streams. Neither the outcrops and drilling records nor our dating methods are good enough to permit detailed interpretations of this kind.

Still another approach is to examine the faulting along the base of the eastern escarpment. The difficulty here is that there are so few datable late Cenozoic rocks involved in the displacements that it is seldom possible to determine the amount and time of any given shift, much less different eposides of movement on the same fault. Even if the vertical component could be measured and dated it would be difficult to say how much of it was caused by uplift of the mountains and how much of it was caused by subsidence of the San Joaquin Valley. Since the valley contains an estimated 8,000 feet of fill in some places it must have subsided appreciably at some time. That there have been numerous displacements along this zone is thus clear, but the details are obscure. A partial exception is indicated by a report that in one place, south of Mono Lake, what is probably an early Pleistocene glacial till has apparently been offset at least 3,500 feet. A small amount of faulting that is definitely very late Pleistocene or Recent was noted in Figure 342.

It will be noted that the indirect evidence for the time of each eposide of uplift or tilting tells only that it took place *after* a given epoch, but not how long after—*post* Eocene, *post* early Pliocene, and *post* early and late Pleistocene. Conceivably, all of these episodes could have taken place very recently—within the last half million years but this is highly improbable, especially if it can be shown that the mountains were high during earlier times. There are conflicting interpretations of the meager evidence on this point.

On the one hand—in favor of low mountains during Miocene-Pliocene time—it has been pointed out that north of Lake Tahoe, in deposits probably as young as late Miocene, there are well-rounded pebbles of rock types apparently foreign to the Sierra block. Their most likely source seems to have been to the east, which means that when they were deposited vigorous gravel-bearing streams may have flowed southwestward across this area from sources in western Nevada. This could have happened only if the Sierra Nevada was then much lower than it is today, or was joined to higher mountains on the east. As further evidence of low mountains, some geologists cite the similarity of fossil plants of probable Miocene and Pliocene age that have been collected both from the west side of the Sierra Nevada and from west-central Nevada. They interpret these as representing similar climates and indicating that there was not then the marked rain shadow which now characterizes the belt immediately east of the range. This would imply that the crest was much lower than it now is.

On the other hand—in favor of high mountains during at least late Pliocene time—is the somewhat stronger argument that some of the Pliocene lavas are surrounded by peaks which today rise at least 4,500 feet above the Pliocene or older surfaces preserved under them. On the assumption that these summits have been undergoing erosion while the surfaces under the lavas were protected, the relief in Pliocene time must have been somewhat greater than 4,500 feet. And since the lavas fill the upper parts of valleys that were at least scores of miles long, and steep enough for their streams to move coarse sediment, it is likely that their floors stood several thousand feet above sea level. This would put the summit of the range at an altitude of at least 6,000 feet.

It is quite possible that different parts of the Sierra Nevada have had different histories and that the concept of a single tilted block, although valuable as a generalization, will prove too simple when enough detailed analyses have been made.

Finally, one wonders how far west this tilted block goes and what is happening there. If it is rising along a fault zone on the east, is it bending, or breaking, on the west? Where is the hinge, if any? Is all of the block rising, or is the western part going down? Data from wells drilled in the eastern part of the San Joaquin Valley show that the crystalline rocks of the Sierra block are buried beneath a wedge of overlapping sediments in the valley. The westward increase in thickness of these deposits, most of which are marine, argues for subsidence of the western part of the block. The changes from marine to nonmarine facies, marking ancient shorelines in these sediments, seem always to have been near or west of the present foothills. The pivot line, therefore, between rising and subsiding portions of the block, has probably always remained somewhere within the eastern half of the San Joaquin Valley.

We may speculate that anything that would mark the edge of such a huge structural block with such a long and active history would probably show up as a major line of disturbance. On this assumption there is no evidence of the west margin of the block across most of the broad San Joaquin Valley, whose thick sedimentary fill is only mildly deformed (Fig. 360, front edge). But along the west margin of the valley and in the adjacent Coast Ranges there is evidence of continuing Cenozoic deformation in the form of localized and shifting sites of sedimentation and considerable folding and faulting of the resulting strata. The San Andreas fault also runs through these hills (Fig. 360), separating unlike rocks along its entire length. Thus it seems reasonable to believe that the Sierra block ends somewhere between the west edge of the San Joaquin Valley and the San Andreas fault, even though we have no direct evidence of its exact position. If it is actually cut off at the San Andreas, its westward continuation, if any, may now be 100 or more miles to the northwest (page 97).

Faulted Ridges in Southern Nevada

The five-mile patch of roughly parallel ridges shown in this aerial view lies between Las Vegas and Lake Mead in southern Nevada. Because of the bright and contrasting colors of some of the strata the local inhabitants refer to it as the Rainbow Gardens.

It can be seen that in all the ridges the strata dip rather uniformly to the right. The sediments range in age from Cambrian at the far left of the photograph to Cenozoic at the far right, and many of them are closely similar to their counterparts in the walls of the Grand Canyon. (The numbers on the wall of the imaginary trench in the drawing correspond to, and extend, the system introduced in Figure 258). Note that the two ridges at the left expose a nearly complete Paleozoic sequence ranging from Tapeats sandstone ([5]) through Kaibab limestone ([13]), all resting nonconformably on Vishnu-like schists ([0]) at the extreme left. A closer view of these strata, here about twice as thick as their equivalents in the Canyon, was shown in Figure 288 (viewed in the opposite direction). Number [14] identifies the Triassic Moenkopi formation and is here placed on a pair of resistant limestones in its otherwise shaly lithology; number [17] is a thick Cretaceous sequence whose basal conglomerate lies unconformably on the older rocks, the intervening erosion having removed various amounts of Jurassic and Triassic strata. The Cretaceous beds also contain an intrusive sill (pattern of v's) and are in turn unconformably overlain by early Cenozoic sediments ([18]).

There are two clear indications of faulting in this scene. One is the abrupt terminations, often involving offsets to right or left, of the well-defined strike ridges. The other is the fact that the same sequence of strata, dipping the same way, is repeated in the same order, several times. It is especially clear in the photograph, for example, that the prominent pair of Triassic limestones ([14]) appear in six unconnected strips, each time overlain to the right by the same thin beds of weaker sediment. This kind of repetition obviously could not have been produced by folding. The gentle westerly dip of some of the fault surfaces is easily determined on the ground: in the photograph it is clearly visible on the near end of the large ridge at the far left.

To thus decipher the major outlines of the structure within a few thousand feet of the surface may help to explain the local landscape, but it can hardly satisfy the geologist interested in diastrophism. *How* and *why* did these blocks become tilted eastward? Is there a larger pattern of which this is only a part? To even begin to answer such questions, it is necessary to know the geology of the surrounding area and to plot all the information accurately on a good map so that the spatial relations can be measured, nearby diastrophism analyzed, and the timing of events in adjacent areas compared.

Viewed in this way, the Rainbow Gardens seem at first to lie between two structural provinces. Thirty miles to the west and northwest is a series of large thrust faults, notable among them the Keystone thrust discussed on pages 322–325, in all of which the overriding rocks have moved relatively eastward. To the east is a belt dominated by blocks tilted, like these, to the east and southeast and separated by gentle to steep faults; sixty miles away these blocks end abruptly at the margin of the Colorado Plateaus (Fig. 37).

We already know that in one place tilted blocks were formed while thrusting was still active, for part of the Keystone thrust plate was broken into such blocks, which were then overridden by the unbroken plate as it continued its eastward movement (Fig. 303). In general the times of thrusting and tilting seem to have overlapped during the Cretaceous period.

Thus, in broad terms, the Rainbow Gardens are part of a larger region of eastward-tilted fault blocks that stops at the edge of the Colorado Plateaus on the east and is complicated by a belt of eastward thrusting on the west. It is obvious that these west-dipping thrust plates once extended farther east and have been eroded back to their present positions. Did they once override the Rainbow Gardens? Are these tilted blocks analogous to those still partly under the Keystone thrust? The possibility gains plausibility when considered in conjunction with an area, explored on the next three pages, that is only 35 miles northeast of Rainbow Gardens.

Fig. 366. Looking northward over Frenchman Mountain (left) and the Rainbow Gardens, 8 to 12 miles east of Las Vegas, Nevada. In the drawing the main faults are shown by light lines; numbers along the imaginary cut conform to, and extend, those used in the Grand Canyon (Fig. 258).

Fig. 367. *Above:* Looking northwestward over the Weiser Bowl and adjacent ridges in the northern
Muddy Mountains, southern Nevada. *At right:* A deep trench has been cut to reveal the
essential structure—a recumbent syncline. Numbers follow the system begun in
Figure 258 and extended in Figure 366.

Combined Folding and Faulting in Southern Nevada

The complexities that have developed from combined folding and faulting in some areas of the crust fascinate, challenge, and sometimes frustrate the structural geologist. Where only sedimentary rocks are involved there is always a reasonable hope that even the most intricate structures can be untangled, because if the relative ages of the strata can be worked out we know not only the original orientation (essentially horizontal) but also the original top and bottom, and therefore which way was up before the disturbance. If igneous intrusion and metamorphism have taken place, the original ages and space relations often become more obscure and the history more difficult to decipher. Here is an example that is entirely free of metamorphism and yields a handsome return on a small investment of observation.

A view northward over part of the northern Muddy Mountains in southern Nevada, 35 miles northeast of the scene on the preceding page, is shown in Figure 367. The feature of greatest interest is the oval depression known as the Weiser Bowl. The regular succession of layers in its floor and walls shows that it is cut into essentially horizontal strata—part of the sedimentary sequence with which we are already familiar. The layer forming a ring of dark patches on the floor of the bowl belongs to the upper part of the Moenkopi formation (14). The lighter-colored, thin, regular strata exposed in the upper wall and rim of the bowl are the equivalent of the limestones in the lower Moenkopi (prominent in Fig. 366). The jagged ridge west of the bowl is Kaibab limestone (13) beyond which are Permian red sandstones (12) equivalent to the Coconino sandstone, Hermit shale, and part of the Supai formation in the Grand Canyon. In other words, the bowl is carved from a nearly horizontal succession that is upside down. This conclusion is further borne out by the fossils in both the Permian and Triassic

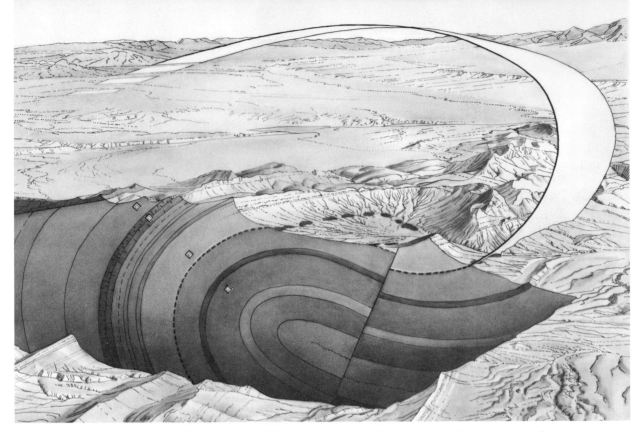

limestones, and by such details of the bedding as ripple marks, cross-bedding, and graded bedding.

The prominent light ridge in the foreground is clearly composed of the same thin-bedded lower Moenkopi limestones that make the rim of the bowl, and near the base of the ridge a row of dark patches represent the same layer in the upper Moenkopi that forms the ring on the floor of the bowl. The westward dip in this ridge would carry these overturned strata beneath their positions in the bowl, so there must be a small fault, upthrown on the west, along the back side of this ridge.

At the far left, where one would expect to find the oldest strata in this overturned sequence, there are instead slabs of limestone and dolomite, representing various Cambrian to Pennsylvanian formations, thrust over the Permian red sandstones. Similar slabs, over fifteen of them in all and covering from less than one acre to several hundred acres each, are scattered throughout this view and to the right of it. These are in fault contact with strata at least as young as Cretaceous, which is interpreted to mean that the whole area was overridden in Cretaceous or later time by a thrust plate, most of which has since been removed by erosion.

To the east, just outside this view, the overturned sequence of the bowl is reversed; the beds dip the same way but are right side up. Still farther east is a faulted steep anticline, overturned to the east.

Thus if we look at the folding alone, the complete structure of one bed would be essentially as shown in Figure 368 (eroded portions shown by broken line).

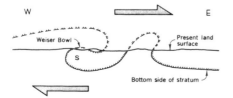

Fig. 368
Generalized sketch of the folding (faulting omitted) in the vicinity of Weiser Bowl. Length of section approximately 5 miles.

397

The only way parts of an originally horizontal blanket of sediments can be inverted while adjacent parts remain upright is through overfolding (Fig. 369) in which the middle limb, common both to the overlying recumbent anticline (A) and to the underlying recumbent syncline (S), is horizontal and inverted. Usually this much deformation also involves some faulting, but faulting as such cannot overturn large masses of strata. It is well to remember, also, that erosion will remove some of the rocks during the deformation, so that the fold, particularly if large, will be far less perfect than a fold in a wrinkled rug on the floor. Nonetheless, the analogy serves as a starting point in understanding this area.

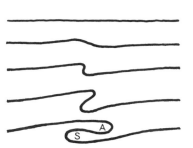

Fig. 369
Diagrammatic representation of the evolution of recumbent folds from originally horizontal strata.
A = recumbent anticline, S = recumbent syncline.

It seems almost certain that there has been east-west compression. The doubly folded single bed sketched in Figure 368 would be 5 miles longer if straightened out to its original unfolded condition. The present extent of the scattered remnants of the thrust sheet is 5 miles east-west by 18 miles north-south; 5 miles is thus also the absolute minimum east-west travel of the thrust plate.

Since pieces of the thrust plate rest on folded strata, it would seem that most of the folding must have taken place before most of the thrust faulting, although there might have been considerable overlap in time. This suggests that the thrusting represents the culmination of deformation that began with the folding and that both folding and thrusting took place in response to persistent forces of east-west compression. This is, indeed, a variety of diastrophism well known in other parts of the world. The Alps are the most famous example, being composed of several series of recumbent folds and thrust plates pushed northward, one on top of another, though the details differ considerably from those of this Nevada scene.

Again, as at Rainbow Gardens, the facts are fairly clear but we lack a full explanation. We are able to work out the geometry of the deformation, but can do little more than speculate on its fundamental causes.

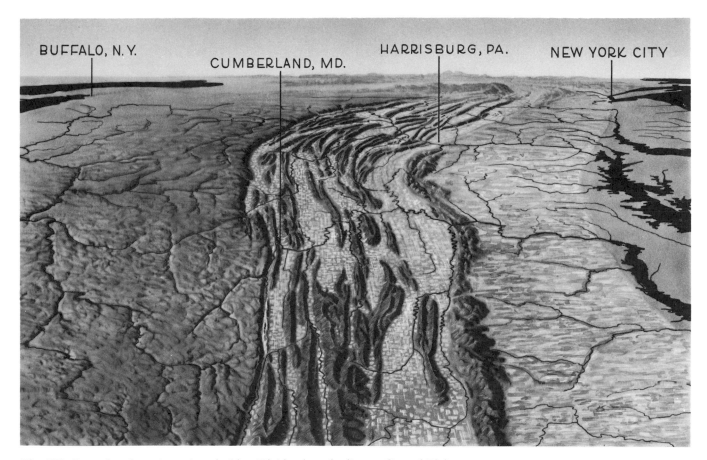

BUFFALO, N.Y. CUMBERLAND, MD. HARRISBURG, PA. NEW YORK CITY

Fig. 370. Central and northern Appalachian Highlands and adjacent Coastal Plain.
Looking northeastward.

The Appalachian Mountain System

The only extensive belt of mountains in the eastern half of the United States is the Appalachian Highlands, an upland that is 75 to 150 miles wide and extends from northern Alabama and Georgia through New England into Canada. The heart of this region, and the part to which we will devote most of our attention here, is the slightly sinuous segment that trends northeast from eastern Tennessee to northeastern Pennsylvania.

A typical cross-section of the Highlands (Fig. 370) consists of three elements. On the northwest is a gently tilted plateau, most of which drains westward into the Ohio River system and whose high southeastern edge is topographically part of the Appalachian Mountains. This dissected upland, whose maximum elevations rarely exceed 4,000 feet, is the Appalachian Plateau (Cumberland Plateau to the south and Allegheny Plateau in Pennsylvania). It ends abruptly in a steep southeast-facing slope, the Allegheny Front of Pennsylvania (Fig. 371), which separates it from a succession of nearly parallel northeast-trending wooded ridges and cultivated valleys. These constitute the second element, the Ridge and Valley province, a sample of which is shown in Figure 372; note the remarkably even crests 500 to 800 feet above the lowlands. The Ridge and Valley belt is bordered on the southeast by a zone of metamorphic and igneous rocks, some of which are part of the Appalachians (e.g., the Blue Ridge of Virginia which widens into the Great Smoky Mountains of Tennessee and North Carolina) and some of which form the Piedmont lowland lying southeast of the mountains. The high parts of this third belt lack the smooth linearity of the ridges west of it and include some of the most rugged topography, and the highest summit (Mt. Mitchell, 6,684 feet) in the Appalachian Highlands.

Fig. 371. The Appalachian Plateau (left), Allegheny Front, and Ridge and Valley province.
Looking northeast toward Lock Haven in central Pennsylvania.

What the Appalachians as a whole lack in loftiness is more than made up in the amount of attention they have received from geologists. Being adjacent to the region first explored by white settlers and ever since accessible from the most densely populated part of North America, a region with a high concentration of universities and enormous appetite for coal and other mineral wealth, they have received more study over a longer period than any other mountains on the continent.

Let us take a closer look at the Ridge and Valley province where it curves across the State of Pennsylvania. The view in Figure 373, looking eastward over the Juniata and Susquehanna rivers, is typical. In the drawing an imaginary trench has been cut across the same scene to reveal the relation between topography and structure.

Fig. 372. Appalachian folds crossed by the Susquehanna River (middle distance) above Harrisburg, Pennsylvania. Cove Mountain, the sharply curved ridge in the foreground, marks a gently plunging syncline that is overturned toward the northwest so that both limbs dip steeply to the right (as may be seen in the river cuts). This syncline is also visible at the far right in the structure section of Figure 373.

Fig. 373. General view eastward over the Juniata River (foreground) and Susquehanna River (background) from a point 30 miles northwest of Harrisburg, Pennsylvania. In the drawing, an imaginary trench has been introduced to reveal the structure. The overturned syncline at the far right may be seen, in a view along its axis, in Figure 372.

The ridges all result from the superior resistance of hard sandstones. Most prominent are the Tuscarora sandstone, at the base of the Silurian, which is responsible both for the ridge in the foreground and the most distant one at the right, and the Mississippian Pocono sandstone, which forms four others shown in the drawing. Somewhat less prominent ridges are formed by thinner sandstones in the Lower Devonian strata; an example (Oriskany sandstone) occurs just beyond the big ridge in the foreground.

The valleys are with equal consistency formed on strata that are predominantly shale and limestone in the Upper Silurian, Upper Devonian, and Upper Mississippian. In this moist climate, limestone, owing to its solubility, is topographically a weak rock; in dry climates the reverse is true, and it often caps the highest peaks. Compare Figure 373, for example, with Figures 366 and 367 (southern Nevada) in which all the major ridge crests are limestone or dolomite and the lowlands are sandstone with some shale.

For obvious reasons (Fig. 373), the Ridge and Valley belt is also known as the "Folded Appalachians."

The geologic map of Pennsylvania, a portion of which is reproduced here (Fig. 374), affords a spectacular means of comparing the structure of the folded belt with that of the areas northwest and southeast of it. Few regions in the world can equal this for demonstrating the relation between outcrop pattern and structure. To wit:

On the plateau at the northwest the pattern of the formations shows broad expanses of the same rock in the flat areas between streams, but along the valley walls closely spaced parallel contacts show that on these slopes the edges of successively lower layers are visible. In other words, the strata are essentially horizontal, the contacts are approximately parallel to the contours (as in the Grand Canyon), and the dominant influence on the outcrop pattern is actually the drainage pattern because it is only on the sides of the valleys that several formations are visible in a small area; on most of the flats between valleys only one formation is exposed at the surface.

Fig. 374. Part of the geologic map of Pennsylvania: Harrisburg is in the center of the lower right quarter. Arrows show the location and direction of views in Figures 371, 372, 373 and 375. (*Reproduced with permission from the* Geologic Map of Pennsylvania, 1960, *issued by the Commonwealth of Pennsylvania, Department of Internal Affairs.*)

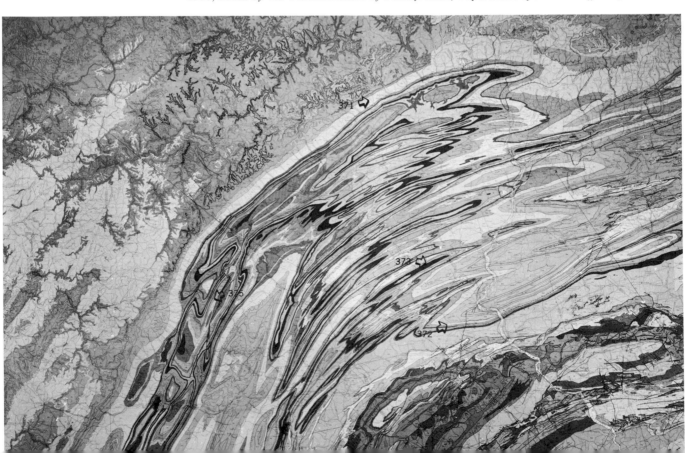

Fig. 375. Looking southwest across Loop Mountain south of Hollidaysburg, Pennsylvania. The Z-shaped ridge is formed by the outcrop pattern of Lower Silurian Tuscarora sandstone which is involved in a syncline (through kink at left) and an adjacent anticline (through kink at right), both plunging gently to the right. Allegheny Plateau in the background.

In the folded belt, on the other hand, the gently plunging anticlines and synclines cause the outcropping edge of each stratum and formation to loop back on itself (compare Figs. 89, 372, and 373). If, for example, a person started at the lower right corner of Figure 375 and walked along the crest of the wooded Z-shaped ridge, the strata would always dip to his right, the overlying younger rocks would always be down the slope to his right, and the old rocks would always dip beneath the ridge from outcrops on his left. Upon rounding the first kink (at the left) he would thus cross the axis or keel of a syncline that plunges gently toward the right, and at the next kink he would go around the nose of an anticline plunging gently the same way. Between the two kinks the strata dip toward us and constitute both the far limb of the syncline and the near limb of the anticline. (This double fold may easily be demonstrated with a sheet of paper.) In contrast to the plateau, the folded belt has an outcrop pattern dominated by the structure of the rocks, and the upturned edges of individual formations are traceable for hundreds of miles as they lace back and forth. It is noteworthy that in both the plateau and the folded belt the traces of individual formations parallel each other; very few are cut off or displaced by faults and, except for a few minor dikes west of Harrisburg, none penetrates or crosses any other. Such a pattern would not be possible if the rocks did not consist of continuous sheets laid one on another and later wrinkled (and eroded) without faulting or widespread igneous intrusion.

The outcrop pattern southeast of the folded belt is fundamentally different from the other two. Instead of the edges of persistent sheets, we see individual rock masses (formations) that occur as twisted pods, many of them cut off by faults, and many invaded by igneous intrusions. *403*

Some idea of the prodigious amount of rock removed since Paleozoic time can be obtained by restoring, in imagination, the eroded upper portions of the folds; some would rise more than 20,000 feet above the present landscape. Indeed, the present Appalachians are only the stumps of a mountain system; how much higher they may have been in the past is unknown. Today their topographic relief is but a small fraction of their structural relief.

This résumé of Appalachian history leaves many fundamental questions unanswered. Why are the same strata that lie essentially horizontal and undeformed in the plateau so folded and faulted in the Ridge and Valley belt? The boundary between the two is much more abrupt than any lateral changes in the sedimentary rocks themselves, implying that the crystalline basement is fundamentally different in the two provinces. If so, what is the difference? And why do the sediments end so abruptly at the southeast, where they are thickest? Did they accumulate in a lop-sided trough with a very steep east side? If not, what has become of their former continuation toward the Atlantic? And how can such a continuation be reconciled with the landmasses that supplied the wedges of clastic sediment?

Full answers to these questions are still being sought. Most of the eastward-thickening blanket of Paleozoic sediments, which consists of limestones, shales, and clean sandstones, must have accumulated in shallow water. This is especially true of the Silurian and younger rocks. The Pennsylvanian coal measures and parts of the underlying Mississippian beds are continental lowland deposits. Toward the southeast, however, a few formations, notably the thick eastern facies of Ordovician shale and sandstone (Martinsburg formation), have many of the characteristics of turbidity-current deposits in deep water.

Shallow-water sediments can accumulate to thicknesses of 5,000 to 30,000 feet only if the site of deposition subsides about as fast as it is filled; thick deep-water sediments require that subsidence keep well ahead of sedimentation. On all counts it would appear that subsidence was greater and more rapid toward the southeast. This puts the most active sinking close to the most active uplifts—those that produced the clastic wedges and probably most of the other clastic sediment. (Water-current structures in the Martinsburg formation indicate that it came from an eastern source—all 8,000 cubic miles of it.) In short, there was evidently a zone on the east in which deformation was more active and more complex than it was in the Appalachians proper.

The existence and nature of this active zone must be deduced chiefly from what is left of it in the present Piedmont belt of crystalline rocks. These may be divided into two groups:

The larger consists of metamorphic rocks that range from slate, quartzite, and marble to schist and gneiss. Among these are metasediments derived from the kind of massive and poorly sorted deposits (like the Martinsburg formation) that characterize deep water not far from land, where debris is abundant but out of reach of most of the currents and other processes that help to differentiate types of sediment; e.g., clay and sand are mixed, instead of separated to form beds of shale and sandstone. Occurring with these are metavolcanics derived from submarine eruptions and a few low-grade metamorphics that strongly resemble those in some of the formations to the west. Most important, there are a few mildly metamorphosed rocks in which recognizable early Paleozoic fossils have been found.

The second group in this belt comprises granitic and other plutonic rocks, which occur as irregular bodies embedded in the metamorphics.

Early workers in the Appalachians naturally supposed that these crystalline rocks of the Piedmont were Precambrian, the roots of the mountains whose erosion had supplied the Paleozoic sediments to the west. But the implications of the fossils and the lithologic similarities mentioned above have been greatly strengthened by modern radiometric age determinations, which seem to indicate that all of the plutonic rocks and most of the metamorphic ones are Paleozoic and that only a relatively small part of the Piedmont crystalline belt is Precambrian. In other words, most of these crystalline rocks formed at the same time as the sedimentary rocks to the west.

It now appears that the Piedmont belt was anything but a narrow highland passively shedding sediments into a narrow trough along its northwest flank; it had a complex and lively history of its own. It probably represents the zone of maximum sedimentary accumulation (estimated at 50,000 feet) in this region. Probably it had already been receiving deposits for many tens of millions of years when, in Middle and Late Cambrian time, the sea that covered it began to spread northwestward onto the continental platform, there to leave the lowest strata of the Appalachian Paleozoic sequence. During early Paleozoic time parts of it apparently remained deep beneath the sea despite infilling deposits, while other parts rose into debris-shedding ridges. In middle and late Palezoic time it was intruded by a variety of granitic and other plutonic rocks which, because bodies of such rock are typically found in the roots and cores of mountains, are probably further indirect evidence of crustal deformation. Apparently this crustal unrest reached a first climax near the middle of the Paleozoic and a second toward the close of the era. In the course of the latter, the whole belt, probably now all above sea level, was for some reason crowded northwestward at least several tens of miles against its own former shallower margin, crumpling part of the thinner sequences there into the Folded Appalachians. Certainly this last event greatly compressed and confused any transition zone that may have existed between the Ridge and Valley province, which was relatively quiet throughout Paleozoic time, and the Piedmont belt, which was quite active during at least the first part of the era.

If this interpretation is at all close to the truth, it means that the events in the geologic history of the Appalachian Plateau and the Folded Appalachians were only a side show that started late and involved much less vigorous crustal deformation than took place in the Piedmont belt. It also makes one wonder if the amount of later deformation and mountain-making may not be related in some way to the amount of earlier subsidence; certainly from northwest to southeast across the Appalachians the progressive increase in depth of the crystalline floor seems to be matched by evidence of progressively stronger deformation. One also wonders where and how all this stopped toward the southeast. The record is obscured today by the overlapping Cretaceous and younger deposits along the coast, but most geologists would argue that there was nothing comparable to the folded belt and plateaus on this side because a short distance offshore there is a fundamental change from continental- to oceanic-type crust.

The Appalachians are the mountains in which it was first observed (over 100 years ago) that mountain-making seems often to involve mainly the thicker parts of a sedimentary sequence —as though the deformation were in some way localized along the site of the former trough whose subsidence accounted for the great thickness. This relationship has since been recognized in many parts of the world. The subsiding sediment-filled trough is a geosyncline (page 290) and the many-faceted concept of its spatial, temporal, and possibly even genetic relation to later mountain-making is one of the great generalizations of geology—one that has been outstandingly fruitful, both as an aid to understanding such very complex structures as the Alps, and as a guide to speculation regarding what happens beneath the crust. We will return to this last problem briefly in Part VI.

VI   Implications

Some fact and speculation bearing
on the deeper problems raised by
observations at the earth's surface.

35 MEASURING CRUSTAL MOVEMENT

Scattered through the preceding sections of this book are many arguments in support of the general conclusion that for hundreds of millions of years the earth's crust has been active. Present-day volcanism, earthquakes, and warping of shorelines are not the only evidence of the crust's activity. Evidence resides in every outcrop of uplifted, tilted, folded, or faulted sedimentary strata. It is recorded in every widespread unconformity, and where, as in the Grand Canyon and at Turtle Bay, one unconformity truncates another, the record is both long and rich. It is implicit in every exposure of originally deep-seated rocks, such as granite or gneiss, for these can be made visible only by the removal of the vertical miles of rock beneath which they formed, and this in turn is possible only where there has been a comparable amount of uplift. Where metamorphosed sediments are seen at the surface a complete cycle of subsidence (with sedimentation) and uplift (with erosion) is spelled out. The very existence of most of the world's mountain ranges and subsiding basins, as well as of its tectonic landscapes, is a kind of general proof that in many areas crustal movements have kept ahead of erosion. It proclaims in grand terms the victory of diastrophism over the processes of erosion, which by persistently taking material away from high places and depositing it in low places would otherwise have long ago reduced the lands nearly to sea level. (Indeed, the average load of the Mississippi River today corresponds to the removal of one vertical foot from its drainage system approximately every 7,000 to 10,000 years, and the average for all United States rivers is believed to be about twice this. At such rates North America could be reduced to sea level in about 12 to 25 million years, but of course the rate would not be maintained as the general level declined and stream gradients and discharges were correspondingly reduced.)

Are diastrophic movements taking place today? Has the crust been as active in the Christian era, or the twentieth century, as it was in the Cenozoic Era or the Pleistocene?

One approach to these questions is archeological. For example, near Naples there is a Roman ruin (Fig. 377) that probably served as a bayside marketplace. Today it stands in several feet of seawater, but old engravings and photographs show it as dry in the seventeenth, eighteenth, and early nineteenth centuries. On its tallest columns, well above present sea level, there are borings of marine organisms and other signs of submergence to a height of 20 feet above its floor. On the reasonable assumption that this building was originally constructed on dry land, its very simplest history includes submergence and then emergence of more than 20 feet between Roman times and approximately the eighteenth century, followed by partial resubmergence since that time. We know that this is largely a result of warping of the land rather than of changes in sea level because other points along the Italian coast do not record the same changes.

Another approach to the question of historical crustal movements is geodetic—by direct measurement. Geodesy is the branch of applied mathematics and engineering concerned with accurate determination of the position, size, and shape of large areas of the earth's curved surface and, ultimately, the size and shape of the earth itself. These aims require precise determinations of latitude, longitude, and sea level, as well as of such factors affecting these measurements as the strength and direction of gravity and of the earth's magnetic field. In determining horizontal positions, triangulation is used. Once the direction and length of the line connecting two points are accurately known, this line can be used as a *baseline* from which the location of a third point can be determined if it forms a roughly equilateral triangle with the first two. The sides of this first triangle may in turn be used as bases for new triangles whose apices are new points, and in this way a *triangulation net* can be extended over hundreds or thousands of square miles. (An example, with the connecting lines removed, may be seen in Fig. 381.) Vertical positions are determined by carrying mean sea level inland along sequences of short level lines whose vertical separation, and therefore elevation, are carefully measured with special rods.

Fig. 377
Columns of the so-called Temple of Jupiter Serapis near Naples, Italy. The flooding with seawater proves submergence and the borings by marine organisms—concentrated in the lower halves of the large columns—indicate emergence. (*Photo courtesy of Bruno D'Argenio, Naples.*)

Permanent brass markers (bench marks) are usually set at the points in a triangulation net and the elevation of each is determined by leveling. In regions where earth movements are suspected, such as those where earthquakes are relatively frequent, the locations and elevations may then be redetermined at appropriate intervals. In some places such measurements have disclosed slow systematic changes in position. But even with such data in hand many questions remain: Has the movement been tectonic, growing out of real activity of the crust, or superficial, like the subsidence that accompanies the compaction of sediments? (Parts of San Jose, California, have subsided over 9 feet since 1912, primarily as a result of groundwater withdrawal; the Wilmington oil field near Long Beach, California, has subsided more than 25 feet in some places because of reduced underground fluid pressures attending withdrawal of oil and gas.) If the movement has been tectonic, did it occur gradually, as a series of jerks, or in some other way?

On the next seven pages we will explore examples in which some questions of this kind can be partly answered.

411

36 DIASTROPHISM AND TIME

The Buena Vista Hills

Large-scale movements in the crust generally take place so slowly that we are unaware of them. We speak of *terra firma*, of the "everlasting hills," and establish legal boundary lines in terms of river courses, shorelines, and rock monuments. To do so causes very little practical difficulty, but it does tend to obscure the fact that the positions of these features are not necessarily permanent.

At the right are scenes in the Buena Vista Hills oil field north of Taft, California. Since 1910, when its first successful well was completed, this field has produced large amounts of gas and over 4 million barrels of oil from a total of more than 1,000 wells.

Like Signal Hill (page 365), the Buena Vista field is located on a northwest trending low ridge that is the surface expression of a gentle anticline in the sediments filling a Cenozoic basin —in this case the southern San Joaquin Valley. As indicated in the diagram at the right, dips on the flanks of the fold reach 20° to 25° in Pliocene strata at depths of 3,500 feet but are more gentle in the Pleistocene, nearer the surface. This change in dip and the fact that some layers are thinner and others missing entirely over the crest of the fold show that the anticline was growing during at least part of Pliocene and Pleistocene time. Has it stopped growing?

In the early years of the oil field's development there were an unusual number of casing failures. By 1932 twenty-three wells had become so crooked as to require repairs. At depths ranging from 76 to 794 feet the steel casings lining the wells had been offset by local bends that displaced their vertical alignment by as much as 15 inches. Tools could no longer be lowered through them. Although the depths of these bends differed from well to well, when all were plotted on a map they defined a nearly plane surface dipping about 25° a little east of north. When this surface was projected to the surface of the land near the crest of the hill it coincided with a number of places where paved roads had broken up, telephone wires had sagged, and, most spectacular of all, pipelines buried in shallow ditches trending up the hill had buckled and slowly risen into arches like that shown in Figure 379 at the right. In short, here was a small, actively moving thrust fault.

Since the 1930's the difficulty with the wells has been reduced by deep excavations and enlargement of their upper parts. Geologists have kept track of the movement through frequent measurements between markers carefully placed for the purpose. These measurements indicate movement at an average rate of 1 foot in about 14½ years. There are also 19 points in a U.S. Coast and Geodetic Survey triangulation net that lie within about a mile of the surface-trace of the fault. Between 1932 and 1959 the 8 points lying north of the fault trace moved southward an average of 9½ inches and the 11 points south of it moved north-northeast an average of 7½ inches. This shows that movement at a similar rate (1 foot in about 19 years) extends considerably beyond the small area in which it was first discovered. Evidently the deformation has involved at least several square miles of terrane in movement that has persisted for at least the past 50 years. The measurement is comparable to the displacement of about 1 foot in 24 years at the winery on the San Andreas fault (page 97).

The Buena Vista Hills thrust fault is evidently a local and minor feature associated with the anticline. Its outcrop can be followed for less than 2 miles along the 15-mile fold, and it is not recognizable deep on the north flank. Combined with the recent discovery (from relevelings between 1957 and 1964) that the surface of the oil field is also subsiding, these observations suggest that at least some of the movement may be a side effect of the withdrawal of oil and gas. Whether or not some is also tectonic is still uncertain.

Fig. 378. Diagrammatic view east-southeastward over the Buena Vista Hills north of Taft, California. Note changes in thickness over the anticlinal crest. The view in Figure 379 is in the same direction, along the outcrop of the fault, just to the right of the wells.

Fig. 379. Detail along outcrop of Buena Vista Hills active thrust fault. Pipe is buckling out of its ditch because of shortening as hill on left advances over lowland on right.

Measured Horizontal Movements in Imperial Valley, 1939-1954

On the evening of May 18, 1940, the regular pattern of orange trees in the scene below was disrupted by movement along a previously unknown fault trending obliquely through the grove. Since the trees are 25 feet apart and the north-south (right-left) shift amounted to about half the interval between rows, this component of movement was about 12½ feet. Along the fault trace the displacement was 14½ feet. Many irrigation ditches and roads, including the highway across the upper right corner of this view, acquired sharp kinks (which have subsequently been straightened out) showing that the sense of movement was right lateral and the maximum slip along the fault trace was about 19 feet. The accompanying earthquake was one of the most severe to strike this area within historic times.

The disruption of the land was not everywhere so marked as in the orange grove; most of the surface breaks were shorter, less straight, showed less displacement, and were only roughly aligned. The explanation for this variety probably lies in the properties of the near-surface rocks. According to seismic surveys, the crystalline basement is about 15,000 feet below the surface in this area. What occurred at the surface was therefore the effect of the shock—after it had passed through a three-mile blanket of weaker rocks and unconsolidated alluvium—produced by whatever happened in the crystalline rocks below. The broad pattern, however, was so simple and systematic that it could only be interpreted as quite faithfully representing real displacement of the crust. Note the following:

A triangulation net covering the Imperial Valley region was completed in 1939. In the spring of 1941 the positions of the points in the most severely shaken part of the valley were redetermined. Figure 381, A, shows, by means of vectors, the direction and magnitude of the changes between 1939 and 1941—most of which must have taken place at the time of the earthquake.

Fig. 380. View westward over orange grove east of Calexico, California, 19 years after it was offset by slippage along a northwest-trending fault.

Two fundamental facts are demonstrated here: (1) All points on one side of the fault moved in the same direction; all points on the other side moved in the opposite direction. (2) The total displacement was greatest close to the fault trace and diminished with distance from it.

Ever since the San Francisco earthquake of 1906 there has been keen interest in horizontal earth movements in California. With increasing frequency and accuracy, the U.S. Coast and Geodetic Survey has resurveyed carefully chosen belts and nets of triangulation in areas of suspected movement. Across the San Andreas fault zone in the central part of the state these have disclosed slow drift at rates of up to one or two inches a year, the coastal part of the state moving northwestward relative to the inland part. As a result of this, a horizontal line drawn at right angles to the fault trace would, as seen on a map, be rotated clockwise about one second of arc every ten years.

Because the San Andreas zone passes into the Gulf of California, it was suspected that similar slow distortions might be affecting the Imperial Valley region, so the triangulation net in this area was surveyed again in 1954, 14 years after the earthquake. During this interval there had been no earthquakes of importance in the Valley, yet, as shown in Figure 381, B, the net had become distorted. If the station in the upper right corner is assumed to have remained fixed, then the maximum shift, at the west, was about 4 feet to the northwest—or again a twisting of about one second of arc per decade.

These observations have greatly strengthened the theory that slipping of the type that took place on the San Andreas fault in 1906 and in the Imperial Valley in 1940 is not caused by a sudden convulsion of the earth but rather by the release of strains that have been slowly building up during the preceding years. It seems certain that there is some kind of slow rock flowage taking place at depth and that the more brittle crust can accommodate itself to the resulting distortions only up to a certain point. When this limit is reached the rocks break and their elasticity snaps them back toward an unstrained position. The resulting vibrations are the quake of the earthquake.

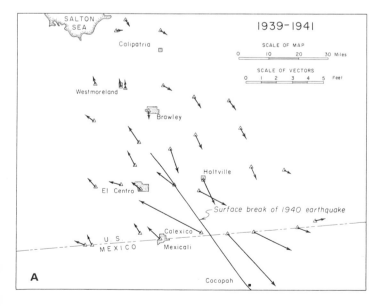

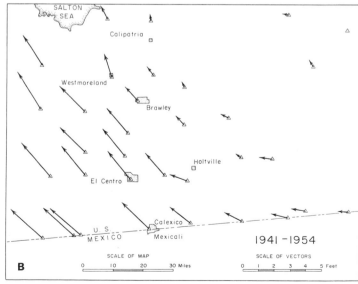

Fig. 381. A portion of the triangulation net in Imperial Valley, California. In A the vectors show the horizontal displacements between 1939 and 1941, the period in which the earthquake occurred. In B, the vectors indicate the shift during the period 1941–1954 on the assumption that the northeast edge of the net had remained fixed. (*From C. F. Richter, Elementary Seismology, W. H. Freeman and Company, 1958.*)

Diastrophism in Time Perspective

We have seen evidence that the long history of the earth's outer crust is the story of a thin skin that for several billion years has been heaving and sinking, folding and fracturing, flowing, melting, and erupting. No part of the crust known to man has escaped. If this concept is difficult to grasp, it is only because of our habits of thought about time—the short perspective afforded by a human lifetime. Were it possible to make a time-lapse motion picture that telescoped the events of a billion years into an hour, the earth would appear as a seething, writhing, erupting sphere. Against this background one must view with wonder—and appreciation—the parallel story of the stuggle of life to survive by adapting to this uncertain footing.

Certainly no quest is more central to geology than the search for an explanation of this crustal restlessness. Before speculating on the deeper mechanism, let us summarize what we are trying to explain—the forms produced by diastrophism, the magnitudes involved, and the rates at which the deformation takes place. We have seen evidence of:

Regional subsidence of 5 to 10+ miles
    Vishnu schist
    Appalachian geosyncline
Regional uplift of 1 to 10+ miles
    Vishnu schist
    Colorado Plateaus
    Appalachian Mountains
Local subsidence of 1 to 5+ miles
    Death Valley
    Los Angeles Basin
    Salton Basin
Local uplift of up to 4+ miles
    Black Hills
    Signal Hill
    Sierra Nevada

Shortening (by wrinkling) of 5 to 10+ miles
    Weiser Bowl area
    Folded Appalachians
Strata completely overturned
    Weiser Bowl area
Vertical offsets of 1+ miles
    Waterpocket monocline
    Hurricane fault
Strike-slip faulting of 10's of miles
    San Andreas fault zone
Thrust faulting of 10+ miles
    Keystone thrust fault
Lava outpouring of 10,000's of cubic miles
    Columbia River basalts

The little we know about rates of diastrophic movement is based on observations covering periods of time that are extremely short when compared with the time required to produce the features just listed. On earlier pages we noted the following rates of deformation (here expressed as *feet per century* without regard to the direction of motion):

Death Valley tiltmeters	1+
San Andreas fault zone	4 to 16
Buena Vista Hills thrust	5 to 7
"Temple of Jupiter Serapis"	2+
Imperial Valley (post-earthquake)	20+

Many others could be added. The elevations of some bench marks along the railroad through Cajon Pass, 50 miles east of Los Angeles, have been determined as many as six times between 1906 and 1961. Using a bench mark about 17 miles northeast of the pass as an arbitrary fixed point, the 26 bench marks between it and the summit have been rising at an average rate of 0.45 foot per century. The corresponding figure for 25 bench marks along 5.7 miles of track on the steeper southwest side of the pass is 2.56 feet per century. The second group comes within 2 miles of the San Andreas fault zone. Tide gage records in Alaska show that parts of the area west of Juneau are rising at rates as high as 12 feet per century, probably in response to un-loading as glaciers melt. The average rate of uplift of points on the east European plain, west of the Urals, is about 1 foot per century with maxima of over 3 feet per century. In Japan rates of uplift range up to about 25 feet per century, the average being about 1.5. In parts of the Persian Gulf uplift is taking place at rates of 1 to 3.3 feet per century.

Figures like those just cited need to be examined geologically as well as mathematically. First, it is necessary to know whether the movement is truly tectonic or only superficial, such as might arise from compaction of recent sediments, shrinkage of clays, or creep. This question can usually be settled by attention to the geology surrounding the site; if a bench mark is mounted on solid crystalline bedrock it probably is not susceptible to such changes. Also, since most surface processes lead to lowering of the land, a bench mark that is rising is probably responding to tectonic movements, although in a few places even this is not certain. Second, any kind of displacement must be measured in relation to some frame of reference; but when the whole earth is considered, what should that reference be? The simplest system, geometrically, for a spinning spheroidal body like the earth, would be to describe points according to their direction and distance from its center, or in relation to its axis of rotation. At the present time both are impractical. We approximate them with latitude and longitude, which we determine by observing the stars, but neither of these is measured directly from earth's center and neither one can be determined to within anything like an inch on the earth's surface. Indeed, accuracy of horizontal position to within a few tens of feet by this method would be considered very good, but actually there is no way of knowing exactly how accurate such a position is; all we can do is to compare it with others nearby, which have about the same inherent uncertainty. Distance from the center of the earth is also impossible to determine very accurately; we use sea level as a datum surface, instead, but this is fraught with complications associated with both position and time (pages 178-179).

(Triangulation using artificial satellites, and distances measured with the geodimeter—which makes use of the speed of light between the instrument and a distant mirror—are now providing improved accuracy in the determination of long distances from point to point on the earth's surface. But the value of these new tools is limited to the refinement of *relative* positions; they help us to know where one point is in relation to others we have established, but cannot provide an independent determination of where we are on the earth.)

In short, the only movements of the crust we can measure directly, and with known precision, are those that involve small areas and relative position at a given time. If the reference point 17 miles from Cajon Pass rose (in relation to sea level) between 1906 and 1961, all the figures given for uplift in that area are too small; if it subsided, all are too large; if sea level changed during that period, a further correction is necessary and we must specify both the location and time of our reference datum. If the triangulation station in the upper right corner of Figure 381, B, moved (in relation to some other point) between 1941 and 1954, then all the vector arrows need to be changed in value and probably also in direction; but is that "other point" any better as a reference?

Note, however, that although the problems of precise location become elusive when we try to reckon absolute positions rather than positions in relation to an arbitrary fixed point, the uncertainties do not invalidate the main conclusions we are drawing from these data. There has definitely been *warping* along 22 miles of track through Cajon Pass, regardless of whether it was superimposed on regional uplift or on subsidence in relation to sea level; and there can be no doubt about the *distortions* within the triangulation net in Imperial Valley between 1939 and 1954, no matter what happened beyond the boundaries of the net.

The examples cited above indicate that deformation rates of from 1 to 5 feet per century are a plausible measure of the speed of diastrophic events in active areas during modern times. If such rates persisted they would yield deformation amounting to 2 to 10 miles in a million years, or 20 to 100 miles since the Miocene, which is more than adequate to account for our interpretation of what has actually happened. In published studies of the San Andreas fault, for example, the largest displacements so far suggested include a shift of about 65 miles since Late Miocene time. The simple anticline of the Kettleman Hills (left foreground, Fig. 360) has risen about a mile and a half since a blanket of probable early Pleistocene sediments was laid across the area, i.e., at a rate of about two miles per million years. Uplift like that now apparently occurring at Cajon Pass could produce a Mt. Everest in 9 million years even if allowance be made for 2 feet of erosional lowering for every 3 of tectonic uplift.

The rates cited above for diastrophism are remarkably similar to those for deformation produced by external loads on the crust. The weight of the water in Lake Mead (about 40 billion tons) has caused crustal subsidence along its shores at the rate of about 6 feet per century. In both Fennoscandia and eastern North America hundreds of thousands of square miles of land, centered in the area where the ice was thickest during the Pleistocene, have risen more than 900 feet since the ice melted, clearly in response to the reduced load. The adjustment has probably taken place in the past 12,000 years or less, or at an average rate of at least 7 feet per century. Similar effects are inferred from the old shorelines around Great Salt Lake (page 361) where the rate is more than 2 feet per century. Whether pushed from above or below the crust is evidently capable of yielding at rates of a few feet per century.

Such rates are not found all over the world, however, nor have they characterized all of the geologic past. Probably nothing so rapid is needed to account for the Cenozoic history of the Appalachian region. Over much of central Canada and the interior of the United States, Paleozoic and younger strata are but little deformed, proving that the region has been relatively stable for most of Paleozoic and later time; well-defined subsiding basins have developed from time to time, but certainly continuous deformation at the rates cited above has not taken place.

From all this it is reasonable to conclude that, in broad perspective, peaks of diastrophic activity have been local both in area and in time, and that so far as we can tell deformation need never have taken place any more rapidly than it is taking place right now in the more active parts of the continents.

37 WITHIN AND BELOW THE CRUST

In Search of a Mechanism

How are changes in the position, attitude, and shape of rock masses actually accomplished? How do rocks yield to deforming stresses? What are the physical properties of the crust? The study of these problems is a complex and rapidly advancing frontier of earth science; let us limit ourselves here to a few generalizations that seem to follow from the observations we have made.

From the samples presented throughout this book it is clear that when the materials of the earth's outer crust crack along joints and faults they behave as brittle solids, much as smaller pieces do when struck with a hammer. Yet the mechanism of slow accumulation of strain followed by sudden release, which has produced earthquakes on the San Andreas fault and in Imperial Valley shows that in larger masses near the surface the same rocks are elastic to a small but important degree. These two properties are not inconsistent; the rocks rupture when their elastic limit is exceeded. (The same is true of many familiar substances. Common glass tubing, or sheets of glass, can be bent quite noticeably, and will spring back with near-perfect elasticity, if deformed only a little: greater distortion leads to breaking.)

It is also clear from the samples presented on earlier pages that there are an equal or greater number of situations in which rocks have been bent and folded (e.g., Waterpocket monocline, Appalachians). In many of these it is possible to prove that there has also been squeezing and flowage in the rock. Some of the limestones involved in the Appalachian folding show marked changes in thickness when traced from the limbs to the crests and troughs of the folds. One, at the top of the Cambrian, contains great numbers of pin-head-size calcareous pellets (oölites) that are essentially spherical in the undeformed rock but have been conspicuously flattened and elongated where the limestone is tightly folded. By measuring the amounts and directions of flattening and elongation of hundreds of oölites from different parts of a fold it has been shown that the deformation of these little pellets is systematically related to their position in the fold and to the thickness of the limestone. Considered as scattered samples of the rock, of known original shape, they prove that the limestone mass has behaved plastically. It has flowed from areas of high pressure toward those of low, and as a result the thickness of the limestone sequence has more than doubled in some places while thinning in others. Many gneisses and schists also contain evidence of plastic deformation. The metaconglomerate shown in Figure 54 is an example of what are sometimes referred to as "stretch-pebble" conglomerates because of the squeezing and flattening of the pebbles. The intricate wrinkles and crenulations and the squeezed and deformed fossils occasionally found in mildly metamorphosed rocks tell the same story.

Plastic flow in the crust is also implied by its slow response to significant changes in surface load. Lake Bonneville, for instance, added its 10,000 billion tons to an area of about 35,000 square miles, depressing the region an unknown amount. Since the large lake disappeared, about 10,000 years ago, the land has risen at least 210 feet in the center of the depression and is still rising. As mentioned on page 418, a similar, but greater recoil has followed the melting of the Pleistocene ice at the centers where it was thickest—near Hudson Bay and in Fennoscandia. Uplift around Hudson Bay has tilted the basins of the Great Lakes as much as 400 feet and raised marine features near Quebec more than 900 feet above sea level; the process is still going on at rates up to 1½ feet or more per century. Uplift near the north end of the Gulf of Bothnia has produced a dome which, as recorded in warped marine shorelines, is now more than 900 feet high and rising about 3 feet per century. The most likely explanation of these adjustments is that at some relatively shallow depth within the earth there is a zone in which the rocks are plastic enough to be displaced by local loads on the surface. In such a zone the rock would be squeezed out from under the load and would flow back when the load was removed. The warping that can be measured today in the examples cited above indicates that the recovery is still taking place; theoretical considerations suggest that recoil from the weight of the Pleistocene ice is more than half complete and slowing down. These observations are among the many in support of the concept that there is a tendency for the larger features of the crust to maintain a flotational balance, called *isostasy* (Greek: *isos*, equal + *stasia*, condition of standing).

What determines whether a rock will yield as a brittle (and elastic) body or as a plastic one? If we consider the earth as a whole and think in terms of very large masses, the three principal factors seem to be (1) the physical properties of the rock, (2) the rate at which it is deformed, and (3) the depth in the crust at which the deformation takes place. Since we are seeking broad generalizations, we may eliminate variations in the first in our speculations about what goes on deep in the crust. All the deep-seated rocks must be crystalline types adjusted to the higher temperatures and pressures of that environment. Such rocks must have physical properties more like those of gneiss and granite than like those of sandstone or limestone, and any differences among them are probably negligible for our purposes. As pointed out earlier, we may reasonably assume that what happens to ice under 200 feet of its own thickness, to rock salt under several thousand feet of overlying sediment, and to limestone involved in folding of the Appalachian type, probably also happens to gneiss at proportionately greater depths, confining pressures, and perhaps deforming stresses.

The importance of the second factor, the rate of distortion, may be readily appreciated from consideration of the propagation of earthquake waves. The two principal types of waves associated with earthquakes are (1) compressional, analogous to sound waves, and (2) transverse, consisting of successive rhythmic displacements at right angles to the direction of advance of the wave—like ripples on water or the waves produced in a hanging rope by a single shake of its lower end. The advance of such shear waves depends on keeping the transverse vibrations alive, which in turn requires the existence of a restoring force—one which tends to return the displaced particles of rock, water, or rope to their original positions. For the water waves and the hanging rope this force is mainly gravity, which has the effect of pulling down the wave crest and of straightening out the rope; without this the first wave trough or crest would simply remain below or above the water surface, the initial kink in the rope would freeze in position, and neither wave would travel. In the earth the restoring force is the elasticity of the rocks, which makes them tend to return to the shape they had before the distorting wave came along. The significant thing is that these crosswise vibrations and resulting transverse waves travel quite readily through the outer 1,800 miles of the earth, proving that to this depth the earth reacts elastically to sudden distortions. Yet this part of the earth also includes all the observable rocks exhibiting evidence of plastic deformation. There is an enormous difference in the corresponding rates of deformation, however; earthquake waves are commonly generated by rock movements of several feet per second whereas plastic deformation, as we have seen, probably does not take place faster than a few feet per century. The rates differ by a factor of about three billion.

The influence of the third factor, depth, on the physical behavior of rocks arises principally from the effects of increased pressures and temperatures. Earlier we considered some of the geological evidence, from migmatites and contorted schists and gneisses, that rocks become plastic with the greater confining pressure and temperature that accompany increased depth. This can be demonstrated in the laboratory. Samples of natural rock that shatter when put in a hydraulic press, can be stretched or shortened in an apparatus designed to maintain the specimen at temperatures and confining pressures comparable to those which exist several miles down in the crust.

Putting these observations together, we conclude that the main constituents of the crust yield as brittle and remarkably elastic materials when at or near the earth's surface, whether the distortion arrives as a sudden earthquake wave or accumulates slowly, as it is now doing along the San Andreas fault and in Imperial Valley. But with increasing depth rocks yield plastically to slowly imposed deformation, although they continue to react elastically to the passage of earthquake waves. We may refer to the property of resisting sudden deformation as *rigidity*, and that of resisting long-continued stress as *strength*. In these terms the earth below the zone of brittle behavior within a few miles of the surface may be said to have high rigidity and low strength. (Such a combination of properties is spectacularly displayed by silicone, ("bouncing") putty which when molded into a sphere bounces better than a golf ball, yet when left on a flat surface, spreads into a thin pancake under its own weight in a few minutes. When struck with a hammer it shatters; with patience it may be poured.)

What are the mechanisms that might account for the formation of folds, geosynclines, faults of large displacement, subsiding basins, and uplifted plateaus and mountains? From the geologic record and by direct observation we know that these are the products of very slow deformation—probably not over a few feet per century. Under the stresses attending such tectonic processes there must then, at all places and times, be a boundary separating an upper brittle zone from a lower plastic zone. This would correspond to the depth of about 200 feet in glacier ice below which the ice flows plastically but above which it cracks and rides along as a brittle carapace. In rock salt this boundary is at a depth of at least several thousand feet. In limestone, to judge from laboratory experiments, it is at a depth of at least several miles. In average plutonic and metamorphic rocks, by analogy, it is presumably still deeper—perhaps somewhere between 10 and 25 miles; it may be a poorly defined zone rather than a sharp boundary.

The term "crust" might seem appropriate for the rocks above this boundary, since they exhibit brittle behavior. But the crust is not so defined because so little is known concerning the properties that would be critical in such a definition. Instead, the crust is currently defined in terms of earthquake waves, whose velocities and paths can be measured fairly accurately. In general, the deeper the path followed by a seismic wave, the faster it travels. But the relationship is not simple and there are several levels at which abrupt changes occur. One is at a depth of about 1,800 miles, below which the speed of compressional waves is suddenly reduced by about 45% and transverse vibrations seem to disappear altogether (Fig. 382). Another occurs at an average of 18 to 25 miles beneath the continents, the depth being greater under some mountains, and much less under the oceans. Upon passing downward through this boundary, known as the Mohorovičić discontinuity, the velocity of seismic waves abruptly increases 15 to 20%. By general agreement these two discontinuities divide the earth into the three principal zones shown in Figure 382, the *core* below about 1,800 miles, the *crust* above the Mohorovičić discontinuity, and the *mantle* between. A third irregularity in seismic velocities is associated with a zone lying between about 40 and 150 miles below the surface; from its effect on transverse seismic waves, this layer within the upper mantle seems to be less rigid than the material immediately above or below it. The top of this "low velocity" or plastic layer may mark the maximum depth of brittle tectonic behavior and therefore be of critical importance in any mechanism of diastrophism.

The rocks, mountains and structures we can see are all features of the crust; whether the mantle is involved in producing them, and if so how and to what extent, are still essentially unanswered questions. We reason that there must be changes in the physical properties of the earth both at the Mohorovičić discontinuity and at the boundary between brittle and plastic behavior, but we know too little about the materials to describe accurately the change at either place. Yet the zone between these two boundaries is almost surely a critical one. Obviously, an important first step in discovering the mechanism of diastrophism would be to learn what goes on just below what is recorded in the rocks we see—in the lower part of the crust and the outer part of the mantle, including the low-velocity layer. This is probably where any turmoil and

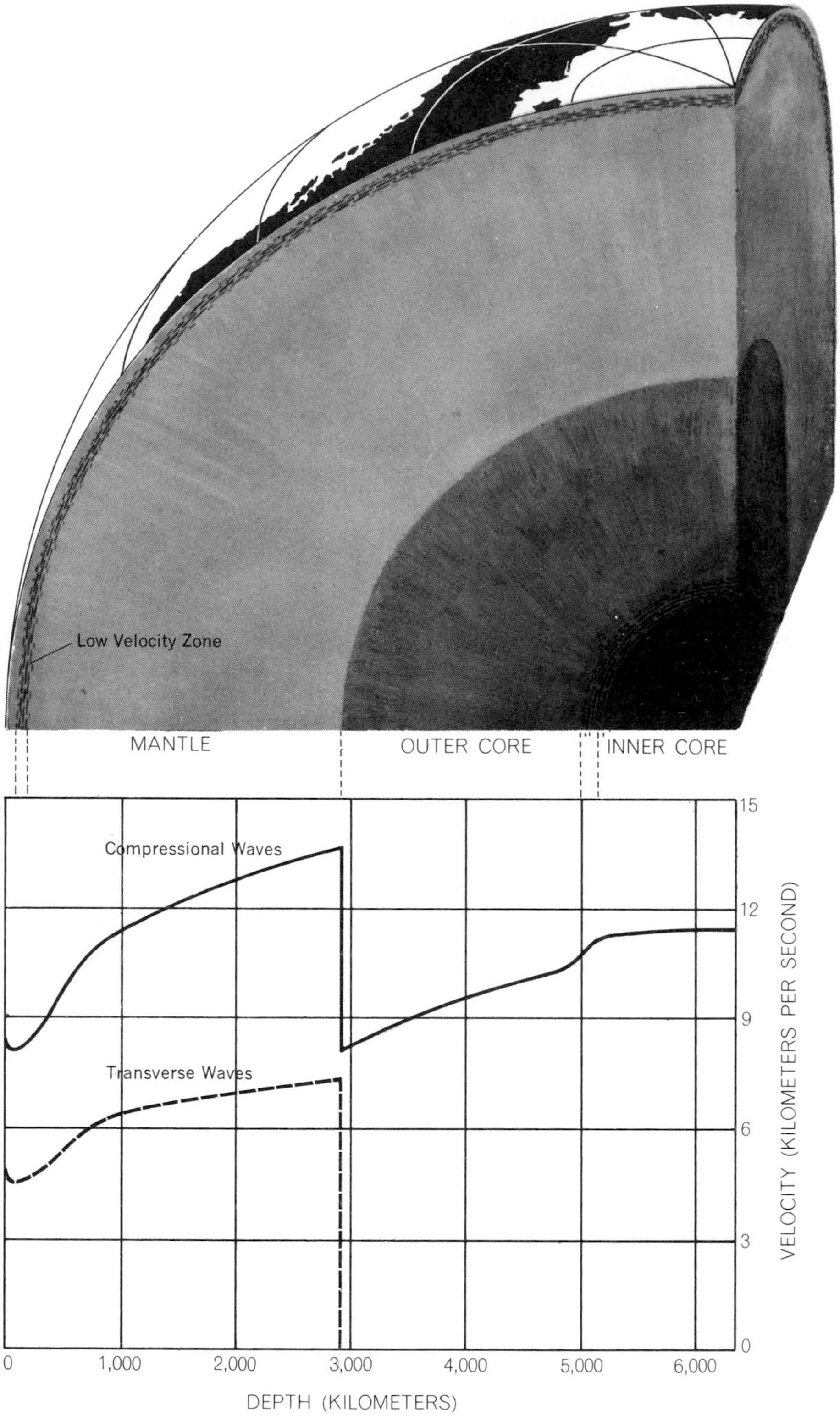

Fig. 382. Section through a segment of the earth, showing the principal depth zones. The graph below shows the change, with depth, in the velocity of seismic waves; the irregularities in these curves are the main basis for the subdivision of the earth's interior. (*After* D. L. Anderson, "The Plastic Layer of the Earth's Mantle," Scientific American, July 1962. Copyright © 1962 by Scientific American, Inc. All rights reserved.

restlessness within the earth is transmitted to the part we ride on. Current research in this field is lively; let us close with a look at some of the associated speculations.

Most geologists look inside the earth for the ultimate driving force of diastrophism; no known exterior forces are sufficiently versatile to account for the variety of deformation we see. From what we have been able to learn so far, it would seem that plastic creep, perhaps in the upper part of the mantle, is the active element, and the brittle crust on which we live is passively riding on this very slow flow. Of course, discernible forces arise from the rotation of the earth, from the tides, and from gravity acting differentially on irregularities in the crust and its surface topography, but these influences probably can do no more than modify and locally complicate what is probably the essential mechanism of crustal deformation—very slow plastic movements at about the level of the upper mantle.

This concept is attractive for many reasons. By postulating different directions of flow in the upper mantle, it is possible to imagine many different kinds of stress being imparted to the lower side of the comparatively passive crust. If the flow involves circulation in three dimensions it must include rising currents in some areas and sinking currents in neighboring ones hundreds or thousands of miles away, as well as horizontal transfer from the first type to the second. Over centers of descending flow horizontal currents would converge, thus providing optimum opportunity for horizontal compression in the outer crust. Over centers of upwelling the currents would diverge, thus producing tension in the outer crust. Between such centers, where horizontal currents would predominate, there could be persistent horizontal stresses, perhaps like those associated with the San Andreas fault zone today. If the degree to which the outer crust is coupled to the plastic interior is also variable—i.e., if the coefficient of friction between the two changes from place to place—then horizontal compression and tension could also be achieved along a single horizontal current; where it had the best "grip" on the overlying crust it would exert the greatest tangential stress, which would tend to produce compression ahead of that area and tension behind it. That the horizontal displacements produced by such a mechanism would probably be much larger than the vertical ones also fits our observations, for it will be recalled that horizontal movements of the crust are often measured in tens, if not hundreds, of miles whereas vertical movements are generally a whole order of magnitude less.

There are several reasons for thinking that the circulation in the deep plastic zone probably involves rising and sinking columns as well as horizontal currents. An important one is simply that the most feasible driving force for such circulation is the combination of gravity with variations in the density of the material. Under the influence of gravity, relatively dense columns (cooler?) of mantle or plastic material will tend to sink while relatively lighter ones (warmer?) tend to rise. The horizontal currents, then, are essentially incidental to these primary changes. Recent measurements of the heat flow from the crust at different points on the continents, and especially on the ocean floors, have supplied unexpected support for the existence of such rising currents. The heat flowing up through the ocean floors is abnormally high in places where tension seems to prevail—where one would postulate a rising column. In fact, it is so high that according to our present understanding of the conductivity of rocks, so much heat could not be delivered to the ocean floor by conduction alone; the amount seems to require the actual rise of masses of rock from hotter regions deeper in the earth.

It is for reasons such as these that some kind of very slow thermal convection—the rise of relatively warm columns and sinking of relatively cool ones—is a favored hypothesis for the ultimate cause of diastrophism.

This is about as far as we should go if we are going to limit ourselves, as promised at the beginning of these speculations, to mechanisms and models that are justified by the outdoor observations that constitute the meat of this book. Of course there are many other lines of evidence that contribute to the thinking about this frontier of knowledge. Chief among them are the results obtained from laboratory work on minerals and rocks subjected to very high temperatures and pressures. It is now possible to simulate the conditions that are believed to exist down to about 200 miles beneath the surface, and it is increasingly clear that this environment is the home of some minerals that are rare or unknown in familiar rocks, and of denser crystalline forms of others, such as coesite and stishovite (page 369), that are new varieties of old friends. Such studies will surely lead to better identification of the materials underlying the crust, and thence to better understanding of their physical properties and perhaps to answers to such questions as: Can the rocky materials in the upper mantle circulate slowly in response to temperature differences?

But let us not forget that before a hypothesis such as that of thermal convection currents in the upper mantle can even be formulated, let alone tested, we need to know the magnitudes and directions and patterns of uplift, subsidence, tension, compression, and shearing in the outer crust. And that before these can be known we must establish the sizes, shapes, and positions of such major features as the little-disturbed Colorado Plateaus and the more-disturbed Appalachian geosyncline, and we must determine the amount and direction and time of the displacements on such great faults as the San Andreas. And that before we can do this we must define and map many formations and study these until we know what facies changes they show and how the rocks in one area match those in another, so that in such features as the walls of the Grand Canyon we see not just an unbroken pile of strata but a succession of floodplains and seas, of peneplains and mountains, of rising and subsiding land masses—recognizing, all the while, that the record in the rocks before us is far from complete.

And finally, before we can do any of these things, we must be able to tell one rock from another—which is just about where we started.

SUGGESTIONS FOR FURTHER STUDY

The following suggestions are offered for those who may not have ready access to a university or college geology department:

GENERAL READING

King, Philip B. (1959), *The Evolution of North America*, Princeton University Press, Princeton. 190 pp.

> The geologic histories of the major regions of the United States and Canada, described with the perspective of an able and mature geologist lecturing to university undergraduates. Emphasis on tectonics. Many line drawings and sketch maps; no photographs. Good lists of references.

Thornbury, William D. (1965), *Regional Geomorphology of the United States*, John Wiley and Sons, New York. 609 pp.

> The area of the 50 States is divided into 27 natural provinces and regions, each of which is analyzed in terms of the rocks, structures, and erosional history responsible for the present landscape. Emphasis on sculpturing processes. Many photographs and useful references to literature on specific areas.

GUIDES TO OTHER SOURCES OF INFORMATION

Matthews, William H., Ed., (1964), *Selected References for Earth Science Courses*, Prentice-Hall, Inc. (Educational Book Division), Englewood Cliffs, New Jersey 07632. 33 pp.

> A guide to selected textbooks, teaching guides and handbooks, reference works, general treatises, periodicals, etc., in the fields of astronomy, earth science, geology, meteorology, oceanography, and physical geography. Designed as an aid to teachers.

Matthews, William H., Ed., (1964), *Sources of Earth Science Information*, Prentice-Hall, Inc. (Educational Book Division), Englewood Cliffs, New Jersey 07632. 48 pp.

> Lists appropriate departments in degree-granting institutions, government agencies (both federal and state), scientific organizations, museums, etc. in the fields of astronomy, geology, meteorology, oceanography, and physical geography. Designed as an aid to teachers; many useful addresses to which inquiries may be sent.

The United States Geological Survey is the largest single source of geological information in North America. Requests for specific information as well as for free lists of available maps and publications may be directed to the *Geological Survey, Washington, D.C. 20242*. There are also regional offices in some cities.

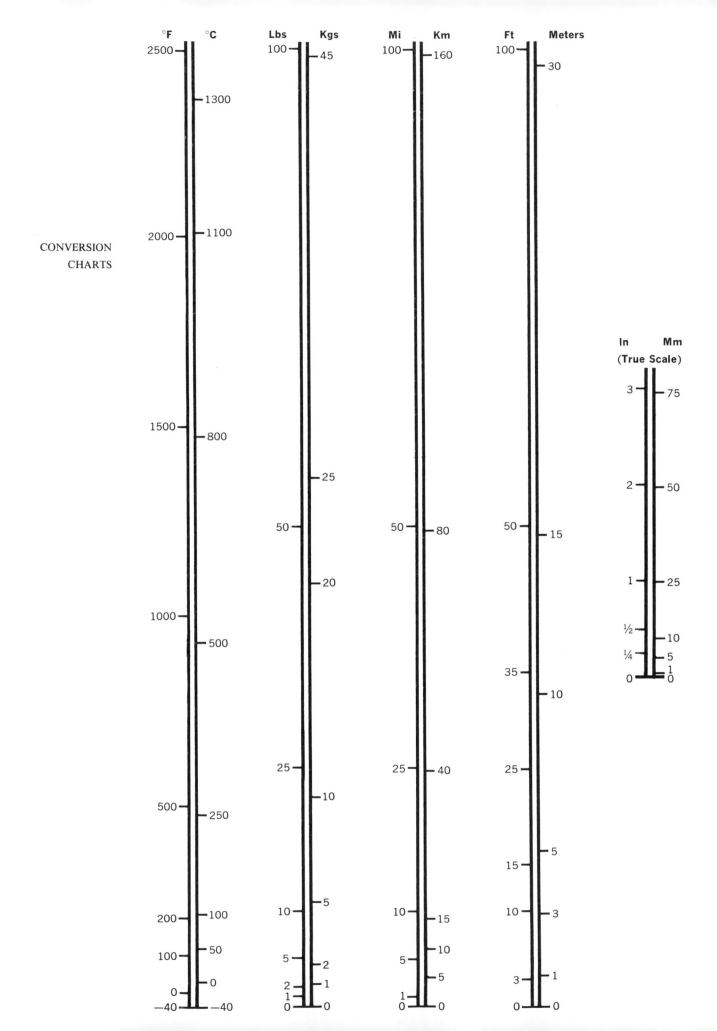

CONVERSION
CHARTS

INDEX